Women-Led Innovation and Entrepreneurship

There is a glaring lack of recognition for the contributions of women to innovation, entrepreneurship, and sustainable development in literature. This book seeks to address this gap by highlighting the often-overlooked accomplishments of women in these crucial fields.

Women-Led Innovation and Entrepreneurship: Case Studies of Sustainable Ventures and Practices features interviews with industry thought leaders and experts, providing exclusive insights into the changing role of women in driving sustainable development through innovation. The book offers a range of resources, including checklists and tools, designed to foster a supportive environment and offer practical advice and strategies for aspiring entrepreneurs and innovators. It outlines actionable steps for implementing sustainable practices and provides valuable insights that can be applied in the real world. Furthermore, this book presents in-depth case studies of women from various industries and regions, showcasing their innovative ventures, strategies, and how they have overcome obstacles to achieve success in entrepreneurship while promoting sustainability.

It serves as a valuable reference for academics, policymakers, and professionals interested in exploring the intersection of gender, innovation, entrepreneurship, and sustainable development.

Women in Industry 4.0: Shaping World Foundations in Engineering, Business, and Sustainability

Series Editors: Tilottama Singh, Richa Goel, and Kayla L. Tennin

According to a World Economic Forum report, only 22% of AI professionals globally are female. This series will bring stories straight from the heart of these innovative women who have established themselves, to inspire millions and create a wave of impact. This will be a series of books and authors that will focus on sustainability through technology and entrepreneurship in Industry 4.0. The series will act as a global platform for bringing together organizations and women sharing their life cases, research, and grounded theory which will be an interesting read for engineers, businesses, government, researchers, and students working on startups and contributing to gender diversity.

Each book in the series is dedicated to illuminating women's research and achievements in key, targeted areas of the contemporary tech world and raising awareness of the pivotal work women are undertaking in areas of importance to our global community toward sustainable development. It will be a step in bringing forth the reality of what overshadows women in workplaces and the strategic attempts of organizations to pave the path for women's representation in companies and fields of engineering.

Books that focus on women in engineering and technology exploring and describing how women have greatly contributed to the world at large through these fields and their many subsections is a focal point of this series.

If you are interested in writing or editing a book for the series or would like more information, please contact Cindy Carelli, cindy.carelli@taylorandfrancis.com.

Women-Led Innovation and Entrepreneurship: Case Studies of Sustainable Ventures and Practices
Edited by Esra Sipahi Döngül, Şerife Uğuz Arsu, Richa Goel, and Tilottama Singh

Women-Led Innovation and Entrepreneurship
Case Studies of Sustainable Ventures and Practices

Edited by
Esra Sipahi Döngül, Şerife Uğuz Arsu,
Richa Goel, and Tilottama Singh

CRC Press
Taylor & Francis Group
Boca Raton London New York

CRC Press is an imprint of the
Taylor & Francis Group, an **informa** business

Designed cover image: Shutterstock

First edition published 2025
by CRC Press
2385 NW Executive Center Drive, Suite 320, Boca Raton FL 33431

and by CRC Press
4 Park Square, Milton Park, Abingdon, Oxon, OX14 4RN

CRC Press is an imprint of Taylor & Francis Group, LLC

© 2025 selection and editorial matter, Esra Sipahi Dongul, Serife Uguz Arsu, Richa Goel, and Tilottama Singh; individual chapters, the contributors

Reasonable efforts have been made to publish reliable data and information, but the author and publisher cannot assume responsibility for the validity of all materials or the consequences of their use. The authors and publishers have attempted to trace the copyright holders of all material reproduced in this publication and apologize to copyright holders if permission to publish in this form has not been obtained. If any copyright material has not been acknowledged please write and let us know so we may rectify in any future reprint.

Except as permitted under U.S. Copyright Law, no part of this book may be reprinted, reproduced, transmitted, or utilized in any form by any electronic, mechanical, or other means, now known or hereafter invented, including photocopying, microfilming, and recording, or in any information storage or retrieval system, without written permission from the publishers.

For permission to photocopy or use material electronically from this work, access www.copyright.com or contact the Copyright Clearance Center, Inc. (CCC), 222 Rosewood Drive, Danvers, MA 01923, 978-750-8400. For works that are not available on CCC please contact mpkbookspermissions@tandf.co.uk

Trademark notice: Product or corporate names may be trademarks or registered trademarks and are used only for identification and explanation without intent to infringe.

ISBN: 978-1-032-84147-2 (hbk)
ISBN: 978-1-032-84059-8 (pbk)
ISBN: 978-1-003-51142-7 (ebk)

DOI: 10.1201/9781003511427

Typeset in Times
by Newgen Publishing UK

Contents

Foreword by Joanna Paliszkiewicz ix
About the Editors xi
Contributors xiii
Introduction xv

Chapter 1 Womanizing Land: The Culture of Women Not Owning
Land – Narratives and Challenges for a Sustainable Planet 1

Monica Mastrantonio

Chapter 2 Psychological Well-Being of Self-Employed Women:
European Perspective 22

Tomasz Skica and Katarzyna Miszczyńska

Chapter 3 Modern Data-Driven Management and Optimization of
Development Opportunities Creation for Women: A Referential
Investigative Focus on the UN SDG Targets for Women 64

Chikezie Kennedy Kalu

Chapter 4 Woman-Led Innovations: Strategies, Barriers, and Solutions
for Women Entrepreneurs in a Changing Landscape 89

Almula Umay Karamanlıoglu and Iper Incekara

Chapter 5 The Role of Multi-Stakeholder Partnerships in Promoting
Women's Leadership in Southeast Asia: Challenges,
Opportunities, and Trends 103

Putri Hergianasari, Michael Koks, and Rizki Amalia Yanuartha

Chapter 6 Unpacking the Effect of Family Support for Women's
Entrepreneurial Success in Pakistan 142

Aemin Nasir and Shajara Ul-Durar

Chapter 7 Unveiling the Factors of Women Entrepreneurs on Social
Media to Achieve Enterprise Sustainability 165

Anshu Rani, Vichitra Somshekar, Ramya U., and Mercy Toni

Chapter 8 Women's Entrepreneurship and Changing Landscape in
Africa: Strategies and Solutions 184

*Benneth Uchenna Eze, Festus Ekechi, Oluyemisi Agboola,
Iyabode Abisola Adelugba, and Adeola Abosede Oyewole*

Chapter 9 An Overall Evaluation of Women's Entrepreneurship 192

Mustafa Altintaş

Chapter 10 An Analysis of Women's Entrepreneurship in Turkey:
Bibliometric Methods 206

Meltem Ince Yenilmez

Chapter 11 Social Work Perspective on Women's Empowerment: Women's
Entrepreneurship 219

Hatice Ozturk

Chapter 12 A Systematic Literature Review on Women and Leadership:
Exploring the Contribution of Women Leaders to Companies'
Sustainability 230

*Durdana Ovais, V. Selvalakshmi, Ravi Chatterjee, and
Ravinder Rena*

Chapter 13 Exploring the Role of Women Entrepreneurs in Advancing
Education and Development in Bangladesh 246

Md. Harun Rashid and Wang Hui

Chapter 14 Preserving Heritage, Inspiring Innovation: The Role of
Indigenous Knowledge in Culinary Entrepreneurship – Case
Study of JhaJi Store and Namakwali 258

Anu Kohli, Neha Tiwari, and Haider Abbas

Chapter 15 Women's Brains: Creative Thinking and Entrepreneurship 274
*Steve Fernando Pedraza Vargas and
Ramón Antonio Hernández de Jesus*

Index 285

Foreword

In today's fast-evolving global landscape, the need for innovation and sustainability has never been more pressing. We face unprecedented environmental, social, and economic challenges, but tremendous opportunity lies within this context. One of our time's most significant and transformative movements is the rising tide of women-led innovation and entrepreneurship. Women are contributing to global economies and reshaping the foundations of business practices with sustainability at the core of their endeavors.

Women-Led Innovation and Entrepreneurship: Case Studies of Sustainable Ventures and Practices is an interesting monograph on the diverse and dynamic ways women are pioneering change in business and society. Through the collaboration of esteemed editors—Dr. Esra Sipahi Dongul, Dr. Serife Uguz Arsu, Dr. Tilottama Singh, and Dr. Richa Goel—this volume sheds light on the unique challenges women face, as well as the groundbreaking solutions and strategies they have developed in response. The book presents a global tapestry of women entrepreneurs who, through their vision and resilience, are driving sustainable ventures that balance economic success with social and environmental responsibility.

This collection of case studies goes beyond simply recounting success stories; it delves into the intersection of gender, innovation, and entrepreneurship with a keen focus on sustainability. It examines the critical role women play in leadership and green business practices, offering readers inspirational accounts, practical insights, and research-backed strategies for fostering women's entrepreneurial success. The case studies from various regions of the world, covering diverse sectors, reveal the profound impact of women's leadership in advancing sustainable development goals, promoting gender equality, and addressing social justice issues.

One of the book's core strengths lies in its multidimensional approach to understanding the factors influencing women's entrepreneurial journeys. The chapters examine a broad spectrum of themes, from psychological well-being and empowerment to data-driven management and the role of family and social media in fostering sustainable enterprises. Doing so provides a comprehensive and nuanced exploration of the socio-cultural, economic, and psychological factors that both challenge and propel women entrepreneurs.

The significance of multi-stakeholder partnerships, innovation capabilities, and creative risk-taking emerges as critical elements for advancing women's leadership. Moreover, the book presents forward-looking strategies for overcoming systemic barriers and highlights the vital role women entrepreneurs play in developing solutions that not only support their ventures but also contribute to broader societal goals.

As we look toward a future defined by uncertainty and transformation, this book serves as both a roadmap and a call to action. As showcased in these case studies, women entrepreneurs' innovative spirit and resilience provide a sustainable development model. Their ability to blend creativity with strategic thinking while addressing some of the most critical issues of our time is a profound source of inspiration for current and future generations.

It is my belief that *Women-Led Innovation and Entrepreneurship: Case Studies of Sustainable Ventures and Practices* will inspire policymakers, educators, investors, and entrepreneurs alike. By spotlighting women's crucial role in driving sustainable change, this book contributes to building a more inclusive, equitable, and sustainable global economy.

Prof. dr hab. Joanna Paliszkiewicz
Warsaw University of Life Sciences

About the Editors

Esra Sipahi Döngül graduated from the Istanbul Gelisim University Business Administration Doctorate Program. Her areas of expertise are management, HRM, data analysis, business, and organization. She has a strong background in research activities in organizational behavior, corporate social responsibility, and management organization areas. She has many published articles in the field of corporate social responsibility and has also attended and presented at numerous international conferences. She is the founder of the human resources platform called flexit24.com. She is currently working at the Aksaray University, Turkey.

Şerife Uğuz Arsu completed her BA in Business Administration at Selcuk University in 2011, her MA in Management and Organization at Aksaray University in 2014, and her PhD in Business Administration in 2015. She is an assistant professor at the Department of Social Work, Faculty of Health Sciences, Aksaray University. She is interested in management organization, organizational behavior, human resources management, strategic management, entrepreneurship, women's entrepreneurship, gender, work psychology, and social work and has conducted studies in these areas, as well as national/international studies on research methods, including qualitative and quantitative research. She also has training and certificates on qualitative research design, qualitative data analysis and computer-aided data analysis, and continues to conduct studies in this field. She is married to Assoc. Prof. Talip ARSU, who is also an academician, and has two sons named Arda and Atlas.

Richa Goel is an accomplished academic with 23+ years of experience in Economics and Management. She is an associate professor at Symbiosis Centre for Management Studies Noida, Symbiosis International University, PUNE. A Gold Medallist in Economics, she holds a PhD in Management with expertise in Diversity Management. She has to her credit more than 20 books and numerous papers published with leading publishers She is involved with training and has conducted many international and global conferences and has been a trainer for many FDPs concerning SDG's.

Tilottama Singh is a certified human resources analyst from IIM and a proficient academic, researcher, and trainer with more than 14 years of experience in the field of Human Resources and Work Dynamics. She is presently the head of Department in the School of Management, Uttaranchal Institute of Management, Uttaranchal University, Dehradun. Her areas of expertise are human resources, entrepreneurship, and strategy, and her teaching concentration includes human resources, strategy, and law. She started her career in the hotel industry with Leela Kempinski in the corporate head office in Human Resources and Training. Later, she joined the education sector with the topmost universities IMS Unison Group and Amity University, the leading university in Pan Asia. She has obtained her master's in human resources and management with distinction. Before that, she graduated in economics with honors and law with a distinction, with an added vocational course in mass communication and analytic certification from IIM. Having an enriched research portfolio with Scopus and international refereed journals, book chapters, and conference presentations, she has been awarded the Best Presentation Award at international conferences. She also serves as a member of AIMA and acts to lead and liaise between the student community and the industry delegates, with a keen interest in training and building sustainable models for business and society.

Contributors

Haider Abbas
University of Technology and Applied Sciences
Oman

Iyabode Abisola Adelugba
Department of Management Sciences
Bamidele Olumilua University of Education, Science and Technology
Nigeria

Oluyemisi Agboola
Department of Entrepreneurship
Ekiti State University
Nigeria

Mustafa Altintaş
Yozgat Bozok University
Çekerek Fuat Oktay Vocational School of Health Services
Turkey

Ravi Chatterjee
Symbiosis International University
United Arab Emirates

Festus Ekechi
Department of Business Administration
Christopher University
Nigeria

Benneth Uchenna Eze
Department of Business Administration
Christopher University
Nigeria

Putri Hergianasari
Satya Wacana Christian University
Indonesia

Ramón Antonio Hernández de Jesus
Secretaria Municipal de Educação
Brazil

Wang Hui
University Putra Malaysia
Malaysia

Meltem Ince Yenilmez
Department of Economics
Izmir Democracy University
Turkey

Iper Incekara
Başkent University
Turkey

Chikezie Kennedy Kalu
Department of Management Science and Engineering
Jiangsu University
China

Almula Umay Karamanlıoglu
Başkent University
Turkey

Anu Kohli
Department of Business Administration
University of Lucknow
India

Michael Koks
Satya Wacana Christian University
Indonesia

Monica Mastrantonio
Arden University
United Kingdom

Katarzyna Miszczyńska
Department of Banking
University of Lodz
Poland

Aemin Nasir
Department of Management
RMIT University
Vietnam

Durdana Ovais
BSSS Institute of Advanced Studies
India

Adeola Abosede Oyewole
Centre for Entrepreneurship
  Development
Yaba College of Technology
Nigeria

Hatice Ozturk
Department of Social Work
Aksaray University
Turkey

Anshu Rani
REVA Business School
REVA University
India

Md. Harun Rashid
University Putra Malaysia
Malaysia

Ravinder Rena
Durban University of Technology
South Africa

V. Selvalakshmi
Velammal College of Engineering and
  Technology
India

Tomasz Skica
Department of Entrepreneurship
University of Information Technology
  and Management
Poland

Vichitra Somshekar
Christ University
India

Neha Tiwari
IILM Academy of Higher Learning
India

Mercy Toni
University of Nizwa
Oman

Ramya U.
REVA Business School
REVA University
India

Shajara Ul-Durar
School of Business Management
University of Sunderland
United Kingdom

Steve Fernando Pedraza Vargas
Universidad Santo Tomás
Colombia

Rizki Amalia Yanuartha
Satya Wacana Christian University
Indonesia

Introduction

The book *Women-Led Innovation and Entrepreneurship: Case Studies of Sustainable Ventures and Practices* delves into the multifaceted experiences of women in leadership, innovation, and entrepreneurship. It sheds light on the challenges and opportunities women encounter in various regions and sectors, particularly with a focus on sustainable development and economic growth. By exploring case studies, data-driven analysis, and the unique strategies employed by women entrepreneurs, this book serves as an essential resource for understanding how women are contributing to shaping a sustainable future.

Each chapter brings a unique perspective, ranging from psychological well-being and empowerment to digital innovation, data-driven management, and leadership in emerging markets. The book emphasizes the critical role of women in advancing the United Nations Sustainable Development Goals (SDGs) and highlights the importance of gender equality in fostering long-term, sustainable economic progress.

In an era where gender equality and economic empowerment are at the forefront of global discussions, this book provides a comprehensive exploration of women's entrepreneurship and empowerment, shedding light on various critical aspects that shape women's roles in the business world.

Building upon this foundational understanding, the book delves into the psychological dimensions of self-employment among women. Self-employment represents a vital pathway to financial independence and personal fulfillment, yet it comes with its own set of mental and emotional challenges. The book explores how women navigate these challenges, balancing work and personal life while managing the stresses and rewards of entrepreneurship. It offers a nuanced view of the psychological impacts of self-employment, including resilience, motivation, and the intricate balance required to sustain both professional and personal well-being.

The exploration continues with a focus on the role of data-driven management in enhancing women's economic opportunities, with particular emphasis on the United Nations SDGs. This section examines how data optimization and technological advancements can be harnessed to accelerate women's economic participation and empowerment. By analyzing global efforts to meet SDG targets and leveraging modern technological tools, the book provides insights into how data can be used to create equitable opportunities for women. Each chapter builds on these themes, offering practical solutions, innovative strategies, and examining multi-stakeholder partnerships that contribute to a holistic understanding of women's entrepreneurship. Through its detailed analysis, the book aims to present a well-rounded perspective on the challenges and opportunities that define women's roles in the business landscape, offering valuable insights for policymakers, entrepreneurs, and advocates alike.

1 Womanizing Land

The Culture of Women Not Owning Land – Narratives and Challenges for a Sustainable Planet

Monica Mastrantonio

1.1 INTRODUCTION

UN Secretary-General Antonio Guterres has urged governments to eliminate the barriers that still exist towards women owning land, and improve the policymaking for women's land ownership worldwide. "To support women and girls to play their part in protecting our most precious resource and to stop land degradation by 2030."

Land ownership is a critical aspect of the economic empowerment and social development of every single country. Historically, women have faced significant barriers in land ownership, which has had profound implications for their well-being, autonomy, and inputs of the global resources. No matter how much effort has been made to address these disparities and promote gender equality in land rights, the distance between the law and its application has remained stable. On the global average, women own around 10–20% of the global farmland, and only some countries do have equal rights and access towards women owning real estate.

Globally, it is estimated that women own approximately 10–20% of the global land, which means the other 80% belongs to men. Ownership is one of the greatest gender gap variables ever, usually overlooked within other discussions (Grabe, 2014). Most of all, it is present in every single continent. In Africa, for instance, the percentage is approximately 10–20%, depending on the region and country. Some countries like Rwanda and Ethiopia have made progress through legal reforms and land titling programs, while others still face significant challenges due to patriarchal norms and legal barriers.

Whereas in Asia, land is owned by around 10–15%, women's land ownership rates are generally low due to traditional inheritance practices, legal restrictions, and cultural norms. Countries like India and Pakistan have even lower rates, while some Southeast Asian countries have seen many improvements due to legal reforms (Chowdhry, 2017).

Although Europe continuously announces laws and policies to cover the gender gap, it is not an example either. The rate is that only 10–30% of its land is women-owned, slowly increasing in recent years. Northern and Western European countries, like Sweden and Norway, have higher rates due to their progressive gender equality policies. However, Southern and Eastern European countries often have lower rates due to traditional norms and inheritance practices.

Apart from the United States and Canada, which have relatively higher rates of women's land ownership, supported by progressive policies and legal frameworks, yet that does not reach the 50%. This means that in the US, about 30% of privately owned farmland is owned by women; in South America, this rate is approximately 15–25%. For example, Brazil and Argentina have seen improvements through land reforms and titling programs, but challenges remain in many areas due to traditional norms, economic, and cultural barriers.

In Oceania, like in Australia and New Zealand, legal frameworks support gender equality, but cultural and economic factors still influence ownership rates. In Pacific Island nations, traditional land tenure systems often limit women's land ownership to the number of only 10–20% land ownership (FAO, 2018).

Despite the variations in these numbers, depending on the region and specific country, either influenced by local laws, cultural practices, or economic conditions, the gender gap is still prevalent all around the globe. And, even though, women are 43% of the global agricultural labour, there is significant discrimination towards ownership, equal pay, decision-making, and access to financial services, among others.

Figure 1.1 shows in which countries women face the highest rates of discrimination towards land ownership. This can be spotted in diverse continents, and there are countries where they own no land, nor any right to ownership, pointing to a culture where women 'do not own anything.'

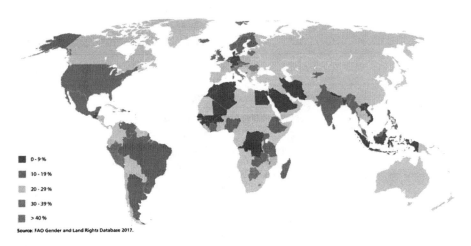

FIGURE 1.1 Distribution of agricultural holders by sex – female.

Source: Food and Agriculture Organization of the United Nations - Global Laws Concerning Property Rights for Women (2017) Women Stats Project. This file is licensed under the Creative Commons Attribution-Share-Print Alike 4.0 International license.

Figure 1.1 is a global map published by the United Nations in 2017 depicting the rate of agricultural landholding by women.

Nevertheless, there are progressive countries where land ownership is much closer to gender equality on equal pay, like Rwanda that went through progressive gender policies and land reforms. The 1999 inheritance law and the 2013 Land Law granted equal rights to land ownership for people, leading to significant increases in women's land ownership across the globe. Ethiopia is not behind with land certification programmes that include joint ownership titles for spouses have increased women's land rights and ownership. Legal reforms in Vietnam, and joint land use certificates for both spouses have helped increase the proportion of land owned by women, and the same happened with Uganda, where policies promoting gender equality and land reforms have allowed greater land ownership by women.

However, there are countries where land rights are indeed non-existent. Saudi Arabia, for instance, has traditional legal restrictions on women's rights and that significantly limits women's land ownership. Cultural norms and ongoing conflict in Yemen have severely restricted women's rights, including land ownership. Despite legal reforms in India, traditional practices and societal norms often limit women's ability to either own or inherit land. Strong patriarchal structures in Pakistan and legal loopholes often prevent women from owning land, despite formal laws guaranteeing their rights.

Europe is neither different. The percentage of women owning land across European countries generally ranges from approximately 10% to 30%. This range varies significantly based on the country, region, and local cultural and legal contexts. There are countries with a higher percentage of ownership like Sweden, known for strong gender equality policies with progressive policies promoting gender equality, including in land ownership. But, there are countries with moderate rights, like Germany and France. Countries like Italy, Spain, and Eastern Europe, which have traditional family structures and inheritance practices, often result in lower rates of land ownership by women. Interestingly, there are only few studies specifically addressing this gender gap in Europe but rather using low hemisphere countries to exemplify the matter.

In the Americas, the percentage of women owning land generally ranges from approximately 15% to 25%. This range encompasses variations across different countries and regions within North, Central, and South America. In terms of real estate, it is one of the most equal in the world in terms of women being able to buy and finance their homes. Canada does have progressive gender equality policies contributing to relatively higher rates of land ownership by women, but only recently had their first rise in this number. About 30% of privately owned farmland in the United States is owned by women, though this includes co-ownership of land within a marital status.

Brazilian legal reforms and land titling programs have improved women's land ownership, but cultural and economic barriers still exist. Mexico is not any different; reforms in agrarian laws have increased women's land ownership, but traditional norms and economic challenges persist. In countries such as Guatemala, Colombia and Honduras, the patriarchal norms and limited legal protections result in lower rates of land ownership by women despite some legal advancements. Even so, it has one of the highest rates compared to other continents.

Figure 1.2 shows women land workers performing agricultural production in greater Santa Fe.

FIGURE 1.2 Women land workers doing agricultural production in greater Santa Fe. This file is licensed under the Creative Commons Attribution-Share Alike 4.0 International licence.

This chapter shall explore why do these disparities still occur? Is it just a matter of law and policy making? What psycho-social structures are present and preventing women from owning land? And what should be done towards closing this gender gap? What other variables might be involved in impeding the equal share?

The objective is to explore the multifaceted causes and implications of laws and behaviours, while mapping the difficulties that women find in applying an equal law. The history of women not owning properties, or managing them, is imbricated by multiple and intersectional aspects that need to be stated and addressed to cover the SDG's goals and promote equal access and management of land and real estate. There are, here and there, initiatives which tackle these issues and promote women's empowerment, but yet, these are not enough to cover the disparity which has a long historical and cultural base.

Besides that, it is understood that owning aligns well with disposition towards women's empowerment like being able to set their own salaries, being able to protect their production, having a safe place against work and domestic violence, and the list of advantages could go on. However, real estate and farmland ownership is directly imbricated in ideological narratives and cultural principles that are still prevalent in most of these societies, either from north or south hemisphere countries. Within the micro-power of cultural and heritage narratives, the forces that still keep women away from owning land are still active and alive.

Taking the concept of micro-power from Foucault whose analyses of discourses reveals how dominant power structures maintain their power through the creation of particular discourses that are constantly reproduced and shared, this chapter shall look at the land gap through multiple agencies and narratives that perpetuate the difficulties that women face towards ownership.

Foucault regards discourse as a central human activity with specific signification and power maintenance, rather than a universal "general tex.". This knowledge and discursive practices about sex, crime, and psychiatry are sifted between epochs, showing that there is the possibility of questioning and change. Owning land is not simply a matter of law and policy making, most countries like Europe, and The USA do have that, but there are other aspects in between the law and its implementation that perpetuates the condition that women are still educated to be "shopping and looking good, whereas men are incentivised to own assets." Why? Because land is power.

Going deeper into the activity of owning a space, this means to have a place, a locality, a geographical positioning, an address, an existence. It has further implications towards autonomy, independence, and more sustainable practices. This means that owning a material space in the geographical space of the globe needs to be connected to owning farmland, a house, an apartment, a tent, a roof, a camp, a farm, a room under a woman's name and law. Having a space of one's own adds to the construction of one's identity and independence. Not only a fundamental right, owning land also means having shelter, protection, and a way of producing a living. Owning land has very rarely been associated with the possibility of a living without violence, a more sustainable production, and a list of other rights guaranteed (Zvokuomba, Kwashirai, & Batisai, 2020).

Thus, owning land and properties is linked to having a space of one's own, being protected, having a border. It is a basic need towards protection, equal rights, and many other possible outcomes that can only be established when the right of ownership is set and practised. The interpersonal, cultural, and social factors that contribute to the perpetuation of gender disparities in land ownership may help to uncover the historical roots and contemporary manifestations of how women are still denied this basic right, and along with it, all the others.

It is important to emphasise the need for a comprehensive and interdisciplinary understanding of the intersectionality of gender, culture, and socio-economic relations on the perpetuation of this gap. This chapter will apply Foucault's theory of micro-power to comprehend the dynamics of land ownership, and which goes much further than simply having a constitution that mentions women's rights as equal, and perpetuates its unevenness (Landesa, 2016). The challenges that women may encounter in guaranteeing and exercising this right do not only include legal barriers but also discriminatory practices, such as prejudice, complaints, accusations, and even verbal and physical violence. This chapter delves into the micro-narratives and power of interpersonal narratives that are mostly discriminatory and insult women's capacities.

Other questions also arise: do women have access to equal mortgage conditions? Do women receive support towards equal inheritance between male and female family members? Are women incentivised from an early age to be capable of owning and running property? Do women's salary gaps bring more burden to the chain of ownership? Can women protect themselves against violent relationships without having a home of their own?

Thusy, to underscore the significance of women's land ownership may act as catalyst for their economic, social, and personal empowerment. This emphasises the need for concerted efforts from governments, civil society organisations, communities, and interactions from youthful age that can dismantle barriers, challenge cultural norms, and ensure equitable access and management to land and properties by women.

By amplifying the visibility towards these narratives of women in relation to land ownership, this study used global data, women's voices, and Foucault's theory of micro-power to understand the interconnection of the variables on land ownership. Unless there is a broader discussion on gender parity for transformative change in promoting women's rights and agency in land ownership, women will still face struggles and suffering. The habits and customs of women not owning land will continue to put at risk all other gender equal rights, which have been perpetuated across the centuries, with small and restricted periods of systematic change.

1.2 HISTORY OF WOMEN NOT OWNING LAND

The historical context for women not owning land worldwide is varied, but it is also rooted in historically limited land possession intertwined with a complex interplay of social, legal, economic, and cultural factors, such as:

- *Patriarchal societies*: Most traditional societies were patriarchal, meaning they were dominated by men in both public and private spheres. In such societies, men held authority over family and community matters, including ownership of land and property. This male dominance was often codified into law and reinforced through social norms.
- *Roman law*: Under Roman law, women were generally under the guardianship of a male relative and had limited property rights.
- *Feudal Europe*: During the feudal era, land was typically controlled by male heads of households. Women could inherit land only in the absence of a male heir, and even then, their control was often limited.
- *Common law*: In English common law, the doctrine of coverture meant that a married woman's legal rights were subsumed under her husband's, severely restricting her ability to own property independently.

The history of women's land ownership rights has evolved significantly over the past 2,000 years, influenced by legal, social, and cultural changes across different historical periods and regions.

In Ancient Egypt (up to 500 CE), women enjoyed relatively high status and had the right to own, inherit, and manage property independently of their husbands or male relatives. Legal documents from this period show women engaging in land transactions.

On the other hand, women's rights to own land in Ancient Greece were severely restricted. In most city-states, land was inherited by patriarchy processes, and women rarely held property rights. Likewise in Ancient Rome, women had limited rights to own property. Under the control of their fathers or husbands, women could not own land independently. However, with the introduction of the dowry system, women did manage some property indirectly.

During the Middle Ages (500–1500 CE) within Feudal Europe, women's land rights were largely determined by feudal laws. Noblewomen could inherit land if there were no male heirs, but their rights were often secondary to those of their male relatives. In some regions, customary laws allowed women to own land, particularly widows who inherited their husbands' estates.

Nonetheless, in Islamic law, established in the 7th century, women were provided with specific inheritance rights, allowing them to own and manage property. Women were entitled to a fixed share of inheritance, which was revolutionary compared to many contemporary cultures.

During the early modern period in Europe (1500–1800), the legal status of women varied widely. In some parts of Europe, such as Scandinavia, legal reforms began to grant women more rights to own and inherit land. Not any differently, the rise of the common law system in England limited women's property rights. Under coverture, a legal doctrine, a married woman's legal rights and obligations were subsumed by those of her husband.

In colonial America, women's land rights were influenced by English common law, which generally limited their property rights. However, some colonies, like Pennsylvania, granted women more autonomy in property ownership.

During the 19th century, there were significant legal reforms in many parts of the world. In England, the Married Women's Property Acts (beginning in 1870) gradually allowed married women to own and control property in their own right. Similar reforms were enacted in the United States, with states passing laws that granted women property rights independent of their husbands.

The 20th century witnessed a global movement towards gender equality. Many countries reformed their legal systems to grant women equal property and inheritance rights.

In the early 1900s, the suffragette movement in Western countries pushed for women's rights, including property rights. Women gained the right to vote and greater legal recognition, which paved the way for property rights reforms. Consequently, countries in Africa, Asia, and Latin America reformed their laws to improve women's property rights.

However, customary practices often continued to restrict these rights. It is specifically these stated rules and commons that create the difficulties to owning and managing land. International conventions, such as the Convention on the Elimination of All Forms of Discrimination Against Women (CEDAW) adopted in 1979, called for the elimination of discrimination against women, including in property and inheritance laws.

During certain historical periods, women were also used as labour in multiple rural tasks without leading to ownership. This is exemplified in Figure 1.3 when women worked in mainly traditionally male roles to cover the gaps during war. However, this does not imply gaining or accessing rights that could change women's path into ownership or the continued use of these new roles after the war was over and they were

FIGURE 1.3 Millicent Fawcett, one of the most influential feminists of the past 150 years. For more information visit https://brightonmuseums.org.uk/discovery/history-stories/millicent-fawcett-one-of-the-most-influential-feminists-of-the-past-150-years. (Creator: Garry Knight. Copyright: Copyright free.)

not needed anymore in the fields. There is a subtle difference in terms of micro-power activities. In the first case, this condition is the consequence of a market need and the lack of men in the fields, and only the second is it an option and a conquest.
There is also a significant difference between farming and ploughing and owning. It is especially important to establish these differences and the focus of attention that ownership is totally different from working the land.

Legal reforms in many countries have aimed to eliminate discriminatory laws and practices. Despite progress, challenges remain, particularly in enforcing these laws and changing deeply rooted cultural practices that continue to discriminate against women.

Many countries have introduced gender-sensitive land reforms. For instance, India's Hindu Succession (Amendment) Act of 2005 granted daughters equal inheritance rights to ancestral property. Equally, Rwanda's 2013 land law explicitly provides for gender equality in land ownership and inheritance, reflecting a broader trend towards legal reforms promoting women's land rights.

Over the past 2,000 years, women's land ownership rights have evolved from near-total exclusion in many ancient societies to increasingly equal rights in modern legal systems. This journey has been shaped by legal reforms, social movements, and international conventions advocating for gender equality. While noteworthy progress has been made, ongoing efforts are needed to fully realise women's land rights globally, especially if the conditions are not the law. More specifically, to provide the means for equal ownership, management, and access towards creating a more equal and sustainable world.

The main difficult factors towards covering this gap have been:

1 *Economic systems* that are often structured in ways that poor women are limited in acquiring and retaining land. Women's roles were typically confined to the domestic sphere, where they performed unpaid labour. This economic dependence on male relatives or husbands meant that women had fewer resources to purchase land.
2 *Cultural and religious beliefs* also play a significant role. Many cultures viewed women primarily as caretakers and homemakers, roles that did not require land ownership. Religious texts and teachings in various traditions often emphasised women's subservient role to men, further justifying legal and social restrictions on women's land rights.
3 *Colonialism* often reinforces or exacerbates these inequalities. Colonial powers imposed their own legal systems, which frequently disadvantaged women. For example, in many parts of Africa and Asia, colonial rulers replaced matrilineal systems, where women had significant property rights, with patriarchal systems that marginalised women's land ownership.

The above has happened all across the world. For example, in pre-colonial India, Hindu law varied but generally restricted women's property rights. British colonial rule further codified these restrictions, limiting women's inheritance rights until reforms in the mid-20th century. Traditional customs in many African societies involved communal land ownership, with male elders controlling land allocation. Colonial administrations often formalised these practices, entrenching male control. Also, Spanish, and Portuguese colonial laws were heavily influenced by Roman Catholic doctrines, which emphasised male authority and restricted women's property rights.

Despite the above, reforms and social movements, especially the ones held during the 19th and 20th centuries, such as the women's suffrage movements, often included demands for property rights. As women gained the right to vote, they could also influence legislation more effectively. Other countries reformed their legal systems to grant women equal property and inheritance rights. For example, the Married

Women's Property Acts in the UK (late 19th century) allowed women to own and control property independently of their husbands, which is a great advance towards being a sole property owner (see Figure 1.4).

Interpersonal, cultural, and economic barriers to women's land ownership still persist indicating that women still own a disproportionately small percentage of land worldwide. According to the United Nations Food and Agriculture Organization (FAO), women account for less than 15% of all landowners globally. This statistic highlights significant gender disparities in land ownership, which are more pronounced in certain regions than others.

Specific examples include women in sub-Saharan Africa (Joireman, 2008) whose property rights and customary law are imbricated in each other. These women often face legal and customary barriers to land ownership, accounting for less than 20% of the land titles in countries like India and Nepal (Mishra & Sam, 2016).

Thus, women rarely own land due to patriarchal inheritance practices and legal barriers. Studies indicate that women in South Asia own less than 10% of the land. Some progress has been made in Latin America; however, women still own a significantly smaller proportion of land compared to men. The disparities also involve other correlated factors. For instance, in the United States, women own about 30% of farmland, but their landholdings are generally smaller and less productive than those owned by men. Amidst these disparities, women may be lacking a set of resources to finance, manage and retain their land and properties, or even being instructed towards the management of the ownership as they, historically, are less experienced than men in the matter.

From the lack of financial resources because women generally tend to have less access to financial resources, such as loans and credit, which are essential for purchasing land to other factors such as informal talks, facing a "machismo" culture, or even being indirectly forced to sell their land, it is important to understand the deep reasons why equal access is not occurring. For example, financial institutions usually require collateral benefits, which women may lack due to their limited asset ownership, to higher poverty rates which hinder their ability to acquire and retain land.

Educational barriers and lower levels of education among women, especially in rural areas, also may contribute to the lack of awareness about legal rights and land acquisition processes. This limits their ability to navigate the legal and bureaucratic systems needed to secure land ownership safely and continuously.

Some administrative hurdles, complex bureaucratic processes, together with high taxes may also add burdens discriminatory practices within administrative systems, where corruption and gender bias can impede their efforts to obtain land titles. Being underrepresented in political and decision-making governmental organisations, they lack influence in land policies and reforms. This results in policies that do not adequately address or prioritise women's land rights (Gaddis et al., 2022). Plus, women who attempt to assert their land rights often face gender-based violence and intimidation, usually within their families and communities. This can deter them from pursuing land ownership, some may fear that even their lives are at risk.

Large-scale land acquisitions by corporations or the state often disproportionately displace women, who have less power to resist such actions. In conflict zones, women

THE

MARRIED WOMEN'S PROPERTY ACT, 1870,

AND THE

Married Women's Property Act, 1870,
Amendment Act, 1874.

ITS RELATIONS TO THE

DOCTRINE OF SEPARATE USE.

With Appendix of Cases, Statutes and Forms.

BY

J. R. GRIFFITH, B.A., Oxon,
OF LINCOLN'S INN, BARRISTER-AT-LAW.

THIRD EDITION.

LONDON:
STEVENS & HAYNES,
Law Publishers,
BELL YARD, TEMPLE BAR.
1875.

FIGURE 1.4   Married Women's Property Act, 1870. For more information visit www.cambridge.org/core/journals/cambridge-law-journal/article/abs/section-17-of-married-womens-property-act-1882does-it-enable-the-court-to-modify-property-rights/4F405118E459D674DFD7538E36695D8E

are more likely to lose their land and have greater difficulty reclaiming it post-conflict due to weaker legal and social standing (Ajala, 2017).

Addressing these setbacks requires a multifaceted complex approach, including not only legal reforms to ensure equal inheritance and land ownership rights, but educational programs to inform women of their rights, financial services tailored to women's needs, and broader societal changes to challenge and transform patriarchal norms including safety measures.

1.3 CASE STUDIES – STRUGGLING WITH LAND OWNERSHIP

Here are some interviews and personal stories of women from distinct parts of the world who have faced challenges in owning real estate or land property. These accounts provide a human perspective on the systemic issues that hinder women's land ownership rights.

1.3.1 INDIA: LAXMI'S STRUGGLE FOR INHERITANCE RIGHTS

Laxmi, a woman from a rural village in India, faced significant legal and familial opposition when she attempted to claim her inheritance rights after her father's death. Despite the Hindu Succession Act granting daughters equal rights, her brothers contested her claim, leveraging local customs and societal biases. With the help of a local NGO, she fought a long legal battle to secure her share of the family land (Landesa's website – India). In her own words, Savita Devi (India) says: "The land is in my husband's name, and despite working on it day and night, I have no legal claim. If something happens to him, his family can take it away from me" (International Land Coalition).

1.3.2 KENYA: ESTHER'S FIGHT AGAINST CUSTOMARY LAW

Esther, a widow from a Maasai community in Kenya, faced eviction from her husband's land by his relatives after his death. The customary laws did not recognize her rights to the land, despite statutory laws to the contrary. Esther's case was supported by the Kenya Land Alliance, which helped her navigate the legal system and eventually secure her rights. Also, the interview with Maureen Achieng (Kenya) clearly points out that, "Even though the law says I have a right to my father's land, my brothers and other male relatives see it as theirs by default. They consider any claim by a woman as a challenge to their authority" (Human Rights Watch Report on Kenya, Kenya Land Alliance, Human Rights Watch) (Otieno & Muga, 2024; Federation of Women Lawyers, 2019).

1.3.3 NEPAL: TARA'S BATTLE WITH PATRIARCHAL NORMS

Tara, a single mother from Nepal, was denied her rightful share of her family's property because of deeply entrenched patriarchal norms. She was initially unaware of her legal rights due to lack of education. After attending a workshop organised by a women's rights group, she gained the knowledge and confidence to assert her rights and eventually won her case in court (The International Land Coalition, the Nepalese Women's Rights Organization).

1.3.4 Uganda: Grace's Journey to Land Ownership

Grace, a farmer from Uganda, faced immense difficulties when her husband died, leaving her to care for their children. His family tried to seize the land she had worked on for years. Through the support of ActionAid and local legal aid clinics, Grace was able to challenge the customary practices and secure her land, ensuring her family's livelihood. (ActionAid's reports).

1.3.5 Guatemala: Maria's Fight against Land Grabbing

Maria, an Indigenous woman from Guatemala, faced forced eviction from her ancestral land due to land grabbing by powerful agricultural corporations. Despite facing threats and violence, Maria, supported by Indigenous rights groups and international organisations, led a community movement to reclaim the land. Her story highlights the intersection of gender and Indigenous rights in land ownership struggles. (Rights and Resources Initiative – RRI, and local Indigenous rights groups). "In our community, land is seen as a man's property. Women can work on it but owning it is almost unheard of. Even when we inherit land, our male relatives often try to take control," says Maria Elena (FAO Report on Gender and Land Rights in Latin America).

1.3.6 Spain: Maria's Legal Battle for Property Rights

Maria, a woman from rural Spain, faced significant obstacles in claiming her inheritance after her father passed away. Due to traditional practices, her male relatives tried to prevent her from receiving her fair share of the family property. Maria had to navigate a complex legal system and endure family disputes to assert her rights. Her story highlights the challenges posed by both legal intricacies and familial resistance. (European Women's Lobby and local Spanish women's rights organisations).

1.3.7 Italy: Giulia's Fight against Gender Bias in Inheritance

Giulia, from southern Italy, was denied her inheritance rights based on the traditional preference for male heirs. Her brothers claimed the family estate, sidelining her despite equal legal entitlements under Italian law. With the support of a local women's legal aid organisation, Giulia fought a prolonged legal battle to secure her inheritance. (Differenza Donna). Marta Kowalska from Poland says, "While the laws have changed, the mindset hasn't. Many still believe land should be passed down to sons, not daughters. Women like me often have to fight just to keep what is rightfully ours," (European Parliament Study on Women and Land).

1.3.8 RUSSIA: OLGA'S STRUGGLE WITH BUREAUCRACY AND DISCRIMINATION

Olga, a single mother in Russia, faced bureaucratic red tape and gender discrimination when trying to purchase property in Moscow. Despite having the financial means, she encountered numerous obstacles, including discriminatory practices from real estate agents and mortgage lenders who doubted her ability to manage property on her own. (Russian Association of Women Lawyers).

1.3.9 UNITED STATES: SARAH'S EXPERIENCE WITH HOUSING DISCRIMINATION

Sarah, an African American woman in the United States, faced racial and gender discrimination when attempting to buy a home. Despite having a stable job and good credit, she was offered unfavourable loan terms compared to her white male counterparts. Her determination to secure a home for her family led her to work with organisations like the NAACP to fight for fair treatment (NAACP, and the National Fair Housing Alliance), "Even with the advancements in gender equality, women farmers face systemic challenges, from securing loans to being taken seriously by suppliers and buyers. Owning and managing land as a woman is still an uphill battle," (National Farmers Union Report).

1.3.10 CANADA: EMILY'S CHALLENGE WITH PROPERTY LAWS AS A WIDOW

Emily, from a rural area in Canada, struggled with property laws after her husband's death. While Canadian laws generally support women's property rights, Emily faced issues with joint tenancy and legal complications that almost led to her losing the family farm. With the help of a local women's legal clinic, she was able to navigate the legal system and retain her property (Canadian Women's Legal Education and Action Fund – LEAF).

These examples illustrate that all women in every single country, either in the South or North Hemisphere are subject to unique challenges related to property ownership, influenced by cultural and family norms, legal frameworks, and socio-economic factors. These personal stories do not only illustrate the specific challenges women face in diverse cultural and legal contexts but also highlight their resilience and the crucial role of advocacy and legal support in overcoming these barriers.

Globally, women face a plethora of barriers in land ownership, from legal constraints to deep-seated cultural biases. These challenges undermine not only their economic stability but also their empowerment and equality. Interviews and conversations with women from different continents reveal a common thread: land ownership remains elusive due to gender-based discrimination. Even where laws are favourable, enforcement is weak and social norms prevail. These citations and interviews reflect the ongoing struggle women face in securing land rights, illustrating the need for continued advocacy and reform to address these systemic challenges.

The concept of micro-power (Foucault, 1991), which refers to the subtle, everyday forms of power and influence exercised by individuals or small groups, can be a useful lens through which to discuss land ownership by women, and help to both map and combat the discrimination these women suffer whether buying, inheriting, or managing properties, farm, and real estate. Foucault's concept (1980) of power emphasises the multiple informal ways in which power operates within society, which is particularly relevant to understanding the barriers and opportunities that women face in owning and managing land (Mubangizi & Tlale, 2023).

Gender-based cultural practices on women's property rights have long-term impact on women's ability to own property – particularly land. In certain cultures, they do not only prohibit women's access to property ownership but also limit any property ownership by constitutional rights. Denying women's rights to property under the guise of culture is discriminatory and offends against fundamental and intrinsic human rights (Cyrus Chu, Po-Hsuan, & Wang, 2023).

1.4 THE NEED FOR INTERSECTIONAL WORK AND MULTIPLE AGENCY APPROACH

By examining land ownership through the lens of micro-power, it is possible to understand the subtle ordinary ways in which women navigate, negotiate, and may challenge the structures that limit their land rights. Empowering women through small-scale, often informal forms of power can lead to significant changes in land ownership dynamics and contribute to broader gender equality in property rights.

Women's land ownership should not only be a right, but a chain of incentives towards this right. It is not only a matter of adjusting the law, when the uncountable difficulties are not ripped out from these women's paths. In many places and structures, it creates a paper law that may be equal, but the implementation of that is not possible. This is in a world where women are still being incentivised to look beautiful and get married as means towards having a civilian status quo recognized by its community (Maliniak, Powers, & Walter, 2013).

Learning to own and manage land skills is much needed among school girls so they can grow up knowing their rights, and implementing them. There is no gender independence without the appropriate economical and spatial allocation of these rights. Working on the land and not owning it, does not bring any progress towards gender equality (Mubangizi & Tlale, 2023).

The topics below are all needed if the world wants to seriously address gender equality and discrimination. Yet, it is common to see women being taught to buy consumable products, and not to invest, much less to make a living out of these investments, to properly handle inheritance land, and to buy their own properties. Additionally, Marks and Phillips (2021) point that there are both legal and non-legal barriers to land ownership by women.

1. Inheritance laws: Ensuring that women have equal rights to inherit land and property is fundamental. Legal reforms that guarantee women the right to inherit land from their parents and spouses are essential for promoting gender equality in land ownership.
2. Property laws: Laws that explicitly allow women to own, buy, sell, and manage land independently of their male relatives are crucial. Reforms should

address and eliminate discriminatory legal provisions that restrict women's land rights.

Marks and Phillips (2021), when analysing non-legal barriers to land ownership by women, point numerous aspects such as:

3. Gender perceptions: Shifting cultural and societal norms that view men as primary landowners is necessary. Promoting the idea that women are equally capable and entitled to own and manage land can help change deep-seated biases (Doss, 2002).
4. Education and awareness: Raising awareness about women's rights to land and educating both men and women about gender equality can challenge patriarchal norms and empower women to assert their rights.
5. Access to financial resources: Providing women with access to credit, loans, and other financial services can enable them to purchase and invest in land. Financial institutions should create products tailored to the needs of women.
6. Supportive policies: Government policies that provide subsidies, grants, or tax incentives for women to acquire land can help level the economic playing field.
7. Simplifying procedures: Simplifying the bureaucratic processes involved in land registration and titling can make it easier for women to secure land ownership. Reducing the complexity and cost of these procedures is important.
8. Combating corruption: Addressing corruption and ensuring fair and transparent processes in land administration can help prevent discrimination against women in land transactions.
9. Support networks: Establishing organisations and support networks that advocate for women's land rights can provide women with the necessary legal assistance and representation to fight for their rights.
10. Legal assistance: Providing free or affordable legal aid to women can help them navigate the legal system and secure their land rights.
11. Inclusive decision-making: Ensuring that women are included in community decision-making processes related to land use and management can strengthen their position and influence over land-related matters.
12. Local initiatives: Supporting local initiatives and community-led projects that promote women's land ownership can foster grassroots change and empower women at the community level.
13. Training programs: Offering training programs on land rights, legal literacy, and agricultural skills can empower women to manage land effectively and advocate for their rights.
14. Awareness campaigns: Conducting campaigns to inform women about their legal rights and the benefits of land ownership can motivate them to pursue land ownership.
15. Government commitment: Governments should demonstrate a commitment to gender equality by enacting and enforcing gender-sensitive policies that promote women's land ownership.

16 Monitoring and evaluation: Implementing mechanisms to monitor and evaluate the impact of land policies on women can ensure accountability and continuous improvement.

By addressing these key factors, societies can work towards achieving greater gender equality in land ownership, empowering women, and enhancing their socio-economic status and overall well-being, mostly creating a more equal and sustainable world.

Some international agreements which have called attention to the importance of women's land and property rights are, (1) the Beijing Platform for Action pointing that women's right to land ownership and inheritance should be recognized, and (2) the Convention for the Elimination of All Forms of Discrimination against Women (CEDAW) pursues equal treatment in land, plus agrarian reform to women's full ownership. Women's property rights are also correlated to the global SDG goals, more specifically Goal 1 and Goal 3.[1]

UN Women advocates for women's land and property rights as part of its core strategy to enhance women's economic security and rights and reduce feminised poverty. There is a strong focus on ensuring that women benefit from equal rights to property under the law, as well as in actual practice at the grassroots level (United Nations, 2013).[2]

1.5 ADVANTAGES WHEN WOMEN OWN LAND

The positive implications of women owning land are multifaceted and significant, impacting not only the women themselves but also their families, communities, and broader societal development, even global sustainability. This is from the Contribution of Land Registration and Certification being aligned to the Implementation of the SDGs in the Amhara Region, Ethiopia (Mengesha et al., 2022), to gender and Food Security, like the NuME Project and quality protein maize in Ethiopia (O'Brien et al., 2016).

The fact is that women can not only own land but also manage it as or even more effectively. By being more sustainable, innovative, or productive (Doss, 2018), they add positively to global farming in so many ways.

Here are some key positive implications:

1 Increased income and productivity: Women who own land are more likely to invest in and manage it effectively, leading to higher agricultural productivity and income. This economic empowerment enables women to support their families and contribute to local economies (Quisumbing et al., 1999).
2 Financial security: Land ownership provides women with a valuable asset that can be used as collateral for loans, facilitating access to credit and enabling further investment in agriculture or other income-generating activities (Deere & Doss, 2006).
3 Enhanced agricultural output: Women who own land are more likely to grow diverse and nutritious crops, improving food security and dietary diversity for their households (Agarwal, 1994).

4 Sustainable farming practices: Women tend to employ sustainable farming practices that preserve soil fertility and biodiversity, contributing to long-term food security (Meinzen-Dick et al., 2011).
5 Increased decision-making power: Land ownership enhances women's status and bargaining power within their households and communities, leading to greater participation in decision-making processes (Agarwal, 1997).
6 Reduction in gender-based violence: Economic independence through land ownership can reduce women's vulnerability to gender-based violence and provide them with more options to leave abusive relationships (Panda & Agarwal, 2005).
7 Better health: With increased income from land, women can afford better healthcare for themselves and their families, leading to improved health outcomes (Doss, 2001).
8 Educational opportunities: Higher income also enables women to invest in their children's education, leading to better educational outcomes and breaking the cycle of poverty (Katz & Chamorro, 2003).
9 Stronger communities: Women landowners tend to invest in their communities, supporting local infrastructure, health, and educational initiatives, which benefits the broader community (Lastarria-Cornhiel, 1997).
10 Social cohesion: Empowering women through land ownership can foster social cohesion and stability, as empowered women are often leaders in community development efforts (Yngstrom, 2002).
11 Sustainable land management: Women are often stewards of the environment, and land ownership can enable them to implement and maintain sustainable land management practices that benefit the environment (Meinzen-Dick et al., 2011).
12 Climate resilience: Women landowners can adopt and promote climate-resilient agricultural practices, contributing to greater resilience against climate change impacts (FAO, 2011).

1.6 CONCLUSION

Gender gaps in land ownership are common across the world, favouring male-headed farmland and households. In countries, where women lack property, inheritance, and credit rights, these disparities contribute even further to gender inequality. Our findings highlight the consequences of one-gender biassed land property ownership of land. Considering gender disparities in crafting agricultural and climate change strategies, it is crucial for governments, civil society, and international organisations to continue their collaborative efforts to address gender disparities and ensure women's equal access towards land ownership.

When analysing gender differences in property and inheritance law, progress in closing gender gaps in laws related to assets has been among the slowest. Although land ownership either individually, collectively, or corporate based has been under discussion through multiple lens and perspectives, the gender gap is still a prevalent one. If the years ahead are meant to be more sustainable and equally shared, this cannot be achieved without the full access of land to women.

As Agarwal (1994: 1455) points out, "the gender gap in the ownership and control of property is the single most critical contributor to the gender gap in economic well-being, social status, and empowerment."

EXTRA

Here are some highly recommended documentaries that explore the issues surrounding women and land ownership, highlighting the challenges and triumphs women face in securing land rights:

This Land is Our Land: The Fight for Land Rights in Kenya (2016). This documentary by the International Land Coalition explores the struggle of women in Kenya to secure their land rights. It highlights the stories of women who have fought against customary laws and traditions to gain legal recognition of their land ownership. Available at: International Land Coalition website

Land Rush (2012). Part of the "Why Poverty?" series, *Land Rush* examines the impacts of land grabbing in Mali, focusing on the effects on local communities, including women. It provides a broader perspective on how global land deals affect women's access to land. Available at: Why Poverty? website

Women and Land: Challenges and Opportunities in Africa"(2015). This documentary produced by the Food and Agriculture Organization (FAO) delves into the struggles and progress of women in various African countries as they fight for their land rights. It highlights the impact of land ownership on women's empowerment and community development. Available at: FAO website.

A Place of Her Own (2013). Directed by Deepa Dhanraj, this film focuses on the lives of women in rural India who fight for their land rights against social and legal challenges. It sheds light on the intersection of gender, caste, and land ownership in India. Available at: Film festivals archives.

Gaining Ground: Women and Land in the Nile Delta (2010). This documentary by the International Center for Agricultural Research in the Dry Areas (ICARDA) explores the gender dynamics of land ownership in Egypt's Nile Delta. It highlights women's efforts to claim land and the socio-economic benefits that follow. Available at: ICARDA website

These documentaries provide insightful and powerful narratives that bring attention to the critical issues of land ownership and gender inequality, offering a deeper understanding of the obstacles and opportunities women face in various parts of the world.

Final statement: there has been no financial support to this chapter. There is no conflict of interest.

NOTES

1 https://asiapacific.unwomen.org/en/focus-areas/women-poverty-economics/women-s-land-property-rights United Nations. (2013). Realising women's rights to land and other productive resources.
2 United Nations. (2013). Realising women's rights to land and other productive resources www.ohchr.org/sites/default/files/RealizingWomensRightstoLand_2ndedition.pdf

BIBLIOGRAPHY

Agarwal, B. (1994). *A Field of One's Own: Gender and Land Rights in South Asia*. Cambridge: Cambridge University Press.

Agarwal, B. (1997). "Bargaining" and gender relations: Within and beyond the household. *Feminist Economics*, 3(1), 1–51.

Agarwal, B. (2010). Gender and land rights revisited: Exploring new prospects via the state, family, and market. *Journal of Agrarian Change*, 10(1), 184–224.

Ajala, T. (2017). Gender discrimination in land ownership and the alleviation of women's poverty in Nigeria: A call for new equities. *International Journal of Discrimination and the Law*, 17(1), 51–66. https://doi.org/10.1177/1358229117700028

Chowdhry, P. (2017). *Understanding Women's Land Rights: Gender Discrimination in Ownership*. California: Sage.

Mengesha, A. K., Mansberger, R., Damyanovic, D., Agegnehu, S. K., & Stoeglehner, G. (2022). The contribution of land registration and certification program to implement SDGs: The case of the Amhara Region, Ethiopia. *Land*, 12(1), 93.

Chu, C. Y. C., Hsu, P.-H., & Wang, Y.-T. (2023). The gender gap in the ownership of promising land. *PNAS*, 120(24), e2300189120. https://doi.org/10.1073/pnas.2300189120

Deere, C. D., & Doss, C. R. (2006). The gender asset gap: What do we know and why does it matter? *Feminist Economics*, 12(1–2), 1–50.

Deere, C. D., & Doss, C. R. (eds.). (2019). *The Gender Asset Gap: Land in Latin America*. Oxford: Oxford University Press.

Doss, C. R. (2001). Designing agricultural technology for African women farmers: Lessons from 25 years of experience. *World Development*, 29(12), 2075–2092.

Doss, C. R. (2002). Men's crops? Women's crops? The gender patterns of cropping in Ghana. *World Development*, 30(11), 1987–2000. https://doi.org/10.1016/S0305-750X(02)00109-2

Doss, C. R. (2018). Women and agricultural productivity: Reframing the issues. *Development Policy Review*, 36(1), 35–50. https://doi.org/10.1111/dpr.12243

FAO. (2011). *The State of Food and Agriculture 2010-2011: Women in Agriculture: Closing the Gender Gap for Development*. Rome: Food and Agriculture Organization of the United Nations.

FAO. (2018). *The Gender Gap in Land Rights*. Rome: Food and Agriculture Organization.

Federation of Women Lawyers, K. (2019). Women's land and property. *Women's Land and Property Rights in Kenya*, 68(c), 1–35. www.fidakenya.org

Food and Agriculture Organization of the United Nations. (2011). Voluntary guidelines on the responsible governance of tenure of land, fisheries, and forests in the context of national food security. www.fao.org/3/a-ia1689e.pdf

Foucault, M. (1991). *Discipline and Punish. The Birth of the Prison*. New York: Random House.

Foucault, M. (1980). *Power/Knowledge*. Selected Interviews and Other Writings. 1972-1977. New York: Pantheon.

Gaddis, I., Lahoti, R., & Swaminathan, H. (2022). Women's legal rights and gender gaps in property ownership in developing countries. *Population and Development Review*, 48, 331–377. https://doi.org/10.1111/padr.12493

Grabe, S. (2014). Land ownership and gender. In: T. Teo (ed.) *Encyclopedia of Critical Psychology*. New York: Springer. https://doi.org/10.1007/978-1-4614-5583-7_536

Gray, L., & Kevane, M. (1999). Diminished access, diverted exclusion: Women and land tenure in Sub-Saharan Africa. *African Studies Review*, 42(2), 15–39. https://doi.org/10.2307/525363

Joireman S. F. (2008). The mystery of capital formation in sub-Saharan Africa: Women, property rights and customary law. *World Development*, 36(7), 1233–1246. https://doi.org/10.1016/j.worlddev.2007.06.017

Katz, E., & Chamorro, J. S. (2003). Gender, land rights, and the household economy in rural Nicaragua and Honduras. Madison, WI: USAID/BASIS CRSP. *Paper Presented at the* International Research Workshop on Gender and Collective Action, *October 17–21, 2003*, Chiang Mai, Thailand.

King, E., & Mason, A. (2001). *Engendering Development: Through Gender Equality in Rights, Resources, and Voice*. Washington, DC: The World Bank.

Landesa. (2016). *Law of the Land: Women's Rights to Land, Infographic by Centre for Women Land Rights*. Landesa. www.landesa.org/resources/property-not-poverty/

Lastarria-Cornhiel, S. (1997). Impact of privatization on gender and property rights in Africa. *World Development*, 25(8), 1317–1333.

Maliniak, D., Powers, R., & Walter, B. F. (2013). The gender citation gap in International relations. *International Organization*, 67(4), 889–922. https://doi.org/10.1017/S0020818313000209

Marks, K., & Phillips, R. (2021). *Analysing Non-Legal Barriers to Land Ownership by Women* (pp. 100–112). CABI. https://doi.org/10.1079/9781789247664.0009

Meinzen-Dick, R., Quisumbing, A., Behrman, J., Biermayr-Jenzano, P., Wilde, V., Noordeloos, M., ... & Beintema, N. (2011*). Engendering Agricultural Research, Development, and Extension*. Washington, DC: International Food Policy Research Institute (IFPRI),.

Mishra, K., & Sam, A. G. (2016). Does women's land ownership promote their empowerment? Empirical evidence from Nepal. *World Development*, 78, 360–371.

Mubangizi, J. C., & Tlale, M. T. (2023). How gender-based cultural practices violate women's property rights and inhibit property ownership: A South African perspective. *Women's Studies International Forum*, 96, 102678. https://doi.org/10.1016/j.wsif.2023.102678

NCAER. (2020). Land records and services index (N-LRSI) report. www.ncaer.org

O'Brien, C., Gunaratna, N. S., Gebreselassie, K., Gitonga, Z. M., Tsegaye, M., & De Groote, H. (2016). Gender as a cross-cutting issue in food security: The NuME Project and quality protein maize in Ethiopia. *World Medical and Health Policy*, 8(3), 263–286. https://doi.org/10.1002/wmh3.198

Otieno, B. A., & Muga, G. O. (2024). Sociocultural and economic barriers to land ownership among widows in rural Western Kenya. *Journal of Land and Rural Studies*, 12(1), 43–60. https://doi.org/10.1177/23210249231196396

Panda, P., & Agarwal, B. (2005). Marital violence, human development and women's property status in India. *World Development*, 33(5), 823–850.

UN Women. (2013). Progress of the world's women 2011-2012: In pursuit of justice, New York www.unwomen.org/en/digital-library/progress-of-the-worlds-women/2011-2012

Yngstrom, I. (2002). Women, wives, and land rights in Africa: Situating gender beyond the household in the debate over land policy and changing tenure systems. *Oxford Development Studies*, 30(1), 21–40.

Zvox Kuomba, K., & Batisai, K. (2020). Veracity of women's land ownership in the aftermath of land redistribution in Zimbabwe: The limits of western feminism. *Agenda*, 34(1), 151–158.

2 Psychological Well-Being of Self-Employed Women
European Perspective

Tomasz Skica and Katarzyna Miszczyńska

2.1 INTRODUCTION

Self-employment, defined as the engagement of employers, self-employed individuals, co-operative producers, and unpaid family workers (OECD, 2023; Euro-found, 2013), constitutes a simple form of economic activity (Skica, 2020) and is the subject of numerous research studies (Arum & Müller, 2009; Dawson et al., 2009; Le, 1999), conducted not only in developing countries (Gindling & Newhouse, 2014) but also in developed countries (Blanchflower, 2000). An increasing amount of attention is also being paid to studies on the factors distinguishing self-employment in terms of gender (Georgellis & Wall, 2005; Boden, 1996). These studies include aspects related to migration (Apitzsch & Kontos, 2003), race and education level (Fairlie, 2005), and gender discrimination (Rosti & Chelli, 2005). However, the topic of the psychological well-being of self-employed individuals has not been adequately researched, particularly in terms of differences between men and women. This issue requires special attention as the social roles of men and women differ from each other (Caputo & Dolinsky, 1998). Similarly, in most cases, the distribution of domestic chores between men and women presents itself differently (Craig, 2012), which does not remain without influence on the motivation and ability of women to engage in entrepreneurial activities (Kalenkoski & Pabilonia, 2022). Self-employment (among the various options for entering the world of business) provides a space for women who wish (or sometimes have to – due to a lack of alternative in the form of permanent employment) to explore entrepreneurial activities, reconciling them with domestic obligations (Sappleton & Lourenço, 2016). The flexibility of this form of entrepreneurship allows women to manage their time effectively to combine both roles (social and economic) (Scandura & Lankau, 1997).

The key question that remains open concerns the psychological well-being of self-employed women. On the one hand, this relates to the expectations of women with regard to this form of economic activity (Lawler, 1973), and on the other hand, it pertains to the consequences (particularly health-wise) of combining domestic responsibilities with self-employment (Gevaert et al., 2018). Bearing in mind that men and women have different expectations with regard to this form of economic activity (Clement, 1987), and at the same time knowing that, in the case of women, business relations are rather "integrated" rather than separate from family, social, and personal relationships (Brush, 1992), the consequences of engaging in economic activity will be reflected in the health of female entrepreneurs. This text aims to shed some light on this issue. It therefore contributes to the ongoing academic debate on the psychological well-being of self-employed women, on the one hand highlighting the factors that determine a woman's decision to engage in self-employment, and on the other, pointing to the differences in expectations regarding their own economic activity with respect to each gender separately. In order to provide as comprehensive a picture as possible of the link between self-employment and psychological well-being, issues such as the relationships of self-employed women with clients (including dissatisfied ones) and employed workers, business decision-making motives and risk-bearing capacity, and exposure to emotionally challenging situations and stress resulting from running a business will be analyzed. A thorough examination of these issues will not only answer questions about the specificity of self-employment as a form of business activity but also about its impact on the psychological well-being of self-employed women.

2.2 LITERATURE REVIEW

Studies suggest that self-employed women encounter distinct challenges in terms of their psychological well-being when compared to male self-employed individuals and women in salaried employment. Research indicates that these women exhibit lower levels of well-being than their male counterparts and women in wage-employment, often experiencing reduced levels of health and life satisfaction (Xiu & Ren, 2022). Self-employment typically fosters greater eudaimonic well-being, which encompasses autonomy and competence, as opposed to traditional employment, largely attributed to the problem-focused coping strategies employed by the self-employed (Nikolaev, 2022). Moreover, employed women, including those who are self-employed, tend to exhibit higher levels of social and environmental quality of life, as well as psychological well-being domains such as environmental mastery and positive relations with others, compared to non-working women (Kapur, 2023). Moreover, factors such as health anxiety, perceived stress, and self-handicapping can significantly impact the psychological well-being of working women, thereby underscoring the importance of addressing these factors for overall well-being (Esmaelli et al., 2020). Finally, credit counselling has been demonstrated to positively influence the psychological well-being of women entrepreneurs, thereby highlighting the significance of financial support and guidance in enhancing well-being among self-employed women (Ndunda, 2020).

The topic of psychological well-being in women is relevant for several reasons (Nawaz et al., 2024). First, studies show that women frequently experience unique challenges and social pressures that can affect their mental health (Rooney et al., 2003). Second, the changing role of women in society and the labor market necessitates an understanding of how these changes impact their psychological well-being (Gonäs & Tyrkkö, 2015). This is clearly illustrated in Gonäs and Tyrkkö's (2015) analysis of the Nordic countries. Additionally, issues related to gender equality and the fight against discrimination have a significant impact on women's mental health (Hosang & Bhui, 2018). The following studies on psychological well-being in women indicate the growing importance and significance of this topic. Moreover, promoting women's mental health has the potential to create more balanced and fulfilling societies, which is a crucial argument for investigating and supporting women's well-being and demonstrating the need for health services to take steps to strengthen gender awareness (Department of Health, 2022).

Although Kessler et al. suggested that mental health issues impact men and women equally (Kessler et al., 2005), they underlined that some conditions, such as major depression and anxiety disorders, show a greater prevalence in women (Seedat et al., 2009). Furthermore, females experience a higher burden of mental illness according to disability-adjusted life years than males (Whiteford et al., 2013).

The literature review presented above underscores the relevance of the subject matter while simultaneously emphasizing its continued openness. Consequently, there remains scope for further research into the diversity of approaches taken by employed women and men in relation to psychological well-being.

2.3 DATA AND METHODS

2.3.1 DATA

The study of the psychological well-being of self-employed females in Europe was conducted using statistical data obtained from Eurofound (EWCTS Eurofound, 2024). The research included over 44,000 individuals in each weaves (5th, 6t^h, and 7th) of the European Working Conditions Survey (EWCS 2010 and 2015) and the European Working Conditions Telephone Survey (EWCTS Eurofound, 2024). The analysis was focused on women who operated as self-employed. The results were compiled with the results for men. The data used in the analysis originated from:

- EWCS 2010 conducted across 31 European countries, including those that are part of the European Union (UE-27) as well as Norway, Croatia, the former Yugoslav Republic of Macedonia, Turkey, Albania, Montenegro, and Kosovo.
- EWCS 2015, which was expanded to 35 European countries, including those that were part of the European Union (UE-28), five candidate countries (Albania, the former Yugoslav Republic of Macedonia, Montenegro, Serbia, and Turkey), as well as Switzerland and Norway.
- EWCS 2021, the study that conducted in 36 European countries, including those that were part of the European Union (UE-27), as well as the United Kingdom, Norway, Switzerland, Albania, Bosnia and Herzegovina, Kosovo, Montenegro, North Macedonia, and Serbia.

Non-parametric tests of significance of differences were used to verify the relationships tested in the study. Depending on the nature of the analyzed variables, these were the

Mann Whitney U-test, chi-square test or Fisher's exact test. The hypotheses posed within the framework of these tests were as follows:

$$H_0 : \mu_1 = \mu_2$$
$$H_0 : \mu_1 \neq \mu_2$$

In addition, in some cases, correlation analysis (Pearson's correlation coefficient and Spearman's correlation) was used for analysis.

2.4 RESEARCH RESULTS

2.4.1 SELF-EMPLOYMENT AND GENDER

The relationship between employment status and gender is prominent and statistically significant, as confirmed by Spearman's rank correlation coefficient. In all years analyzed, the correlation was negative, that is, it confirmed the correlation that the self-employed group is dominated by men. Over the years spanning from 2010 to 2021, the percentage of self-employed women was noticeably lower than that of men in the workforce (see Figure 2.1).

As illustrated in Figure 2.1, there is a clear downward trend in the share of self-employed individuals in both genders in 2021 compared to 2015. This trend can be attributed to the occurrence of the COVID-19 pandemic and the tendency of women to engage in self-employment. The final fact confirms the results of the research conducted by GEM (2021). Considering that women in recent years have shown a stronger tendency towards engaging in self-employment than men (GEM, 2022), the observed decrease in self-employment among women (28.8 percentage points) is weaker than that among men (45.6 percentage points).

2.4.2 PSYCHOLOGICAL WELL-BEING OF SELF-EMPLOYED WOMEN

The statistical analysis of the psychological well-being of self-employed women was conducted according to group of factors affecting their well-being. The analysis was

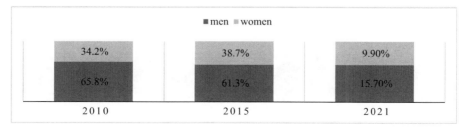

FIGURE 2.1 Self-employed women and men between years 2010, 2015 and 2021. (Own elaboration based on Eurofound database.)

conducted in two stages. In the first stage, self-employment was analyzed from the perspective of gender, age, the number of employees in their companies, the number of clients they were dealing with, and personal preferences of being self-employed. In the second stage, the analysis was deepened to include aspects of psychological well-being, which were also related to gender. We have endeavored to compare the trends in self-employment among women with those of men to provide a more comprehensive analysis. This approach has also enabled us to make more detailed comparisons with socio-economic phenomena that have occurred over the years under the study.

To begin the analysis, it is worth examining how the age distribution among self-employed women and men has evolved (see Figure 2.2). In both men and women, the highest percentage of self-employed individuals was found in the age range of 45–55 years in all analyzed years. Results are consistent with Carrasco's (1999) findings, which indicate that the probability of self-employment is higher among middle-aged individuals and significantly lower among those aged 55 and over. One explanation is that older individuals face problems with accumulating capital necessary to start a business (Blanchflower & Oswald, 1991; Evans & Leighton, 1989). The obtained result also reflects demographic changes, evident in the clear increase in the percentage of older individuals (aged 55 and over) in the labor force, making self-employment as a possible career path available to a broader spectrum of the aging population (Minola et al., 2016). Research on individual orientation towards personal development (in the entire life span) shows that people tend to orient their life decisions towards personal growth in middle adulthood. It plays a less significant role in early adulthood and late adulthood (Ebner et al., 2006). Moreover, the lifespan curve of self-employment motivation (Super, 1980) defines young adulthood as a period of search, middle adulthood as a period of growth and maintenance, and later stages as a period of decline. Since middle adulthood is a phase of both personal development and starting a business, self-employment is best suited to this developmental stage.

Interestingly, in 2021, the difference in the participation of women and men in the age range of 45–55 years was 3.8 percentage points in favor of women. Consequently, the percentage of self-employed women in 2021 in the age range of 45–55 years was

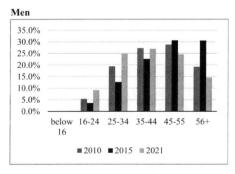
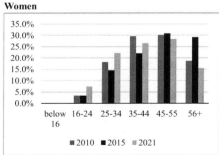

FIGURE 2.2 The share of working age self-employed people in Europe [years 2010, 2015, 2021]. (Own elaboration based on Eurofound database.)

28.4%. Research indicates that although women's participation in the labor market was more severely impacted during the pandemic, it has rebounded at a faster pace than that of men (Botelho & Neves, 2021). Women opted for self-employment to gain greater flexibility in scheduling professional and family obligations (Kalenkoski & Pabilonia, 2022). Simultaneously, industries that were particularly hard-hit by the COVID-19 crisis and that feature a higher proportion of employed women were also forced to switch to self-employment (Graeber et al., 2021).

2.4.3 Self-employment: Employees and Customer Relations

As one examines the realm of business management, it is essential to consider two questions presented in the EWCS database. The first query pertains to the presence of employees within the organization, while the second inquiries about the existence of multiple clients or customers. In the case of the first question, it is worth noting a significant increase in the share of self-employed women who hired employees in their company. This behavior cannot be confirmed among men. Despite the lack of data for 2010, it can be stated that women were more inclined to hire employees during the pandemic (see Table 2.1).

Among self-employed women, the participation of those who have more than one client decreased by 1.7 percentage points from 2015 to 2021. This behavior cannot be confirmed in the case of men, where there was an increase of 1.1 percentage points. The result reflects the specific industry characteristics (Graeber et al., 2021). Cosmetic services, hairdressing, and other personal services are areas of economic activity dominated by women. Considering that these industries have been subject to the most extensive restrictions, women in the observed period were characterized by a higher rate of leaving self-employment than men (Parker, 2018). This is further confirmed by the findings of Abubakar et al. (2023), which indicate that compared to men, hostile COVID-19 lock-down policies are more likely to have negative effects on women's self-employment rates. The proposed position is also consistent with the results of Table 2.2, which shows that the self-employment of women and men have different characteristics (Kowalska, 2023; Georgellis & Wall, 2005). The COVID-19 pandemic has only exacerbated these differences (OECD/European Commission, 2021).When asked whether you deal with more than one customer, more than half of the men answered in the affirmative. There was a downward trend from 2010 to 2015, while in 2021 the percentages were significantly higher than in 2015. The percentage

TABLE 2.1
The share of self-employed who have employees (years 2015 and 2021)

Characteristics		2015 Yes	2015 No	2021 Yes	2021 No
Sex	Men	70.30%	59.10%	71.40%	58.80%
	Women	29.70%	40.90%	28.20%	40.90%

Source: Own elaboration based on Eurofound database.

TABLE 2.2
Regarding your business, do you generally, have more than one client or customer? (years 2010, 2015 and 2021)

Characteristics	2010		2015		2021	
	Men	Women	Men	Women	Men	Women
Yes	64.0%	36.0%	63.5%	36.5%	63.8%	35.9%
No	60.8%	39.2%	55.5%	44.5%	60.5%	39.2%

Source: Own elaboration based on Eurofound database.

of affirmative responses from women increased between 2010 and 2015, only to decrease again in 2021, in contrast to the findings among men.

The impact of the pandemic has been more pronounced on self-employed women, particularly those engaged in healthcare and social assistance services (PARP, 2011, p. 24). Women are most likely to establish service companies within these sectors. The COVID-19 restrictions have sequentially limited direct human interactions, reducing the scale and scope of services offered, as evidenced by the data in Table 2.2.

2.4.4 Self-employment Motivation

The aspect of employee motivations for self-employment was covered only by data from EWCS study (from 2015). The analysis showed that, both in the case of women and men, it was dictated by individual preferences. However, an alarming 32.3% of women stated that they had no other alternatives, while among men, this percentage was 24.1% (see Figure 2.3 and Table 2.3).

Analyzing the data compiled in the form of cross tabulations, it should be noted that women are more likely than men to take up self-employment out of so-called necessity arising from the lack of other alternatives. It is also confirmed by the literature (Minniti et al., 2006). What is more, based on GEM (2006), 44% of women entrepreneurs worldwide choose to become entrepreneurs due to financial necessity, compared to 31% of men. In this study, in the 2015 wave, for 32.1% of women it was a necessity to start a business, while for men it was only 24.1%. We also noted that it is more often men than women who choose this form of work from the perspective of their own beliefs (in 2015 it was a ratio of 65.8% to 34.2% for men). Such a situation is also confirmed in the literature and was supported by studies Bergmann and Sternberg (2007), Giacomin et al. (2007) and Wagner (2005). The situation stating that women are more likely to choose self-employment out of necessity and for men it is more of an opportunity was checked by Spearman rank correlation (see Table 2.4). The investigation allowed us to confirm the assumption that there is statistically significant a relationship between gender and the motive for deciding to become self-employed. Research has shown (confirming the results of other studies) that women choose self-employment out of necessity, and men choose out of personal preference.

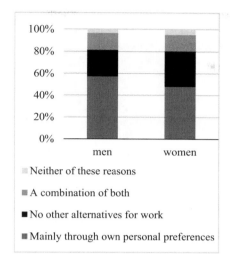

FIGURE 2.3 Personal preferences of being self-employed [year 2015]. (Own elaboration based on Eurofound database.)

TABLE 2.3
Self-employment, was it your own personal preference or you had no better alternatives for work? (year 2015)

Characteristics		Sex		Total
		Men	Women	
Self-employed, was it your own personal preference or you had no better alternatives for work?	Mainly through own personal preferences	57.2%	47.8%	53.6%
	No other alternatives for work	24.1%	32.1%	27.2%
	A combination of both	15.2%	14.7%	15.0%
	Neither of these reasons	3.5%	5.4%	4.2%
Total		**100.0%**	**100.0%**	**100.0%**

Characteristics		Sex		Total
		Men	Women	
Self-employed, was it your own personal preference or you had no better alternatives for work?	Mainly through own personal preferences	65.8%	34.2%	100.0%
	No other alternatives for work	54.8%	45.2%	100.0%
	A combination of both	62.4%	37.6%	100.0%
	Neither of these reasons	51.4%	48.6%	100.0%
Total		**61,7%**	**38.3%**	**100.0%**

Source: Own elaboration based on Eurofound database.

TABLE 2.4
Spearman rank correlation results (year 2015)

Characteristics			Self-employed, was it your own personal preference or you had no better alternatives for work?
rho Spearman's	Sex	Correlation coefficient	.108**
		Significance (two-sided)	<0.001
		N	6169

Source: Own elaboration based on Eurofound database.

2.4.5 Psychological Well-being of Self-employed Women

The psychosocial well-being of self-employed women is frequently discussed in literature from various perspectives. In our study, we decided to focus on the aspects related to, on the one hand, the realities of running one's own business (see McGowan et al., 2012; Crainer & Dearlove, 2000), and on the other hand, psychological well-being (see Gevaert et al., 2018). Table 2.5 presents the questions upon which further analysis were conducted.

2.4.5.1 Day-to-Day Functioning of Own Business

Day-to-day business functioning will be analyzed through five different aspects corresponding with questions provided in Table 2.5. Taking into account long-term illness, a greater proportion of women (over 55% of women) than men (over 50% of men) feel financial insecurity in such a situation. Among women, only 11.6% felt financially secure, which is 3.2 percentage points less than among men (see Figure 2.4 and Table 2.6). Studies show that financial uncertainty is more frequently experienced by self-employed individuals who start their own businesses. The reason is the lack of an alternative to self-employment (Tammelin, 2019). Moreover, work–family conflicts are linked to long-term sickness absence in the working population, in a gender-specific manner. In the case of men, excessive work obligations and, in the case of women, extensive family obligations disrupt the balance between work and family and increase the risk of long-term sick leave (Lidwall et al., 2010).

Research has indicated that individuals who experience financially induced uncertainty as a result of long-term illness are more likely to be those who made their business decision voluntarily. Conversely, those who started a business due to a lack of alternative options (e.g., lack of job opportunities) are significantly less likely to report that their long-term illness would not have led to financial uncertainty (Spasova & Wilkens, 2018). The reasoning behind these findings is quite clear. Individuals who start a business due to current financial uncertainty resulting from a lack of income sources (similar arguments are introduced by Dencker et al., 2021) are driven to find a way to maintain their livelihood by engaging in entrepreneurial activities. Almost 30% of necessity entrepreneurs indicate a negative impact of prolonged absences on their financial situation, while only around 15% claim to be financially secure in the

Psychological Well-Being of Self-Employed Women

TABLE 2.5
Questions used in the study

Perspective	Questions
Day-to-day functioning of own business	1. To what extent you agree or disagree with the following statements? a) I make the most important decisions on how the business is run. b) I enjoy being my own boss c) It is easy for me to find new customers d) I find it hard bearing the responsibility of running my business e) I make the most important decisions on how the business is run
Psychological well-being	2. Does your main paid job involve: a) Dealing directly with people who are not employees at your workplace? b) Handling angry clients, customers, patients, pupils etc.? c) Being in situations that are emotionally disturbing for you? 3. Last month, has it happened at least once that you had less than 11 hours between 2 working days? [Yes, No] 4. Please indicate for each of the five statements which is the closest to how you have been feeling over the last two weeks: a) I have felt cheerful and in good spirits b) I have felt calm and relaxed c) I have felt active and vigorous d) I woke up feeling fresh and rested e) My daily life has been filled with things that interest me 5. How satisfied are you with working conditions in your main paid job? [Very satisfied, Satisfied, Not very satisfied or Not at all satisfied]

Source: Eurofound database.

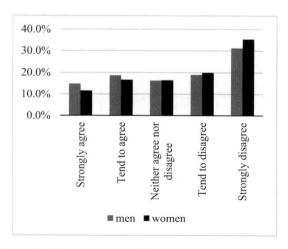

FIGURE 2.4 Share of self-employed women and men confirming that they are financially secure in the event of a long-term sickness [year 2015]. (Own elaboration based on Eurofound database.)

TABLE 2.6
Self-employment, was it your own personal preference or you had no better alternatives for work? Vs. Agree or disagree: If I had a long term sickness, I would be financially secure? (year 2015)

Characteristics		Agree or disagree – If I had a long term sickness, I would be financially secure?		
		Yes	No	Total
Self-employment, was it your own personal preference or you had no better alternatives for work?	Mainly through own personal preferences	66.8%	48.0%	53.9%
	No other alternatives for work	14.8%	32.9%	27.2%
	A combination of both	13.2%	16.0%	15.1%
	Neither of these reasons	5.3%	3.1%	3.8%
Total		**100.0%**	**100.0%**	**100.0%**

Source: Own elaboration based on Eurofound database.

event of a long illness. On the other hand, individuals who view their business entry decision as a chance, usually have already accumulated a certain capital base that will help them survive the period of ongoing illness-related absences. Therefore, almost 70% of them feel financially secure, even in the face of health-related risks. This stability also results from the easier accessibility (if necessary) to external financing (Frid et al., 2016). Generally, both men and women are satisfied or very satisfied with running their own businesses (see Figure 2.5). Consistent with empirical research findings, according to which self-employed individuals enjoy greater job satisfaction than those employed on a salary (Benz & Frey, 2004; Naughton, 1987). Presenting the answer to this question in the Y/N formula, we note that it is more often men than women who are satisfied with being self-employed. On the other hand, the difference is small. Testing whether there are differences in perceived satisfaction among women and men, a chi-square test was used (see Table 2.7). Based on the formation of the responses, it was hypothesized that there are differences in the perception of satisfaction among men and women. In addition, that satisfaction with being one's own boss is higher among men than among women.

Based on the test of significance of differences, a p-value <0.01 was obtained, which means that we reject hypothesis H0 in favor of hypothesis H1. Thus, there are differences in satisfaction among men and women, and men are more often satisfied than women. The interpretation of the obtained result should be approached with caution, as the difference in the results obtained is relatively small. First, previous studies suggest that the difference in job satisfaction between self-employed men and women is not due so much to their gender, as to the composition of the sector in terms of gender (Sappleton & Lourenço, 2016). Second, there is evidence to suggest that, while higher pay increases job satisfaction among men, it does not do

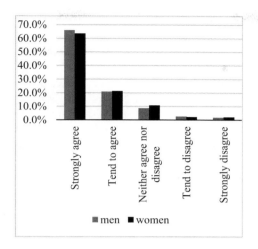

FIGURE 2.5 Share of self-employed women and men who enjoy being their own boss [year 2015]. (Own elaboration based on Eurofound database.)

TABLE 2.7
Agree or disagree: I enjoy being my own boss? (year 2015)

Characteristics		Agree or disagree: I enjoy being my own boss?		Total
		Yes	No	
Sex	Men	64.8%	59.7%	64.2%
	Women	35.2%	40.3%	35.8%
Total		**100.0%**	**100.0%**	**100.0%**

Characteristics	Agree or disagree: I enjoy being my own boss?	
N	14336	
Chi-kwadrat	17,546	
Asymptotic significance	<.001	
Yates' fix for continuity	Chi-square	17,322
	df	1
	Asymptotic significance	<.001

Grouping variable: Sex

Source: Own elaboration based on Eurofound database.

so among women (Sloane & Williams, 2000). Instead, higher job satisfaction among self-employed women arises from their lower expectations of the job market (Lawler, 1973), and the nature of the work they perform (Sappleton & Lourenço, 2016), as well as the flexibility of their hours of work (Kuranga, 2020; Scandura & Lankau, 1997), shorter working hours (Clark, 1997), and greater autonomy (Benz & Frey,

2004). Therefore, the reasons for job satisfaction among both genders are different, and their unequivocal explanation requires further testing. Further research aimed at understanding this situation could also take into account the sector of the economy and the nature of the work being performed.

Only 45% of all women and 47.6% of all men believe that it will be easy for them to find new clients. Interestingly, almost 30% of both women and men do not have an opinion on the matter (see Figure 2.6). To test whether there are statistically significant differences in perceptions of the ease of acquiring new customers, a Mann Whitney U-test of significance of differences was conducted (see Table 2.8).

Based on the results of the Mann Whitney U-test, where p=0.026 and was greater than the assumed significance level of alpha=0.05, there was no basis for rejecting hypothesis H0. Thus, there are no differences in the difficulty or ease of acquiring new customers among men and women. The outcome confirms the position of Kundera (2022) and the results of PARP (2011). The cited works demonstrate that acquiring new clients for business owners is a challenge of equal difficulty for both men and women.

Both men and women tend to agree or strongly agree that it is easy to run their own business, in terms of responsibility (see Figures 2.7–2.8). Consequently, they are not generally deterred by the responsibility associated with self-employment. In the case

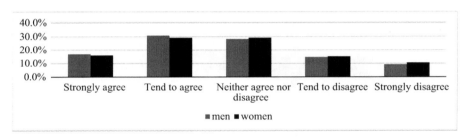

FIGURE 2.6 Percentage of self-employed women and men who agree with the statement that it is easy for them to acquire new clients [year 2015]. (Own elaboration based on Eurofound database.)

TABLE 2.8
Agree or disagree: It is easy for me to find new customers? (year 2015)

Characteristics	Agree or disagree: It is easy for me to find new customers?
Mann Whitney U	5429104.500
W Wilcoxon	14883830.500
Z	-2.222
Asymptotic significance (two-sided)	.026
Grouping variable: Sex	

Source: Own elaboration based on Eurofound database.

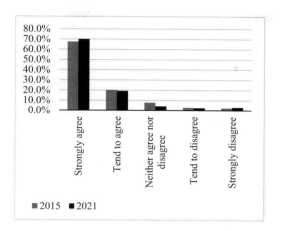

FIGURE 2.7 The share of self-employed men who find it difficult to bear the responsibility of running their own business [year 2015, 2021]. (Own elaboration based on Eurofound database.)

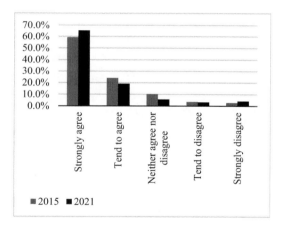

FIGURE 2.8 The share of self-employed women who find it difficult to bear the responsibility of running their own business [year 2015, 2021]. (Own elaboration based on Eurofound database.)

of women, an increasing trend was observed in the proportion of these responses, which could not be said about men in 2015.

According to Zaheer and Hussain's (2015) perspective, women with a defensive character cannot play an offensive game, and therefore are not capable of effectively taking responsibility for business. This view is rooted in the existing literature's stance that women are averse to risk, which transfers to business (see Buratti et al., 2017; Nelson, 2015; Brindley, 2005). The obtained result may suggest differences in the mechanisms observed in the case of self-employment and complex forms of economic activity such as limited liability companies (Skica, 2020). Given that the size of the enterprise and the accompanying change in its organizational form affect the way the company is managed, individual personalities of entrepreneurs (appropriate for self-employment and different for men and women) affecting the way the company is managed are replaced by a combination of traits of people forming complex collegial bodies (e.g., the board of directors). This situation is not only about the way decisions are made but also about the perception of business risk (Fetisovová et al., 2012).

Comparing the responses of men and women (in 2015 and 2021) regarding their attitude towards taking responsibility for the consequences of running a business does not show significant differences (see Table 2.9 and Table 2.10). Both men and women do not perceive this aspect of economic activity as problematic. Furthermore,

TABLE 2.9
Agree or disagree: I find it hard bearing the responsibility of running my business? (year 2015)

Characteristics		Strongly agree	Tend to agree	Neither agree nor disagree	Tend to disagree	Strongly disagree	Total
Sex	Men	12.1%	18.4%	18.0%	19.6%	32.0%	100.0%
	Women	12.3%	19.5%	19.8%	19.3%	29.1%	100.0%

Characteristics		Strongly agree	Tend to agree	Neither agree nor disagree	Tend to disagree	Strongly disagree	Total
Sex	Men	62.6%	61.5%	60.8%	63.3%	65.1%	63.0%
	Women	37.4%	38.5%	39.2%	36.7%	34.9%	37.0%
Total		100%	100%	100%	100%	100%	100%

Source: Own elaboration based on Eurofound database.

TABLE 2.10
Agree or disagree: I find it hard bearing the responsibility of running my business? (year 2021)

Characteristics		Strongly agree	Tend to agree	Neither agree nor disagree	Tend to disagree	Strongly disagree	Total
Sex	Men	19.1%	16.9%	9.4%	18.4%	34.5%	100.0%
	Women	15.5%	17.3%	10.0%	19.9%	34.8%	100.0%

Source: Own elaboration based on Eurofound database.

TABLE 2.11
Test for significance of differences (year 2015)

Agree or disagree: I find it hard bearing the responsibility of running my business?	2015	2021
Chi-square	5.982	0.062
Asymptotic significance	0.014	0.803
Grouping variable: Sex		

Source: Own elaboration based on Eurofound database.

the distance between the responses of men and women has decreased from 2.90 to 0.30 percentage points between 2015 and 2021, and the percentage of respondents indicating difficulties in taking responsibility in business has increased from 32.00% and 29.10% for men and women, respectively, to nearly 35.00% for both genders. In order to test whether there are statistically significant differences between men and women in their attitudes toward risk-bearing in business, test for significance of differences was conducted.

Based on the results of chi-square-test, where p-values were higher than 0.05, so were, in both years, greater than the assumed significance level of alpha=0.05, there were no grounds to reject the H0 hypothesis. Thus, there are no differences among men and women in terms of risk aversion, including business risk. The details are presented in Table 2.11.

The outcome is not surprising. Sebastian and Moon (2017) demonstrate that there are no significant differences between genders in most leadership styles. Men and women are equally adept at handling responsibility in business management. Radu et al. (2017) state directly that the ability of both genders to bear responsibility in business is similar. This is also supported by Polston-Murdoch (2013). In terms of self-employment, Rico and Cabrer-Borrás (2018) and Hagqvist et al. (2018) present a similar perspective. Furthermore, studies of the self-employed in the EU show that both men and women report similar experiences regarding work-related obligations. Despite differences in types of businesses and sectors, these differences do not translate into significant differences in the perceived burden of responsibility between genders (OECD/European Commission, 2021).

2.4.5.2 Psychological Well-being

Psychological well-being was analyzed through ten different questions. Among the self-employed women, almost 56% reported that they spend at least the majority of their work time dealing directly with individuals who are not employees of their workplace. This is a higher proportion than among men, which may indicate a different character of enterprises run by women (see Figure 2.9 and Figure 2.10).

In order to confirm the existence of significant differences between genders in terms of the variable analyzed, and thus indicate that there are differences in the nature of the business between men and women, a chi-square test for significance of differences was conducted (see Table 2.12).

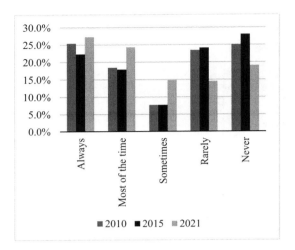

FIGURE 2.9 The share of self-employed men saying that paid job involves direct contact in the workplace with people who are not their employees [years: 2010, 2015, 2021]. (Own elaboration based on Eurofound database.)

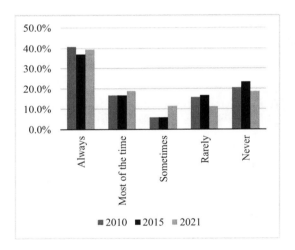

FIGURE 2.10 The share of self-employed women saying that paid job involves direct contact in the workplace with people who are not their employees [years 2010, 2015, 2021]. (Own elaboration based on Eurofound database.)

Based on the results of the chi-square -test, where p<0.001 and was lower than the assumed significance level of alpha=0.05, we rejected the H0 hypothesis in all analysed years. There are differences between male and female entrepreneurs in terms of the type of economic activity they engage in. These findings are consistent with the literature. Female entrepreneurs primarily work in the service sector (Mitchelmore

TABLE 2.12
Testing results (years 2010, 2015, 2021)

Characteristics		2010	2015	2021
Chi-square		837.602	610.770	1196.959
Asymptotic significance		<.001	<.001	<.001
Yates' fix for continuity	Chi-square	837.602	610.770	1196.409
	Asymptotic significance	<.001	<.001	<.001
Grouping variable: Sex				

Source: Own elaboration based on Eurofound database.

and Rowley, 2013; Sullivan and Meek, 2012), especially in highly skilled areas (Bates, 1995). Thus, advanced education and professional experience are strong predictors of the presence of women in these sectors. This position is also supported by studies on national contexts (Agussani, 2020; Byrne et al., 2019; Ghouse et al., 2019; Mitchelmore and Rowley, 2013; Bruni et al., 2004), which show the existence of female-oriented areas of economic activity. On the other hand, the situation is different in the agriculture and industry sectors. Both sectors are dominated by male entrepreneurs (Panić, 2022).

The importance of collaborating with clients, the frequency of such collaboration, and their nature are not without significance for the mental well-being of the self-employed. These factors can be potential stressors that negatively impact the health of entrepreneurs (Nikolova, 2019). In order to determine whether self-employed men or women more often collaborate with dissatisfied clients as part of their businesses, two tabular listings were prepared (see Figure 2.11 and Figure 2.12), separately for each gender.

The aforementioned tabulations (taking into consideration the small fluctuation in variable values) do not provide a univocal answer to the question of whether women as frequently as men collaborated with unwilling clients. In order to resolve this doubt, a cross-tabulated table (see Table 2.13) was used, for which a Chi-square test of significance was conducted.

Based on the analysis of cross-tabulation (see above), a hypothesis was postulated that:

H0: Women as frequently as men had client engagements
H1: Women more frequently than men had client engagements

To validate the hypothesis, a significance test of the Chi-square difference was conducted, based on the results of which, at a significance level of p-value <0.01 and a level of significance $\alpha=0.05$, the null hypothesis was rejected in favor of the alternative hypothesis. Thus, it was established that women more frequently than men had client engagements in each of the years under review. Research indicates that a larger proportion of self-employed individuals in the service sector (including, in particular, shop and restaurant owners) are women (Park et al., 2021), and their work is

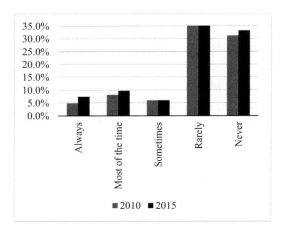

FIGURE 2.11 The share of self-employed men who said their paid job involved dealing with angry customers, clients, patients, students, etc. [years 2010, 2015]. (Own elaboration based on Eurofound database.)

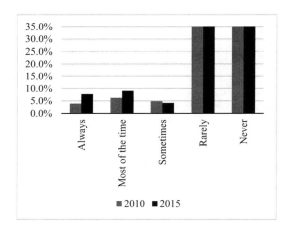

FIGURE 2.12 The share of self-employed women who said their paid job involved dealing with angry customers, clients, patients, students, etc. [years 2010, 2015]. (Own elaboration based on Eurofound database.)

characterized by long working hours, a greater number of days worked per week, and more frequent contact with dissatisfied customers. These factors translate to poorer mental well-being among self-employed women. Studies on the conditions and health of self-employed women reveal that they have higher psychological demands than other professional groups (Bernin, 2002). Furthermore, studies show that self-employed women exhibit behaviors that may increase the risk of illness. Additionally, when compared to men (in managerial positions), self-employed women face greater

TABLE 2.13
Does your main paid job involve – Dealing directly with people who are not employees at your workplace? (years 2010, 2015)

		2010					
Characteristics		Always	Most of the time	Sometimes	Rarely	Never	Total
Sex	Men	3.6%	7.1%	5.0%	42.3%	42.0%	100.0%
	Women	4.2%	7.4%	5.1%	44.2%	39.0%	100.0%
Total		**3.9%**	**7.3%**	**5.1%**	**43.2%**	**40.6%**	**100.0%**

a. Year of survey = 2010

		2015					
Characteristics		Always	Most of the time	Sometimes	Rarely	Never	Total
Sex	Men	5.6%	9.0%	5.8%	40.5%	39.1%	100.0%
	Women	8.1%	10.2%	6.5%	42.9%	32.2%	100.0%
Total		**6.9%**	**9.6%**	**6.2%**	**41.7%**	**35.7%**	**100.0%**

a. Year of survey = 2015

Source: Own elaboration based on Eurofound database.

difficulties in relaxing outside of work (Sevä et al., 2016). These findings correspond with other research indicating that many self-employed women view stress related to work-life balance as one of the biggest problems (Harte, 1996). On the other hand, Gevaert et al. (2018) show that self-employed men have better mental well-being than self-employed women (which is also evident from the present findings). However, self-employed men had lower mental well-being scores in the situation where the decision to start a business was due to a lack of alternative (necessity entrepreneurship). A similar situation occurred among men with poor work-life balance and those who always had contact with irritated customers (Park et al., 2021). On the whole women tend to show greater propensity in their work to participate in emotionally disturbing situations. The details of this situation are shown in cross tabulations (see Figure 2.13 and Figure 2.14).

Thus, in order to test whether there are significant differences in the propensity in their work to participate in emotionally disturbing situations among men and women, and thus (after analyzing the data) whether women's propensity is greater than men's, a chi-square test was conducted. Based on it, with a p-value <0.001, the H0 hypothesis was rejected in each of the analyzed years (2015 and 2021). Thus, it can be concluded that women have a higher propensity to experience emotionally disturbing situations in their work. The conclusion is supported by research conducted by Suh and Punnett (2022). The more frequently individuals are exposed to harmful emotional factors, the greater the risk of exacerbating the mental well-being of self-employed women.

Another aspect we focused on in our study was stress at work. As can be seen from the charts and cross tabulations provided, it was more often women than men who experienced stress at work (see Figure 2.15 and Table 2.14).

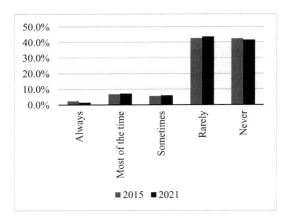

FIGURE 2.13 The percentage of self-employed men said that their paid work involves being in situations that are emotionally disturbing [years 2010, 2015]. (Own elaboration based on Eurofound database.)

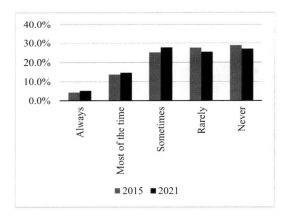

FIGURE 2.14 The percentage of self-employed women said that their paid work involves being in situations that are emotionally disturbing [years 2010, 2015]. (Own elaboration based on Eurofound database.)

Therefore, it was decided to test whether there were significant differences in experiencing stress at work between men and women. For this purpose, the Mann Whitney U-test was used, according to the results of which, with p>alpha, there were no grounds to reject the H0 hypothesis. Thus, there are no significant differences in the experience of stress between the two sexes (see Table 2.15).

The research indicates that there is no significant difference in the experience of stress between men and women. However, it is important to remember that women entrepreneurs more frequently than men find themselves in unsettling situations

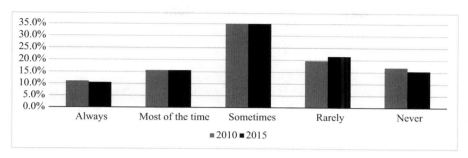

FIGURE 2.15 Which best describes your work situation – You experience stress in your work? [years 2010, 2015]. (Own elaboration based on Eurofound database.)

TABLE 2.14
Which best describes your work situation – You experience stress in your work? (years 2010, 2015)

Characteristics		2010					
		Always	Most of the time	Sometimes	Rarely	Never	Total
Sex	Men	11.0%	15.6%	36.8%	19.7%	16.9%	100.0%
	Women	10.4%	15.4%	39.5%	19.0%	15.7%	100.0%
Total		**10.8%**	**15.5%**	**38.1%**	**19.3%**	**16.3%**	**100.0%**
Characteristics		2015					
		Always	Most of the time	Sometimes	Rarely	Never	Total
Sex	Men	10.5%	15.6%	36.9%	21.5%	15.4%	100.0%
	Women	9.7%	16.7%	38.9%	19.5%	15.2%	100.0%
Total		**10.1%**	**16.1%**	**37.9%**	**20.5%**	**15.3%**	**100.0%**

a. Year of survey = 2015

Source: Own elaboration based on Eurofound database.

(Cooper & Davidson, 1982). The resulting outcome suggests that statistics should not be considered from the perspective of stress experience alone, but rather in terms of the frequency of its occurrence and differences in exposure to stress for both genders. This perspective is in line with the findings presented above by Harte (1996) and Gevaert et al. (2018).

Among self-employed women, 39.4% confirmed that it happened at least once that they had less than 11 hours between 2 working days. In the case of men the share was far higher (60.6%) (see Table 2.16).

TABLE 2.15
Testing results (years 2010, 2015)

Characteristics	2010	2015
Mann Whitney U-test	233048604.500	230283530.00
W Wilcoxona	451129774.500	459488285.00
Z	-1.675	-2.547
Asymptotic significance (two-sided)	.094	.011
a. Year of survey = 2010		
b. Grouping variable: Sex		

Source: Own elaboration based on Eurofound database.

TABLE 2.16
Last month, has it happened at least once that you had worked less than 11 hours between 2 working days? (year 2015)

Characteristics		Sex		Total
		Men	Women	
Response options	Yes	60.6%	39.4%	100.0%
	No	47.4%	52.6%	100.0%
Total		**50.3%**	**49.7%**	**100.0%**

Source: Own elaboration based on Eurofound database.

Thus, to confirm the hypothesis that men are more likely to work more than women, a chi-square test was performed (see Table 2.17). Thus, the following hypothesis was established:

H0: the length of time men and women work is not significantly different
H1: men's working time is greater than women's

Thus, the null hypothesis was rejected in favor of the alternative hypothesis, that is, the length of time men work is greater than women.
Analyzing the results obtained, it is worth beginning with the fact that self-employed individuals generally work longer hours than salaried employees (Park et al., 2020; Hyytinen & Ruuskanen, 2007). Longer work hours may indicate that the company is performing well, which in turn is associated with higher levels of job satisfaction among the self-employed (Millán et al., 2013). This could be a potential reason why, in both cases, among women and men, there are hardly any differences in the share of the respective response groups in terms of gender. Most women and men stated

TABLE 2.17
Testing results (year 2015)

Characteristics	Value	df	Asymptotic significance (two-sided)	Precision relevance (two-sided)	Precision relevance (one-sided)
Chi-square Pearson	524.042[a]	1	<.001		
Continuity correction	523.514	1	<.001		
Reliability quotient	527.345	1	<.001		
Fisher's exact test				<.001	<.001
Test of linear relationship	524.030	1	<.001		
N valid observations	43321				

Source: Own elaboration based on Eurofound database.

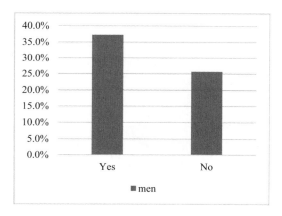

FIGURE 2.16 The share of self-employed men who stated that during the last month they worked less than 11 hours between 2 working days at least once [year 2015]. (Own elaboration based on Eurofound database.)

that they are cheerful most of the time (about 40%), however, this was slightly less in 2021 than in 2015. On the other hand, a larger percentage of men than women were cheerful all the time. Here, for both genders, there was an increase in the percentage of participants (see Figure 2.16 and Figure 2.17).

Based on the analysis presented on Figure 2.18 and Figure 2.19 as well as cross-tabulation (see Table 2.18), a hypothesis was postulated that:

H0: Women as frequently as men felt cheerful and in good spirits over the last two weeks.

H1: Men more frequently than women felt cheerful and in good spirits over the last two weeks.

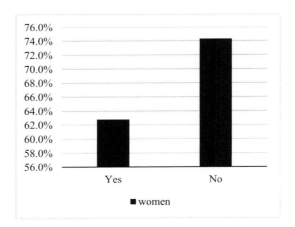

FIGURE 2.17 The share of self-employed woman who stated that during the last month they worked less than 11 hours between 2 working days at least once [year 2015]. (Own elaboration based on Eurofound database.)

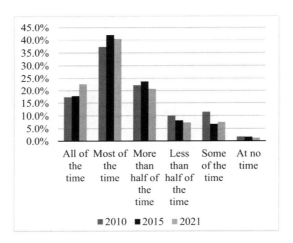

FIGURE 2.18 Share of self-employed men asked how often in the last two weeks they felt cheerful and in good spirits [years 2010, 2015, 2021]. (Own elaboration based on Eurofound database.)

In both years, based on the test performed (see Table 2.19) we rejected H0 stating that there are no differences by gender. Therefore, it can be assumed that this factor differentiate responses. Women and men derive do not equally satisfaction from their business activities. The crucial aspect necessitates taking into account the nature of the industries and sectors being considered. Some of them are, in fact, dominated by women, while others are dominated by men. Findings indicate that in sectors dominated by men, working hours do not influence the level of job satisfaction among

Psychological Well-Being of Self-Employed Women 47

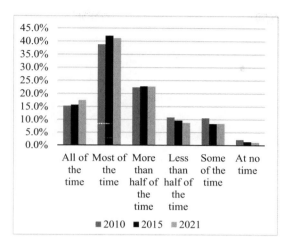

FIGURE 2.19 Share of self-employed woman asked how often in the last two weeks they felt cheerful and in good spirits [years 2010, 2015, 2021]. (Own elaboration based on Eurofound database.)

TABLE 2.18
How have you been feeling over the last two weeks: I have felt cheerful and in good spirit? (years 2015, 2021)

Characteristics		2015		2021	
		Men	Women	Male	Women
Response options	All of the time	64.9%	35.1%	58.4%	41.2%
	Most of the time	62.4%	37.6%	51.6%	48,2%
Total		**63.1%**	**36.9%**	**36.1%**	**29.9%**

Source: Own elaboration based on Eurofound database.

TABLE 2.19
Tests results (years 2015, 2021)

	2015		2021	
Characteristics	Value	Asymptotic significance (two-sided)	Value	Asymptotic significance (two-sided)
Mann Whitney U	10.958	0.04	372.683	<0.001

Source: Own elaboration based on Eurofound database.

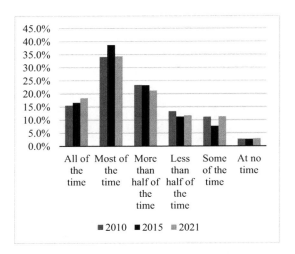

FIGURE 2.20 Percentage of self-employed men describing how often they felt calm and relaxed in the last two weeks? [years 2010, 2015, 2021]. (Own elaboration based on Eurofound database.)

men and women. Conversely, no correlation was observed between working hours and job satisfaction among women in sectors dominated by women (Sappleton & Lourenço, 2016). However, as Bögenhold and Klinglmair (2015) claim, well-being and happiness are higher in female-led companies and areas of activity. Furthermore, Bender and Roche (2016) argue that self-employed women are happier than their male counterparts due to their greater ability to attain their desired outcomes in self-employment.

In all the years that were analyzed, both men and women generally felt calm, relaxed, rested and active at least most of the time (see Figures 2.20–2.21). It is important to note, however, that these values differ significantly in the case of 2021 responses. In most cases, the percentages, regardless of gender, were lower (in 2021 compared to previous years) for positive responses. On the other hand, they were higher for negative responses. Therefore, it is possible to argue that the pandemic had a negative impact (examined in this particular area) on the psychological condition of employed individuals.

Based on the analysis presented in Figure 2.20 and Figure 2.21, as well as crosstabulation (see Table 2.20), a hypothesis was postulated that:

H0: Women as frequently as men felt cheerful and in good spirits over the last two weeks.
H1: Men more frequently than women felt cheerful and in good spirits over the last two weeks.

A Mann Whitney U-test of significance of differences was conducted, based on the results of which, at p-value <0.01 and significance level $\alpha=0.05$, the hypothesis H0

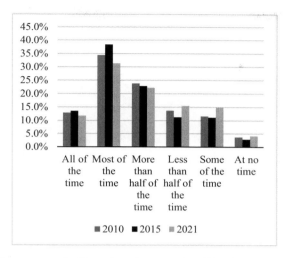

FIGURE 2.21 Percentage of self-employed woman describing how often they felt calm and relaxed in the last two weeks? [years 2010, 2015, 2021]. (Own elaboration based on Eurofound database.)

TABLE 2.20
How have you been feeling over the last two weeks: I have felt calm and relaxed? (years 2015, 2021)

Characteristics		2015		2021	
		Men	Women	Male	Female
Response options	**All of the time**	64.9%	35.1%	58.4%	41.2%
	Most of the time	62.4%	37.6%	51.6%	48.2%
Total		**63.1%**	**36.9%**	**36.1%**	**29.9%**

Source: Own elaboration based on Eurofound database.

was rejected in favor of the alternative hypothesis in 2010-2021. Thus, in terms of this analyzed variable, based on the test conducted, gender is the differentiating variable and it is men who feel more calm. According to the viewpoint of Xiu and Ren (2022), male independent workers enjoy better health than male employees, whereas female independent workers have worse health than part-time female workers. Moreover, self-employed women have lower self-rated health and lower job satisfaction than men after controlling for individual and occupational characteristics (Xiu & Ren, 2022).

In the next step, the analysis was conducted (dividing by gender) to determine whether the self-employed individuals had felt fresh and rested in the past two weeks. In other words, did they have the opportunity to recharge (both physically and mentally) and gain the strength needed to carry out their business activities (see Figure 2.22 and Figure 2.23). Consideration of the gender criterion is important here,

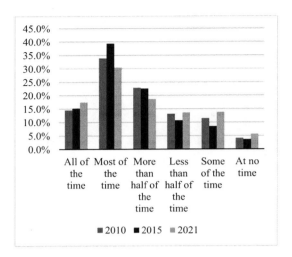

FIGURE 2.22 The share of self-employed men describing how often they felt fresh and rested in the last two weeks [years 2010, 2015, 2021]. (Own elaboration based on Eurofound database.)

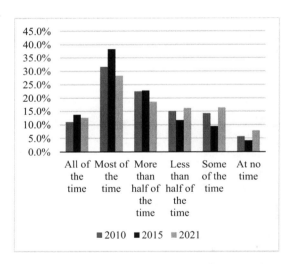

FIGURE 2.23 The share of self-employed woman describing how often they felt fresh and rested in the last two weeks [years 2010, 2015, 2021]. (Own elaboration based on Eurofound database.)

as Sapleton and Lourenço (2016) state that men work the longest hours in male-dominated sectors. This result is significantly higher than the average for those employed full-time in Europe. Cabrita and Ortigão (2011) report that the average weekly work hours for full-time employees in Europe is 38 hours. The average weekly work hours for self-employed women in male-dominated sectors are 46 hours, while for self-employed men in female-dominated sectors, it is 45 hours per week (again, these results are above the European average). At the same time, women in female-dominated sectors work approximately 33 hours per week (Sapleton & Lourenço, 2016).

Based on an analysis of the cross-tabulations (see Table 2.21), the hypothesis is that:

H0: Men feel as rested as women do.
H1: Men are more likely to feel more rested than women.

A Mann Whitney U-test of significance of differences was conducted, based on the results of which, at p-value <0.01 and significance level α=0.05, the hypothesis H0 was rejected in favor of the alternative hypothesis in 2010-2021. Thus, men are more likely to be rested than women.

Despite evidence presented by Cabrita and Ortigão (2011) and later by Sapleton and Lourenço (2016) that self-employed men work longer hours than women, they are statistically more likely to be under-restrained than their male counterparts, as previously demonstrated. The observed result can be easily explained. Women, apart from working on their own accounts, are additionally burdened with domestic responsibilities (Craig et al., 2012), which, even with lesser hours associated with self-employment, significantly exceed the burden of men, who are generally less burdened with domestic chores (see Lunau et al., 2014). Therefore, it can be concluded that men, despite working statistically longer hours, lose to women when considering domestic responsibilities that are equally time-consuming as their professional obligations (Hagqvist et al., 2015).

In most cases, a similar number of women and men were engaged in activities that interested them. However, here we also see a negative impact of COVID-19 – in the case of women. The decline was more pronounced than in the case of men (see Figure 2.24 and Figure 2.25). Perhaps this can be attributed to the fact that men more often choose self-employment as a form of professional self-realization, while for women, the flexibility offered by self-employment in combining professional obligations with domestic ones is a driving factor (Kalenkoski & Pabilonia, 2022). Therefore, the negative impact of COVID-19 on the ability to pursue one's professional aspirations and preferences was felt more strongly by men than by women.

Referring to the statistics presented in Table 2.22, it is worth considering the results of the research conducted by Orser and Riding (2004). They indicate statistically significant differences between genders in the values that individuals assign to different dimensions of success. Women entrepreneurs statistically demonstrate higher and more significant (than men) criteria for success related to personal fulfillment and balance. Furthermore, while the weights of these two factors (i.e., fulfillment and balance) were positive for women, the scores for men in these dimensions were negative. These findings are associated with observations regarding the impact of COVID-19 on self-realization across both genders. Female entrepreneurs attach greater importance to maintaining client relationships and personal relationships, good mental well-being, and balance between work and private life (see Table 2.2). These results also support the argument that women entrepreneurs value success

TABLE 2.21
Been feeling over the last two weeks – I woke up feeling fresh and rested? (years 2010, 2015, 2021)

2010		All of the time	Most of the time	More than half of the time	Less than half of the time	Some of the time	At no time	Total
Sex	Men	14.5%	33.9%	22.9%	13.1%	11.5%	4.1%	100.0%
	Women	11.0%	31.6%	22.5%	15.0%	14.2%	5.7%	100.0%

a. Year of survey = 2010

2015		All of the time	Most of the time	More than half of the time	Less than half of the time	Some of the time	At no time	Total
Sex	Men	15.2%	39.3%	22.6%	10.7%	8.5%	3.6%	100.0%
	Women	12.1%	37.2%	23.0%	12.5%	10.4%	4.9%	100.0%

a. Year of survey = 2015

2021		All of the time	Most of the time	More than half of the time	Less than half of the time	Some of the time	At no time	Total
Sex	Men	17.8%	30.3%	18.6%	13.6%	13.7%	5.7%	100.0%
	Women	12.7%	28.5%	18.5%	16.1%	16.2%	7.8%	100.0%

a. Year of survey = 2015

Source: Own elaboration based on Eurofound database.

Psychological Well-Being of Self-Employed Women

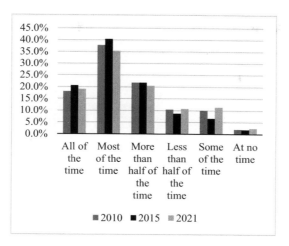

FIGURE 2.24 Share of self-employed men answering the question how often in the last two weeks they felt that everyday life was filled with things that interested them [years 2010, 2015, 2021]. (Own elaboration based on Eurofound database.)

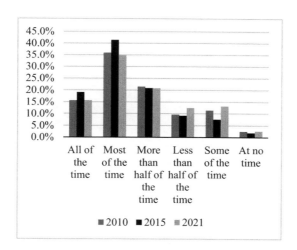

FIGURE 2.25 Share of self-employed woman answering the question how often in the last two weeks they felt that everyday life was filled with things that interested them [years 2010, 2015, 2021]. (Own elaboration based on Eurofound database.)

criteria related to the personal dimension more frequently than men. Therefore, based on a cross-tabulation analysis (see Table 2.22) and the substantive justification derived from the results of the Orser and Riding (2004) study, the hypothesis is that:

H0: Men are as likely as women to feel self-fulfillment due to self-employment.
H1: Men are more likely than women to feel self-fulfillment due to self-employment.

TABLE 2.22
Been feeling over the last two weeks: My daily life has been filled with things that interests me (years 2010, 2015, 2021)

2010		All of the time	Most of the time	More than half of the time	Less than half of the time	Some of the time	At no time	Total
Q2a. Sex	Men	18.1%	37.8%	21.8%	10.4%	9.9%	2.0%	100.0%
	Women	15.6%	38.1%	21.4%	10.8%	11.9%	2.3%	100.0%

a. Year of survey = 2010

2015		All of the time	Most of the time	More than half of the time	Less than half of the time	Some of the time	At no time	Total
Q2a. Sex	Men	20.6%	40.4%	21.8%	8.7%	6.6%	1.9%	100.0%
	Women	19.1%	41.4%	20.9%	9.1%	7.6%	1.9%	100.0%

a. Year of survey = 2015

2021		All of the time	Most of the time	More than half of the time	Less than half of the time	Some of the time	At no time	Total
Q2a. Sex	Men	19.8%	35.4%	20.2%	10.4%	11.2%	2.5%	100.0%
	Women	16.3%	34.9%	20.8%	12.1%	12.8%	2.6%	100.0%

a. Year of survey = 2021

Source: Own elaboration based on Eurofound database.

Psychological Well-Being of Self-Employed Women

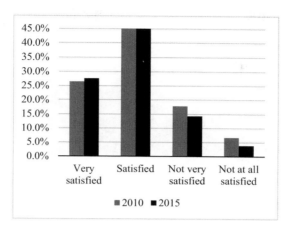

FIGURE 2.26 The share of self-employed men satisfied with working conditions in their main paid job? [years 2010, 2015, 2021]. (Own elaboration based on Eurofound database.)

To verify the hypothesis, a chi-square difference significance test was conducted, based on the results of which, with a p-value <0.01 and a significance level of α= 0.05, the H0 hypothesis was rejected in favor of the alternative hypothesis (in all years analyzed). Thus, men are more likely than women to experience self-fulfillment due to self-employment. The outcome validated the position of Kalenkoski and Pabilonia (2022), expounding the justifications for engaging in professional activities through self-employment, which is linked to the results presented in Figure 5 of their study.

Between 2010 and 2015, both men and women were satisfied with working conditions of their job. In 2015, a slightly smaller proportion of women reported being very satisfied compared to 2010. In contrast, the trend among men was reversed (see Figures 2.26–2.27). The results obtained from research conducted by Pita and Torregrosa (2020) have been useful in clarifying observed regularities, indicating a higher level of job satisfaction among women than men, which supports the gender/job satisfaction paradox.

Based on an analysis of the cross-tabulations (see Table 2.23), a hypothesis has been put forward stating the following.

H0: Women are as likely as men to feel satisfaction with working conditions.
H1: Women are more likely than men to feel satisfaction with working conditions.

In order to verify the hypothesis, a Mann Whitney U difference significance test was conducted, based on the results of which, with a p-value <0.01 in 2010 and p<0.004 in 2015 and a significance level of α=0.05, the H0 hypothesis was rejected in favor of the alternative hypothesis. Women are more likely to experience satisfaction with their working conditions, which supports the views of Pita and Torregrosa (2020). This satisfaction with self-employment as a career path is higher among women than men, as shown by the sum of responses "very satisfied" and "satisfied." This confirms previous findings that the motives for self-employment and the general nature of the

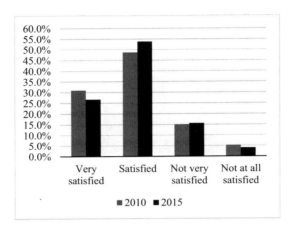

FIGURE 2.27 The share of self-employed woman satisfied with working conditions in their main paid job? [years 2010, 2015, 2021]. (Own elaboration based on Eurofound database.)

TABLE 2.23
On the whole, satisfied with working conditions in your main paid job? (years 2010, 2015, 2021)

2010		Very satisfied	Satisfied	Not very satisfied	Not at all satisfied
Sex	Men	22.1%	56.7%	16.6%	4.6%
	Women	23.6%	57.4%	15.6%	3.3%
2015		Very satisfied	Satisfied	Not very satisfied	Not at all satisfied
Sex	Men	24.5%	58.9%	13.4%	3.2%
	Women	25.5%	58.7%	13.0%	2.9%

Source: Own elaboration based on Eurofound database.

conditions that accompany it, as well as the way it is perceived, are independent of gender.

2.5 CONCLUSIONS

The topic of psychological well-being in women is relevant for several reasons, one of which is the higher burden of mental illness experienced by females according to disability-adjusted life years than males (Whiteford et al., 2013). The changing role of women in society and the labor market necessitates an understanding of how these changes impact their psychological well-being (Gonäs & Tyrkkö, 2015). Research suggests that self-employed women face distinct challenges in terms of their psychological well-being compared to male self-employed individuals and women in salaried employment. Studies indicate that these women exhibit lower levels of well-being than their male counterparts and women in wage-employment, often experiencing reduced levels of health and life satisfaction. However, employed women, including those who are self-employed, tend to exhibit higher levels of social and environmental quality of life, as well as psychological well-being domains such as environmental mastery and positive relations with others, compared to non-working women (Kapur, 2023). Nonetheless, women also frequently experience unique challenges and social pressures that can affect their mental health (Rooney et al., 2003). Additionally, issues related to gender equality and the fight against discrimination have a significant impact on women's mental health (Hosang & Bhui, 2018).

Studies have shown that self-employment, as a simple form of entering the business world, has a different character when considered through the lens of gender. Women and men have different (for each gender separately) motivations for engaging in entrepreneurial activity and expectations regarding its effects. However, the most important value added (captured during the course of research) is the fact that self-employment has a different impact on the psychological well-being of men and women. As shown by empirical research, self-employed women typically engage in activities in the service sector, their undertakings are characterized by a high hourly load, and therefore a proportional greater exposure to disgruntled clients and employees. Combining these facts with the difficulties in relaxing after work (e.g., due to family obligations), their health risk load is disproportionately greater than that of men.

BIBLIOGRAPHY

Abubakar, Y. A., Saridakis, G., Litsardopoulos, N., Torres, R. I. M., Sookram, S., & Hosein, R. (2023). Impact of Covid-19 policies on women self-employment rates: An integrated conceptual framework. *International Review of Entrepreneurship*, 21(1), 1–34.

Agussani, A. (2020). Are woman the dominant entrepreneurs in Indonesia. *The International Journal of Social Sciences and Humanities Invention*, 7(5), 5935–5947. https://doi.org/10.18535/ijsshi/v7i05.01

Apitzsch, U., & Kontos, M. (2003). Self-employment, gender and migration. *International Review of Sociology/Revue Internationale de Sociologie*, 13(1), 67–76. https://doi.org/10.1080/0390670032000086989

Arum, R., & Müller, W. (eds.). (2009). *The reemergence of self-employment: A comparative study of self-employment dynamics and social inequality*. Princeton University Press, Princeton, NJ.

Bates, T. (1995). Self-employment entry across industry groups. *Journal of Business Venturing*, 10(2), 143–156. https://doi.org/10.1016/0883-9026(94)00018-P

Bender, K. A., & Roche, K. (2016). Self-employment and the paradox of the contented female worker. *Small Business Economics*, 47, 421–435. https://doi.org/10.1007/s11187-016-9731-z

Benz, M., & Frey, B. S. (2004). Being independent raises happiness at work. *Swedish Economic Policy Review*, 11(2), 95–134.

Bergmann, H., & Sternberg, R. (2007). The changing face of entrepreneurship in Germany. *Small Business Economics*, 28(2–3), 205–221. https://doi.org/10.1007/s11 187-006-9016-z

Bernin, P. (2002). *Managers' working conditions stress and health* (Doctoral thesis). Karolinska Institutet, Stockholm, Sweden.

Blanchflower, D. G. (2000). Self-employment in OECD countries. *Labour Economics*, 7(5), 471–505. https://doi.org/10.1016/S0927-5371(00)00011-7

Blanchflower, D., & Oswald, A. (1991). What makes a young entrepreneur?. *NBER Working Paper 373*. London School of Economics, Centre for Labour Economics.

Boden Jr, R. J. (1996). Gender and self-employment selection: An empirical assessment. *The Journal of Socio-Economics*, 25(6), 671–682. https://doi.org/10.1016/ S1053-5357(96)90046-3

Bögenhold, D., & Klinglmair, A. (2015). Female Solo self-employment-features of gendered entrepreneurship. *International Review of Entrepreneurship*, 13(1), 47–58.

Botelho, V., & Neves, P. (2021). The impact of the COVID-19 crisis on the euro area labour market for men and women. *Economic Bulletin Issue*, 4. www.ecb.europa.eu/ press/economic-bulletin/focus/2021/html/ecb.ebbox202104_04~686c89e9bb.en.html (Accessed: 31.05.2024).

Brindley, C. (2005). Barriers to women achieving their entrepreneurial potential: Women and risk. *International Journal of Entrepreneurial Behavior & Research*, 11(2), 144–161. https://doi.org/10.1108/13552550510590554

Bruni, A., Gerardi, S., & Poggio, B. (2004). Entrepreneur-mentality, gender and the study of women entrepreneurs. *Journal of Organization Change Management*, 17(3), 256–268. https://doi.org/10.1108/09534810410538315

Brush, C. G. (1992). Research on women business owners: Past trends, a new perspective and future directions. *Entrepreneurship Theory and Practice*, 16(4), 5–30. https://doi.org/ 10.1177/104225879201600401

Buratti, A., Cesaroni, F. M., & Sentuti, A. (2017). Does gender matter in strategies adopted to face the economic crisis? A comparison between men and women entrepreneurs. In Ladislav Mura (ed.) *Entrepreneurship-Development Tendencies and Empirical Approach*, 393–412. http://dx.doi.org/10.5772/intechopen.70292

Byrne, J., Fattoum, S., & Diaz Garcia, M. C. (2019). Role models and women entrepreneurs: Entrepreneurial superwomen has her say. *Journal of Small Business Management*, 57(1), 154–184. https://doi.org/10.1111/jsbm.12426

Cabrita, J., & Ortigão, M. (2011). Working time developments–2010. Eurofound. Ireland. https://policycommons.net/artifacts/1833807/working-time-developments/2575314/ (Accessed: 09.06.2024).

Caputo, R. K., & Dolinsky, A. (1998). Women's choice to pursue self-employment: The role of financial and human capital of household members. *Journal of Small Business Management*, 36(3), 8–17.

Carrasco, R. (1999). Transitions to and from self-employment in Spain: An empirical analysis. *Oxford Bulletin of Economics and Statistics*, 61(3), 315–341. https://doi.org/10.1111/ 1468-0084.00132

Clark, A. E. (1997). Job satisfaction and gender: Why are women so happy at work? *Labour Economics*, 4(4), 341–372. https://doi.org/10.1016/S0927-5371(97)00010-9

Clement, S. (1987). The self-efficacy expectations and occupational preferences of females and males. *Journal of Occupational Psychology*, 60(3), 257–265. https://doi.org/10.1111/ j.2044-8325.1987.tb00258.x

Cooper, C. L., & Davidson, M. J. (1982). The high cost of stress on women managers. *Organizational Dynamics*, 10(4), 44–53. https://doi.org/10.1016/0090-2616(82)90028-6

Craig, L., Powell, A., & Cortis, N. (2012). Self-employment, work-family time and the gender division of labour. *Work, Employment and Society*, 26(5), 716–734. https://doi.org/10.1177/09500170124516

Crainer, S., & Dearlove, D. (2000). *Generation entrepreneur: Shape today's business reality, create tomorrow's wealth, do your own thing*. Financial Times Press, London.

Dawson, C., Henley, A., & Latreille, P. L. (2009). Why do individuals choose self-employment?. *IZA Discussion Papers, No. 3974*. Institute for the Study of Labor (IZA), Bonn. https://nbn-resolving.de/urn:nbn:de:101:1-20090210136; www.econstor.eu/bitstream/10419/35711/1/592883760.pdf (Accessed: 10.06.2024).

Dencker, J. C., Bacq, S., Gruber, M., & Haas, M. (2021). Reconceptualizing necessity entrepreneurship: A contextualized framework of entrepreneurial processes under the condition of basic needs. *Academy of Management Review*, 46(1), 60–79. https://doi.org/10.5465/amr.2017.0471

Department of Health. (2022). Embedding Women's Mental Health in Sharing the Vision (November). www.gov.ie/en/publication/5f65e-embedding-womens-mental-health-in-sharing-the-vision/ (Accessed: 10.06.2024).

Ebner, N. C., Freund, A. M., & Baltes, P. B. (2006). Developmental changes in personal goal orientation from young to late adulthood: From striving for gains to maintenance and prevention of losses. *Psychology and Aging*, 21(4), 664–678. https://doi.org/10.1037/0882-7974.21.4.664

Esmaeili, S., Ghanbari Panah A., & Koochak-Entezar, R. (2020). Prediction of psychological well-being based on health anxiety and perceived stress with the mediating role of self-handicapping in married women working in the school of Nursing and Midwifery, Tehran University of Medical Sciences in 2018. *Iranian Journal of Nursing Research*, 14(6), 45–52. https://doi.org/10.21859/ijnr-140606

Eurofound. (2013). Self-employed or not self-employed? Working conditions of 'economically dependent workers'. *Background Paper*. www.eurofound.europa.eu/en/publications/2013/self-employed-or-not-self-employed-working-conditions-economically-dependent (Accessed: 09.06.2024).

EWCTS, Eurofound. (2024). European Working Conditions Telephone Survey, 2021. [data collection]. 3rd Edition. UK Data Service. SN: 9026, DOI: http://doi.org/10.5255/UKDA-SN-9026-3 (Accessed: 09.06.2024).

Evans, D., & Leighton, L. (1989). Some empirical aspects of entrepreneurship. *The American Economic Review*, 79, 519–535.

Fairlie, R. W. (2005). Entrepreneurship among disadvantaged groups: An analysis of the dynamics of self-employment by gender, race, and education. In: Parker, S. C., Acs, Z. J., & Audretsch, D. R. (eds.), *Handbook of entrepreneurship* (pp. 437–478). Kluwer Academic Publishers, Dordrecht.

Fetisovová, E., Hucová, E., Nagy, L., & Vlachynský, K. (2012). *Aktuálne problémy financií malých a stredných podnikov, Vydavateľstvo EKONÓM Bratislava*. Bratislava.

Frid, C. J., Wyman, D. M., Gartner, W. B., & Hechavarria, D. H. (2016). Low-wealth entrepreneurs and access to external financing. *International Journal of Entrepreneurial Behavior & Research*, 22(4), 531–555. https://doi.org/10.1108/IJEBR-08-2015-0173

GEM (Global Entrepreneurship Monitor). (2006). *Global entrepreneurship monitor. Summary report 2005/2006*. GEM, London.

GEM (Global Entrepreneurship Monitor). (2021). *Global entrepreneurship monitor 2020/21 women's entrepreneurship report: Thriving through crisis*. GEM, London.

GEM (Global Entrepreneurship Monitor). (2022). *Global entrepreneurship monitor 2021/22 women's entrepreneurship report: From crisis to opportunity*. GEM, London.

Georgellis, Y., & Wall, H. J. (2005). Gender differences in self-employment. *International Review of Applied Economics*, 19(3), 321–342. https://doi.org/10.1080/02692170500119854

Gevaert, J., De Moortel, D., Wilkens, M., & Vanroelen, C. (2018). What's up with the self-employed? A cross-national perspective on the self-employed's work-related mental well-being. *SSM-Population Health*, 4, 317–326. https://doi.org/10.1016/j.ssmph.2018.04.001

Ghouse, S., McElwee, G., & Durrah, O. (2019). Entrepreneurial success of cottage-based women entrepreneurs in Oman. *International Journal of Entrepreneurial Behavior & Research*, 25(3), 480–498. https://doi.org/10.1108/IJEBR-10-2018-0691

Giacomin, O., Guyot, J. L., Janssen, F., & Lohest, O. (2007, June). Novice creators: personal identity and push pull dynamics. In: *52nd International Council for Small Business (ICSB) World Conference* (pp. 1–30). http://hdl.handle.net/2078/18311

Gindling, T. H., & Newhouse, D. (2014). Self-employment in the developing world. *World Development*, 56, 313–331. https://doi.org/10.1016/j.worlddev.2013.03.003

Gonäs, L., & Tyrkkö, A. (2015). Changing structures and women's role as labor force. *Nordic Journal of Working Life Studies*, 5(2), 89–108. https://doi.org/10.19154/njwls.v5i2.4795

Graeber, D., Kritikos, A. S., & Seebauer, J. (2021). COVID-19: A crisis of the female self-employed. *Journal of Population Economics*, 34, 1141–1187. https://doi.org/10.1007/s00148-021-00849-y

Hagqvist, E., Toivanen, S., & Bernhard-Oettel, C. (2018). Balancing work and life when self-employed: The role of business characteristics, time demands, and gender contexts. *Social Sciences*, 7(8), 139. https://doi.org/10.3390/socsci7080139

Hagqvist, E., Toivanen, S., & Vinberg, S. (2015). Time strain among employed and self-employed women and men in Sweden. *Society, Health & Vulnerability*, 6(1), 1–16. https://doi.org/10.3402/shv.v6.29183

Harte, S. (1996, July). Women who work it out. *Atlanta Journal Constitution*, C1, 29.

Hosang, G. M., & Bhui, K. (2018). Gender discrimination, victimisation and women's mental health. *British Journal of Psychiatry*, 213(6), 682–684. https://doi.org/10.1192/bjp.2018.244

Hyytinen, A., & Ruuskanen, O. P. (2007). Time use of the self-employed. *Kyklos*, 60(1), 105–122. https://doi.org/10.1111/j.1467-6435.2007.00361.x

Kalenkoski, C. M., & Pabilonia, S. W. (2022). Impacts of COVID-19 on the self-employed. *Small Business Economics*, 58(2), 741–768. https://doi.org/10.1007/s11187-021-00522-4

Kapur, S. (2023). An exploratory factor analysis examining psychological correlates among females. *Bangladesh Journal of Medical Science*, 22, 106. https://doi.org/10.3329/bjms.v22i20.66318

Kessler, R. C., Berglund, P., Demler, O., Jin, R., Merikangas, K. R., Walters, E. E. (2005). Lifetime prevalence and age-of-onset distributions of DSM-IV disorders in the National Comorbidity Survey Replication. *Archives of General Psychiatry*, 62(6), 593–602. https://doi.org/10.1001/archpsyc.62.6.593

Kowalska, M. (2023). Impact of the unconditional basic income on the professional situation of women. *Financial Internet Quarterly*, 19(2), 18–25. https://doi.org/10.2478/fiqf-2023-0009

Kumar, P. (2015). A study on women entrepreneurship in India. *International Journal of Applied Science & Technology Research Excellence*, 5(5), 43–46.

Kundera, C. (2022). Facilitating female entrepreneurship: A case study on how to design tools using Service Design. *Open Innovation and Lean Startup*. www.theseus.fi/bitstream/handle/10024/747482/Kundera_Cecylia.pdf?sequence=2 (Accessed: 01.06.2024).

Kuranga, M. O. (2020). Work life balance and service delivery among women entrepreneurs in South-Western Nigeria. *Financial Internet Quarterly*, 16(4), 24–34. https://doi.org/10.2478/fiqf-2020-0025

Lawler, E. E. (1973). *Motivation in work organizations*. Brooks/Cole, Belmont, CA.

Le, A. T. (1999). Empirical studies of self-employment. *Journal of Economic Surveys*, 13(4), 381–416. https://doi.org/10.1111/1467-6419.00088

Lidwall, U., Marklund, S., & Voss, M. (2010). Work–family interference and long-term sickness absence: A longitudinal cohort study. *European Journal of Public Health*, 20(6), 676–681. https://doi.org/10.1093/eurpub/ckp201

Lunau, T., Bambra, C., Eikemo, T. A., van Der Wel, K. A., & Dragano, N. (2014). A balancing act? Work–life balance, health and well-being in European welfare states. *The European Journal of Public Health*, 24(3), 422–427.

McGowan, P., Redeker, C. L., Cooper, S. Y., & Greenan, K. (2012). Female entrepreneurship and the management of business and domestic roles: Motivations, expectations and realities. *Entrepreneurship & Regional Development*, 24(1–2), 53–72. https://doi.org/10.1080/08985626.2012.637351

Millán, J. M., Hessels, J., Thurik, R., & Aguado, R. (2013). Determinants of job satisfaction: A European comparison of self-employed and paid employees. *Small Business Economics*, 40, 651–670. https://doi.org/10.1007/s11187-011-9380-1

Minniti, M., Bygrave, W., & Autio, E. (2006). *Global entrepreneurship monitor: 2005 Executive report*. London Business School, London.

Minola, T., Criaco, G., & Obschonka, M. (2016). Age, culture, and self-employment motivation. *Small Business Economics*, 46, 187–213. https://doi.org/10.1007/s11187-015-9685-6

Mitchelmore, S., & Rowley, J. (2013). Entrepreneurial competencies of women entrepreneurs pursuing business growth. *Journal of Small Business and Enterprise Development*, 20(1), 125–142. https://doi.org/10.1108/14626001311298448

Mitchelmore, S., & Rowley, J. (2013). Growth and planning strategies within women-led SMEs. *Management Decisions*, 51(1), 83–96. https://doi.org/10.1108/00251741311291328

Naughton, T. J. (1987). Quality of working life and the self-employed manager. *American Journal of Small Business*, 12(2), 33–41. https://doi.org/10.1177/104225878701200203

Nawaz, M. A., Naveed Ranjha, A., Noor, S., & Almas, I. (2024). Social response and problems faced by self employed skillful women in District Bahawalpur. *Harf-O-Sukhan*, 8(2), 718–731. https://harf-o-sukhan.com/index.php/Harf-o-sukhan/article/view/1393 (Accessed: 01.06.2024).

Ndunda, N. L., Mark, N., Mutinda, M., Owen, Ngumi, O., & Gachohi, N. (2020). Factors influencing the psychological well-being of women enterprise fund beneficiaries in Njoro Sub-County, Kenya. *Psychology*, 11(11). https://doi.org/10.4236/psych.2020.1111109

Nelson, J. A. (2015). Are women really more risk-averse than men? A re-analysis of the literature using expanded methods. *Journal of Economic Surveys*, 29(3), 566–585. https://doi.org/10.1111/joes.12069

Nikolaev, B., Lerman, M., Boudreaux, Ch, J., & Mueller, B. (2022). Self-employment and eudaimonic well-being: The mediating role of problem- and emotion-focused coping. *Entrepreneurship Theory and Practice*, 47(6), 2121–2154. https://doi.org/10.1177/10422587221126486

Nikolova, M. (2019). Switching to self-employment can be good for your health. *Journal of Business Venturing*, 34(4), 664–691. https://doi.org/10.1016/j.jbusvent.2018.09.001

Nordenmark, M., Vinberg, S., & Strandh, M. (2012). Job control and demands, work-life balance and wellbeing among self-employed men and women in Europe. *Vulnerable Groups & Inclusion*, 3(1), 1–18. https://doi.org/10.3402/vgi.v3i0.18896

OECD. (2023). *OECD employment outlook 2023: Artificial intelligence and the labour market.* OECD Publishing, Paris. https://doi.org/10.1787/08785bba-en

OECD/European Commission. (2021). *The missing entrepreneurs 2021: Policies for inclusive entrepreneurship and self-employment.* OECD Publishing, Paris. https://doi.org/10.1787/71b7a9bb-en

Orser, B., & Riding, A. (2004, June). Examining Canadian business owners' perceptions of success. In *Canadian Council for Small Business and Entrepreneurship Conference* (pp. 1–24).

Panić, D. S. (2022). Qualitative features of women's entrepreneurial activity in the Republic of Serbia: The sectoral distribution perspective. *Facta Universitatis-Economics and Organization*, 19(2), 83–94.

Park, J., Han, B., & Kim, Y. (2020). Comparison of occupational health problems of employees and self-employed individuals who work in different fields. *Archives of Environmental & Occupational Health*, 75(2), 98–111. https://doi.org/10.1080/19338244.2019.1577209

Park, J., Kim, H., & Kim, Y. (2021). Factors related to psychological well-being as moderated by occupational class in Korean self-employed workers. *International Journal of Environmental Research and Public Health*, 19(1), 141. https://doi.org/10.3390/ijerph19010141

Parker, S. (2018). *The Economics of Entrepreneurship.* 2nd edition. Cambridge University Press, Cambridge.

PARP. (2011). *Przedsiębiorczość Kobiet w Polsce.* Polska Agencja Rozwoju Przedsiębiorczości, Warszawa. www.parp.gov.pl/storage/publications/pdf/12839kobiety.pdf (Accessed: 31.05.2024).

Pita, C., & Torregrosa, R. J. (2020). Re-evaluating job satisfaction. www.researchgate.net/profile/Ramon-Torregrosa-2/publication/339067736_Re-evaluating_Job_Satisfaction/links/5e3b9ea5458515072d831355/Re-evaluating-Job-Satisfaction.pdf (Accessed: 09.06.2024).

Polston-Murdoch, L. (2013). An Investigation of path-goal theory, relationship of leadership style, supervisor-related commitment, and gender. *Emerging Leadership Journeys*, 6(1), 13–44.

Radu, C., Deaconu, A., & Frasineanu, C. (2017). Leadership and gender differences: Are men and women leading in the same way. In: A. Alvinius (ed.), *Contemporary Leadership Challenges* (pp. 63–82). IntechOpen, London.

Rico, P., & Cabrer-Borrás, B. (2018). Gender differences in self-employment in Spain. *International Journal of Gender and Entrepreneurship*, 10(1), 19–38. https://doi.org/10.1108/IJGE-09-2017-0059

Rooney, J., Lero, D., Korabik, K., Whitehead, D. L., Abbondaza, M., Tougas, J., Boyd, J., & Bourque, L. (2003). *Self-employment for women: Policy options that promote equality and economic opportunities.* Status of Women, Ottawa, Canada.

Rosti, L., & Chelli, F. (2005). Gender discrimination, entrepreneurial talent and self-employment. *Small Business Economics*, 24, 131–142. https://doi.org/10.1007/s11187-003-3804-5

Sappleton, N., & Lourenço, F. (2016). Work satisfaction of the self-employed: The roles of work autonomy, working hours, gender and sector of self-employment. *The International Journal of Entrepreneurship and Innovation*, 17(2), 89–99. https://doi.org/10.1177/1465750316648574

Scandura, T. A., & Lankau, M. J. (1997). Relationships of gender, family responsibility and flexible work hours to organizational commitment and job satisfaction. *Journal of*

Organizational Behavior, 18(4), 377–391. https://doi.org/10.1002/(SICI)1099-1379(199 707)18:4<377::AID-JOB807>3.0.CO;2-1

Sebastian, J., & Moon, J. M. (2017). Gender differences in participatory leadership: An examination of principals' time spent working with others. *International Journal of Education Policy and Leadership*, 12(8), 1–16. http://journals.sfu.ca/ijepl/index.php/ijepl/article/view/792 (Accessed: 01.06.2024).

Seedat, S., Scott, K. M.., Angermeyer, M. C., Berglund, P., Bromet, E. J., Brugha, T. S., et al. (2009). Cross-national associations between gender and mental disorders in the WHO World Mental Health Surveys. *Archives of General Psychiatry*, 66(7), 785–795. https://doi.org/10.1001/archgenpsychiatry.2009.36

Sevä, J. I., Vinberg, S., Nordenmark, M., & Strandh, M. (2016). Subjective well-being among the self-employed in Europe: Macroeconomy, gender and immigrant status. *Small Business Economics*, 46, 239–253. https://doi.org/10.1007/s11187-015-9682-9

Skica, T. (2020). *Wpływ polityki gmin na rozwój lokalny: cele strategiczne, polityki budżetowe oraz instrumentalizacja wsparcia*. Wyższa Szkoła Informatyki i Zarządzania z siedzibą w Rzeszowie oraz Oficyna Wydawnicza ASPRA, Rzeszów-Warszawa.

Sloane, P. J., & Williams, H. (2000). Job satisfaction, comparison earnings, and gender. *Labour Economics*, 14(3), 473–502. https://doi.org/10.1111/1467-9914.00142

Spasova, S., & Wilkens, M. (2018). The social situation of the self-employed in Europe: labour market issues and social protection. In: B. Vanhercke, D. Ghailani & S. Sabato (eds.), *Social policy in the European Union: State of play* (pp. 97–116). European Trade Union Institute & European Social Observatory, Brussels.

Suh, C., & Punnett, L. (2022). High emotional demands at work and poor mental health in client-facing workers. *International Journal of Environmental Research and Public Health*, 19(12), 7530. https://doi.org/10.3390/ijerph19127530

Sullivan, D. M., & Meek, W. R. (2012). Gender and entrepreneurship: A review and process model. *Journal of Managerial Psychology*, 27(5), 428–458. https://doi.org/10.1108/02683941211235373

Super, D. E. (1980). A life-span, life-space approach to career development. *Journal of Vocational Behavior*, 16(3), 282–298. https://doi.org/10.1016/0001-8791(80)90056-1

Szaban, J. (2018). Self-employment and entrepreneurship: A theoretical approach. *Central European Management Journal*, 26(2), 89–120.

Tammelin, M. (2019). The solo self-employed person and intrinsic financial security: Does the promotion of self-employment institutionalise dualisation? *Journal of Poverty and Social Justice*, 27(2), 219–234. https://doi.org/10.1332/175982719X15535215192741

Wagner, J. (2005). Nascent necessity and opportunity entrepreneurs in Germany: evidence from the regional entrepreneurship monitor (REM). *Discussion paper No. 1608*. Institute for the Study of Labor. http://ftp.iza.org/dpl608.pdf (Accessed: 31.05.2024).

Whiteford, H. A., Degenhardt, L., Rehm, J., Baxter, A. J., Ferrari, A. J., Erskine, H. E., Charlson, F. J., Norman, R. E., Flaxman, A. F., Johns, N., Burstein, R., Murray, Ch, J. L., & Vos, T. (2013). Global burden of disease attributable to mental and substance use disorders: findings from the Global Burden of Disease Study 2010. *Lancet*, 382, 1575–1586. https://doi.org/10.1016/S0140-6736(13)61611-6

Xiu, L., & Ren, Y. (2022). Gain or loss? The well-being of women in self-employment. *Frontiers in Psychology*, 13, 986288. https://doi.org/10.3389/fpsyg.2022.986288

Zaheer, R., & Hussain, S. (2015). Globalization, media access and role of woman entrepreneurship in sustainable economic development of Pakistan. *Journal of History and Social Sciences*, 6(1), 57–81.

3 Modern Data-Driven Management and Optimization of Development Opportunities Creation for Women
A Referential Investigative Focus on the UN SDG Targets for Women

Chikezie Kennedy Kalu

3.1 INTRODUCTION

In 2015, the member states of the United Nations (UN) collectively agreed to pursue and implement the vision of sustainable development across countries globally; and they set out 17 Sustainable Development Goals (SDGs) to guide development efforts at various levels. One of the fundamental principles adopted in order to achieve adequate sustainability and sustainable developments is that of *inclusiveness*, which means society must leave no one behind in the process of development. Women, in particular, should benefit from opportunities for business, for learning and for and connecting to social and economic networks. Information and communication technologies (ICTs) are playing a crucial role in modern day development, and they are now deeply embedded as the platform through which information is gathered, opportunities are shared, skills training are delivered and can also contribute greatly to women's empowerment and development (Hussain, 2016). As defined by the UN Women's Empowerment Principles in 2011, "Empowerment means that people, both women and men can take control over their lives: gain skills (or have their own skills and knowledge recognized, set their own agendas, solve problems, increase self-confidence and develop self-reliance". Therefore, the process of change that increases choice (resources) and enhances the capacity to make choices favorable to oneself and to society in general, can be largely defined as empowerment (Balk, 1996). To empower women is ensuring that they adequately and effectively participate fully in economic, social and political life, and this is very much needed in our world.

With the varying and evident challenges to women's empowerment, cutting across countries globally, innovative measures are continually needed to overcome these challenges and create processes, systems, tools, and cultures that can provide cost effective, efficient, intelligent, and better management and optimization of women empowerment, the creation of better opportunities for women and women's empowerment at large. This analytical data driven research analysis is therefore aimed at exploring the variables that more partly and directly influences women's empowerment and developmental opportunities for women across countries, with reference to the UN SDG targets in relation to socio-economic development, using new or innovative approaches, with the aid of machine learning engineering, data science and technological tools.

3.2 LITERATURE REVIEW

The SDGs provide new impetus to the push for policies and systems to improve gender equality at a global level; though the global progress in reaching these developmental targets has been uneven. Gender equality, for example, in many domains is still a far reach in many countries, despite progress (even somewhat impressive) in enrolling girls in primary education (Queisser, 2016). As stated by Signal (2003), even in developing countries, women are now gradually being considered as the development core, thereby further indicating that women, gender and cooperation are crucial subjects in development (Singhal, 2003). Generally, the Millennium Development Goals (MDG) is mainly about socio-economic development, measured in terms of jobs and incomes, and includes improvements in education, health, human development, and opportunities (Mohammad and Razmi, 2012). In 2000, the United Nations (UN) declared the MDG with the aim of sparing no effort to free humanity from abject and dehumanizing conditions of poverty, which more than a billion people are subjected to globally (Ackon, 2010). However, there are still numerous areas where gender imbalances exist due to diverse conditions of women's work, cultural beliefs, value systems, and attitudes (Akinola and Tella, 2013). Many scholars have also stated theories in relation to women empowerment and its challenges, as a testament of the importance of the development of women to our societies. Malhotra (2005) describes

> the term empowerment as a process by which women obtains larger command over intellectual and material resources, which help them to raise their self-dependence and increase them to laid stress on their rights and dare the philosophy of patriarchy and the gender based prejudice against women.
>
> (Malhotra and Schuler, 2005)

Mason and Smith (2003) stated that women's empowerment absolutely imposes that in all human societies the men commands women, or to be further specific, males control at least small segment of the women of their societal strata, especially of their families and households (Mason, Oppenheim & Smith, 2003). Boender et al. (2002) said that "Women make up half of the world's population, and form a cross-cutting group that overlaps all other groups in society" (Boender, Malhorta & Schuler, 2002). Jehan (2000) states that impediments on women strengthening start from social practices and convictions. Patriarchal-based control is legitimized and practiced on the premise of society, a composite of custom, and religion which decides and characterizes the endorsed part of women in the public arena, their versatility and their admission to financial assets, social, and political power too (Jehan, 2000).

Globally, about 50% of all women and about 75% of all men are noted to participate and contribute to the labor force in the world; while also the literacy rate for all females is 82.7% and 90% for all males (UN Statistics, 2015; Central Intelligence Agency, 2013). Such noted observations and data makes it pertinent for all concerned sectors to concentrate on activities, developments, and policies geared to improve the percentage bars by mainly promoting women's active participation and contribution in professional work force. The SDGs continuous work to make the right choices in improving loves in a sustainable way for future generations is carried out in a spirit of partnership and pragmatism. The SDGs are an inclusive agenda to provide clear guidelines and targets for all countries to adopt and integrate with their own priorities and the environmental challenges of the world at large. The SDGs are positioned to tackle the root causes of poverty and challenges of humanity; and unite us together to make positive changes for both people and planet (UNDP Report, 2016). In the changing world of work, several needed measures have been framed to ensure women's economic empowerment, which includes bridging the gender gap which stands at a global score of 24%; addressing the gender gaps in leadership, entrepreneurship and access to social protection. Furthermore, there needs to be more done in terms of ensuring gender responsive economic policies for poverty reduction, job creation, sustainable, and inclusive growth (UNDP Report, 2016). The adoption of the concept of women engagement and its advocacy of global goals should be done by not only the governments but also at the work places. Women are vital parts of creation and there is therefore no need for any special significance to be given but honor and encourage their participation in the various activities of society to create a better place for living (Anurag, 2017)).

3.3 METHODOLOGY

This research study is very much data driven, as it involves the synthesis of certain socio-economic factors and some related UN SDG target variables for countries around the world from approved databases and portals. The research activities revolved around the various data analysis, data analytics involving statistical social science terminologies, data science and machine learning processes to carry out the necessary studies and obtain results. In addition to the theoretical and practical concepts and parameters already known, new metrics are derived to capture and measure and confirm key tests and analysis.

Optimization of Development Opportunities Creation for Women 67

Furthermore, for such various methodical analyses, data acquisition, data cleaning, data wrangling are among key preliminary steps to ensure that the required data is importantly used for such analysis, following the key important steps:

Step 1: Acquire data from validated open access data stores and live data web portals.
Step 2: Clean the data, label it appropriately, wrangle it and make it fit for purpose.
Step 3: Store the data and partition them accordingly for use.
Step 4: Feed the data into the particular analysis tool, model and process as required.
Step 5: Prepare and specify how results will be reported.

Furthermore, since the set datasets obtained are from reliable and approved data sources, stores, and portals, approved as global statistical benchmarks, therefore, the datasets are assumed to be reliable, credible, and acceptably correlated.

3.3.1 Conceptual Model of Influencing Factors

Considering the drivers or factors that influence women's empowerment and/or opportunities available for women's development and empowerment, certain measurement metrics for our analysis are derived and defined to specifically define the socio-economic potential of women can be measured and managed with reference to certain UN SDG target metrics; and also decision-making processes can be optimized for more efficient and effective management of women's empowerment and women's opportunities for development.

From the literature and earlier described principles, we build a conceptual framework to describe and analyze the link between the (dependent variable): *women's development capacity* ($W\psi$) and the key factors (independent variables) affecting water development or the creation of development opportunities for women. These independent variables or set of target UN SDG metrics used as a reference of measurement are: ***access to electricity as percentage of population (var1), gross tertiary education enrolment (var2), proportion of youths and adults with information technology skills (var3), legal frameworks for gender equality (GE*) achievement (var4), women population (var5), country GDP per capita (var6),*** and ***females with mobile phone within a population (var7).*** By incorporating these independent variables, a more robust predictive model for $W\psi$ can be obtained which can aid effective and efficient policy decision-making, resource forecasting, systems design, better monitoring and efficient management.

3.3.2 Research Questions and Hypothesis

Having established a good theoretical background and link between the factors that influence women's development management in a given region or country, a model was built (Figure 3.1) to show how all these variables connect together and utilized to carry out analyses. Analyses on how these factors influence the metric women's

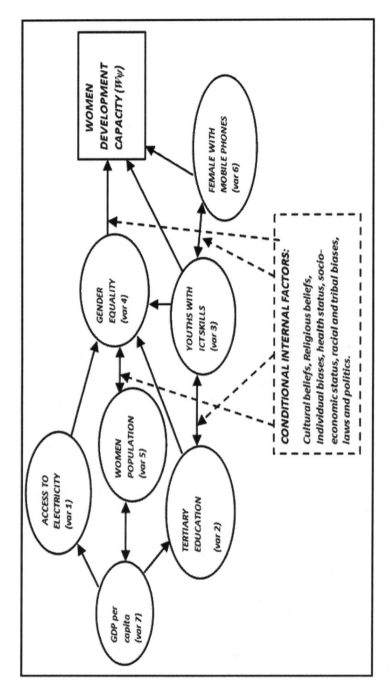

FIGURE 3.1 Conceptual model for the determination and prediction of women's development capacity ($W\psi$). (Developed by author.)

Optimization of Development Opportunities Creation for Women

development capacity ($W\psi$) or the measure of the socio-economic potential of women in a given region as a key metric for effective and efficient management, creation and optimization of development opportunities for women empowerment were to be carried out. The research questions and accompanying hypotheses are thus:

Q1: Can the socio-economic potential of women or women's development capacity ($W\psi$) in a given region be determined or measured and/or predicted?
Hypothesis 1(H1): Women's development capacity ($W\psi$) in a country can be measured and also predicted.
Q2: Is women's development capacity ($W\psi$) for region/country affected by many factors and not just one like the GDP per capita of a country?
Hypothesis 2(H2): Internal factors or independent variables within a region/country can positively or negatively affect how the socio-economic potential and opportunities available to women ($W\psi$).
Q3: Which of the factors most influences women's development capacity ($W\psi$) for a country?
Hypothesis 3(H3): Some factors have greater influence than others on the women's development capacity ($W\psi$) in a region/country.

3.3.3 Mathematical and Analytical Descriptions for Women's Development Capacity ($W\Psi$)

For this data-driven analytical mathematical approach to the management and optimization of the creation of socio economic opportunities for women's empowerment, the focus is to be able to precisely analyze, model and measure women's development capacity or a measure of the socio-economic potential of women ($W\psi$) relative to the creation of opportunities for women with reference to the UN SDG target goals for sustainable development.

Therefore, for this research work, the analytical back ground to determining the derived women's development capacity ($W\psi$) is described:
From Definition via UNDP (United Nations Development Program):

$$HDI = \sqrt[3]{LEI * EI * II} \qquad (3.1)$$

Where: *HDI* is the Human Development Index; *LEI* is the Life Expectancy Index; *EI* is Education Index; *II* is the Income Index; all within a given country. *HDI* usually combines multiple dimensions of development to provide a comprehensive measure of population well-being.

Since *HDI* involves the income earnings of a population, which is indirectly also influenced by how rich or poor a country is, it can thereby also be correctly postulated that:

$$HDI \propto GDP_c \qquad (3.2)$$

Where: GDP_C is a country's GDP (gross domestic product) per capita.

Furthermore, and also from definitions, it can be postulated that the development of women in a country will no doubt be affected by the *HDI* of the country and also unique factors that affect women's development in the socio-economical, legal, and political spaces.

Therefore, it can mathematically be postulated that:

$$W\Psi \propto HDI \text{ and also; } W\Psi \propto V_W \tag{3.3}$$

Where: V_W is defined as unique factors of reference or variables that affects women's empowerment in a country. For the focus of this research work; V_W will therefore be the set of UN SDG target goals that affects women's empowerment.

With such multiple direct proportional relationships, the mathematical partial or part variation principle is then applied. Here $W\psi$ varies partly with *HDI* and also varies partly with influencing variables (V_W).

Therefore for every country we have:

$$W\Psi = HDI + Wp(We * Wt * Wm * Wg * ...) \tag{3.4}$$

Where: *Wp* is the percentage proportion of women in the population. *We* is the percentage population having access to basic resources(in this case: electricity). *Wt* is the percentage gross enrolment of the population (both sexes) into tertiary education. *Wm* is the percentage of females having mobile phones. *Wg* is the percentage proportion of legal frameworks for gender equality achievements within a country.

Since these variables vary with time, it can then be defined that during a period of *T* number of years; the annual women's development capacity ($W\psi$) is defined as:

$$W\Psi = \left[HDI + Wp(We * Wt * Wm * Wg * ...)\right] * T \tag{3.5}$$

Thus, for a single year dataset, as applied in this research work, T = 1.

Therefore, calculating the women's development capacity ($W\psi$) as a sum across *T* number of years is defined as:

$$W\Psi = \sum_{1}^{T}([HDI + Wp(We * Wt * Wm * Wg * ...)] * T) \tag{3.6}$$

Furthermore, in the case of changes within years; the women's development capacity ($W\psi$) is defined as a first-order differential equation, given as:

… Optimization of Development Opportunities Creation for Women

$$\frac{dW\,\Psi}{dT} = \left[HDI + Wp(We * Wt * Wm * Wg * \ldots) \right] T + CW\,\Psi \qquad (3.7)$$

Where: C is a constant of factors that do not change with time or experience very minimal changes with time; for example: bio-data, race, identity, etc.

3.3.4 Decision Trees Algorithm

A decision tree has leaves, branches, and nodes. Nodes are where a decision is made. A decision tree consists of rules that we use to formulate a decision (or prediction) on the prediction of a data point.

Every node of the decision tree represents a feature, while every edge coming out of an internal node represents a possible value or a possible interval of values of the tree. Each leaf of the tree represents a label value of the tree (So et al., 2020).

An important metric used to measure the degree of randomness in a variable is called **entropy**. In information theory, entropy measures how randomly distributed the possible values of an attribute are. The higher the degree of randomness is, the higher the entropy of the attribute. The formula for calculating entropy is given as:

$$H(\text{distribution}) = \sum_{i=0}^{n} -p_i \log_i p_i \qquad (3.8)$$

Where: p_i represents the probability of one of the possible values of the target variable occurring. So, if this column has n different unique values, then we will have the probability for each of them ($[p1, p2, \ldots, pn]$) and apply the formula.

Also, **information gain** is the entropy values at the different levels of the decision tree at different branch nodes. It is used to decide which branch nodes become part of the decision tree from the root node.

There is also another metric used to measure the degree of randomness of a variable's distribution, and it is called **Gini impurity**. The formula for calculating Gini impurity is given as:

$$Gini(\text{distribution}) = 1 - \sum_{i=0}^{n} p_i^2 \qquad (3.9)$$

Where: p_i here represents the probability of one of the possible values of the target variable occurring.

3.3.5 K-nearest Neighbors (KNN) Algorithm

K-nearest neighbors (KNN) is a supervised machine learning algorithm that can be used for either regression or classification tasks. Since the KNN algorithm does not make assumptions about the underlying distributions of the data, it is thereby known to be non-parametric in nature.

To perform the algorithmic procedure, we choose **k**, which is the number of nearest neighbors, and **p** is the number of classes or number of data columns.

Where: $\mathbf{k} = \sqrt{Nt/2}$; Nt = Number of training data.

Also, **k** should be an odd numbered value and must not be a multiple of the number of classes.

The distance between the sample data points is described as the *Minkowski* distance (**Dm**) is calculated using the formula:

$$\left(\sum_{i=1}^{n} |x_i - y_i|^p \right)^{1/p} \tag{3.10}$$

Where: where X and Y are specific data points or data nodes, **n** is the number of dimensions, and **p** is the Minkowski power parameter.

3.4 RESULTS AND DISCUSSION

This section outlines the results obtained and the various associated interpretations, and discussions on the results obtained in harmony with theories derived and literature obtained.

Figure 3.2 shows the scaled profile charts of how certain socio-economic factors affect women's development in relation to key target UN SDG women factors for selected countries around the world, using Table 3.1. The derived development capacity or the measure of socio economic potential of women per country (**Wψ**) is expectedly shown to increase with the corresponding increase in a country's HDI, GDP per capita, and observed UN SDG services like number of females having a mobile phone and those having access to electricity within a population. Such a relationship between these observed variables also occurs vice versa, as shown in Figure 3.2. Furthermore, and also interestingly, the women's development capacity for the UN SDG Target service observed in Figure 3.2 shows a greater increase and decrease in relation with the number of females having phones than for the case of the percentage of a population having access to electricity for each country. While there is shown to be an estimated 10–15% increase or decrease in (**Wψ**) for the case of the number of people with access to electricity, there is about a corresponding 20–30% increase or decrease in (**Wψ**) in relation to the increase or decrease in the number of females with mobile phones, respectively. Therefore, from the observations, it supports the socio-economic factors that the richer or greater the GDP per capita is for a country; the more opportunities there are for women develop and increase their potentials, and vice versa. Additionally, the development capacity of women is not only boosted by the income or GDP of a country but must also be sustained by critical and strategic developmental projects, to enable a corresponding increase in HDI and also service available to the women and greater population. It has been defined that economic development of societies and a country has a direct relationship with women

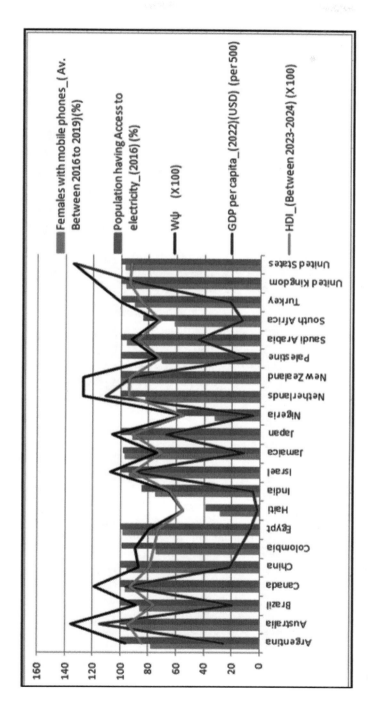

FIGURE 3.2 Plot of socio-economic factors and UN SDG variables (infrastructure) affecting women's development. (Developed by author.)

empowerment (Duflo, 2011). A report by McKinsey in 2015 also indicated that the global annual GDP could see an additional rise if women play an importantly identical role to men in the labor markets (Malik, 2018). Furthermore, the greater effect of the number of females using a mobile phone than the number of people having access to electricity on the socio-economic potential of women once again sheds light on the powerful influence of ICT and information technologies on the growth of society and women in particular, especially in our modern world. This does not in any way reduce the need for basic services like electricity that contribute immensely to the development of women and society; but there it uniquely contributes to a body of knowledge that the need to communicate, conduct business, and create opportunities using modern technologies via mobile phones can more hugely and rapidly influence/affect the socio-economic potential of women in a country and the society at large. There is no doubt that ICT continues to play a vital role in women's employment, opportunities for global well paid jobs, female entrepreneurs' facilitation, online job creation, and economic empowerment (Prasad & Sreedevi, 2007; Nikulin, 2016). Furthermore, adequate internet access, computers, and mobile phones have been shown to be key tools for women in E-business as an enabler of self-sufficiency and such tools are considered important instruments for women to advance as entrepreneurs and in organizations (Davidson, 2012; Afrah & Fabiha, 2017; Lutfunnahar, 2022).

In Figure 3.3, scaled profile charts show how certain socio-economic factors affect women's empowerment in relation to key target UN SDG women factors (in this case tertiary education enrolment and legal frameworks that boost gender equality) for selected countries around the world, using Table 3.2. The derived measure of socio-economic potential of women per country ($W\psi$) which has earlier been shown to increase or decrease with how wealthy or poor a country is respectively; is also positively or negatively affected by the corresponding increase or decrease with tertiary education enrolment and legal support for gender equality within a country; as also shown in Figure 3.3. Furthermore, the women's development capacity for the UN SDG target service observed in Figure 3.3 shows a greater increase and decrease in relation with the tertiary education gross enrolment for both sexes than for the case of the percentage amount of legal frameworks for gender equality achievement for each country. While there is shown to be an estimated 20–30% increase or decrease in $W\psi$ for the case of the legal frameworks for gender equality; there is slightly higher corresponding 35–40% increase or decrease in $W\psi$ in relation to the increase or decrease in the grow tertiary education enrolment for both sexes, respectively. Also, it is observed that the legal frameworks amount is affected by the tertiary education enrolment; and these frameworks for gender equality are shown to more greatly affect the women's development capacity ($W\psi$) for more developed or richer countries. Therefore, from the observations in the charts of Figure 3.3, the development capacity of women or the potential for the creation of more developmental opportunities for women within a country increases with the corresponding increase in tertiary education enrolment and the pertinent, practical legal frameworks that encourage and institute gender equality in societies. It has also been observed that women are more likely to be poor, less educated, malnourished, and overworked relative to men; as a negative effect of gendered economic opportunities which affects women negatively

Optimization of Development Opportunities Creation for Women

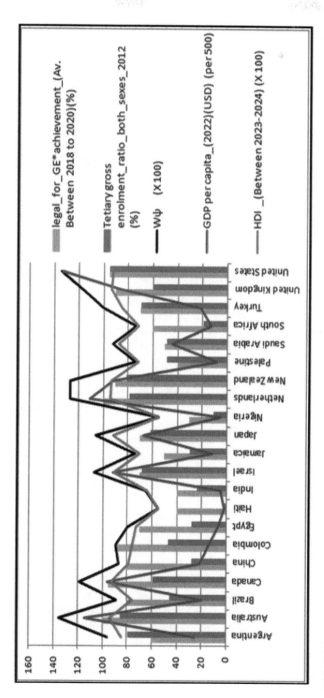

FIGURE 3.3 Plot of socio-economic factors and UN SDG variables (equality and education) affecting women's development). (Developed by author.)

and makes women and men on average occupy different class positions (Davis, 1981; Benerõa & Roldan, 1987; Deere, 1990; Wright, 1996). Furthermore, the positive influence of tertiary education enrolment on the legal frameworks for gender equality and cumulatively on the development capacity of women underscores the importance of education in societies as a major contributor to the creation of more opportunities for women, as it also supports a solid critical foundation of frameworks that encourage gender equality within the society, businesses, and organizations for a greater utilization of the huge potentials of women in countries and societies at large. The need therefore for robust, encompassing strategies that take into account promotion of continuous education for women and enabling laws to promote gender equality for the good of society and the rapid development and empowerment of women, becomes paramount. Earlier postulations have also established that the effect of gender earnings differentials that raise access to new technologies on growth and foreign exchange earnings depends on the level of education since skilled labor is more conducive to the effective and efficient absorption of new technologies (Nelson & Phelps, 1966; Seguino, 2000).

3.4.1 Decision Trees Classification Analysis

From the methodologies and theories for the machine learning engineering decision trees classification algorithm described earlier, the resultant decision tree for certain socio-economic factors affecting women's empowerment, with respect to key UN SDG targets for women for selected countries around the world using data from Table 3.1 and Table 3.2, is shown in Figure 3.4. Using the listed tabulated variables and its associated attributes as described, the decision tree was calculated and determined by the classification algorithm using node gain or node entropy as the figure of merit (Agarwal, 2016). From the algorithmic calculations using the python programming language, the number of females having mobile phones within a country attribute has the highest gain or entropy (3.807), as it also has the most homogenous branches in the tree; and so it becomes the decision or root node. Furthermore, for the successive hierarchal branch nodes from the root node, the algorithm intelligently calculates that GDP per capita and tertiary education enrolment measures within a country has the highest gains (2.807 respectively) and thereby both forms the branch nodes and leads to their individual leaf nodes which indicates the end points of the tree. Additionally, the end point of the tree indicates the process has run out of attributes or all the data available has been classified perfectly, thereby resulting in the final decision tree shown in Figure 3.4.

From the target value of the root node (i.e. <= 94.245 or > 94.245); it is calculated that an estimate of about 14 samples or countries meet each for both criteria, which is then followed up by its branch nodes having seven countries each meeting the corresponding GDP per capita and gross amount of enrolment in tertiary education criteria being influenced from the root node initial target; and then the process runs out of attributes stemming from the root node calculated target criteria. These values means that for the **decision rule** and target reference from the root node, when value is <= 94.245, the branch node chosen will be tertiary education

Optimization of Development Opportunities Creation for Women 77

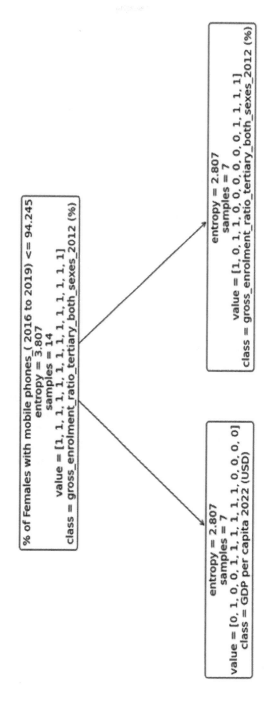

FIGURE 3.4 Decision tree plot of socio-economic factors and UN SDG variables affecting women's development. (Developed by Author).

Notes:

For UN SDG metadata: https://unstats.un.org/sdgs/metadata/?Text=&Goal=1&Target=

For demographic data: World Bank Data Portal: https://data.worldbank.org/indicator/NY.GDP.PCAP.CD

enrolment, and therefore, countries selected will be countries who should take tertiary education enrolment as the topmost priority to boost women's development capacity. Alternatively, when the value is > 94.245, the branch chosen will be the GDP per capita; and countries selected under this branch will need to take boosting or maintaining their GDP per capita as a topmost priority to boost women's development capacity. This is a brief demonstration of an integrated intelligent process using decision trees to manage, organize, optimize and create better strategic monitoring and implementation of key UN SDG goals with respect to women's empowerment. For even further analyses, Table 3.1 and Table 3.2 could be increased or reduced to accommodate more or fewer factors or attributes respectively, to analyze the effects of diverse target variables using the decision tree algorithm. Therefore, from the observations of the analyzed decision tree, this indicates that for the set of available data analyzed, the number of females having mobile phones has the strongest influence on the development of women and the creation of developmental opportunities. This further supports research-based arguments, that information, the internet, and communication technology have very great potential in growing an empowered women community in countries around our modern world. As highlighted during the review of progress in the implementation of the platform of action in the twenty-third special session of the UN General Assembly, based on knowledge and emerging experiences, it was recognized that ICT had created various new opportunities for women and contributed to electronic commerce, networking and knowledge sharing activities (UN, 2005). Additionally, this decision tree analysis indicates that the GDP per capita of a country and enrolment in tertiary education will also strongly affect the number of mobile phone available for females in a country, which will ultimately influence the development capacity for women. Therefore, a robust strategy and/or modern tools and systems to address, manage and optimize developmental opportunities for women has to factor in better creation of wealth and educational growth, needed to enhance the productivity of women and also bring them up to speed with the rapid technologically driven modern society, which has great potentials for individuals, businesses, organizations, societies, and countries globally. As Moser (1989) puts it, five approaches are key in analyzing women and development relationships, and most of these approaches appear simultaneously. He stated that these approaches should include welfare approach, justice approach, anti-poverty approach, empowerment approach; and these approaches should be considered as a special and undependable approach (Moser, 1989).

3.4.2 Prediction of the Women's Development Capacity ($W\psi$) from Some Key Referential UN SDG Target Variables Using the KNN Analysis

Furthermore, with the aim to thoroughly analyze the influence of the key target UN SDG variables water sources for women's empowerment and opportunities creation across countries and in the world, a study, analysis, and prediction is also made in Figure 3.5 for the average development capacity or the **measure of socio economic potential of women ($W\psi$)** Globally, which is also the dependent variable, using the KNN Machine Learning probabilistic Model from independent variables: *% access*

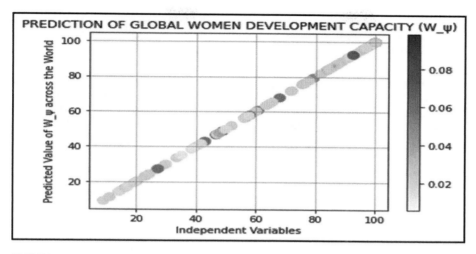

FIGURE 3.5 Predicted global women's development capacity ($W\psi$) using a KNN analysis machine learning model. (Developed by author.)

Notes:
For UN SDG metadata: https://unstats.un.org/sdgs/metadata/?Text=&Goal=1&Target=
For demographic data: World Bank Data Portal: https://data.worldbank.org/indicator/NY.GDP.PCAP.C

to electricity as percentage of population (2016), % gross tertiary education enrolment (2012), % proportion of youths and adults with information technology skills (2019), % legal frameworks for Gender equality (GE*) achievement (2020), % of Women in population (2022) and % of females with mobile phone within a population (2017). These independent variables available datasets were obtained for each country and for specific years where complete, and globally as recorded by the UN SDG portals. Additionally, the analysis was based on the definition of Equation 3.5; but here the HDI was excluded (i.e. HDI = 0), to enable the analysis of the linear relationship between the derived women's development capacity metric and the set of listed independent variables, to further understand how women's development capacity ($W\psi$) is affected more directly by the set of target UN SDG variables. This analysis is also pertinent to the development of efficient and modern processes, systems and management strategies for better management, and development of opportunities for women globally.

For the KNN internal flow prediction analysis, the model was built, as defined in the methodology with the following parameters: (K Nearest-Neighbors = 6, Number of Classes or Columns = 6, Total Data Sample Number (N) = 261, Training Data Sample size = 208) shows a **f1-score** of close to 1 (100%) which is excellently optimal and indicates the model has done an excellent job of predicting the global development capacity ($W\psi$) giving other UN SDG independent global variables as earlier listed. For a real life situation as this case, the f1-score is of most importance and a better metric due to the presence of imbalance class distributions which usually associated with real life datasets. Furthermore, the almost perfect alignment of the linear prediction KNN plot in Figure 3.5 is an indication of its **precision**, which also had a near perfect score of about of 1 (100%), which indicates the excellent intuitive ability of the model not to label as positive a data sample that is negative and vice versa. Additionally, the close to perfect prediction plot is further confirmed by also the **recall** score of 1.00 (100%), which is the best and highest recall value and indicates the intuitive ability of the model to find all positive samples, which also means the ability of the model to predict positive outcomes out of actual positives. Also, a very important classification metric of the KNN predictive model if the confusion metrics shown as:

Confusion Matrix: [[0 0]
[0 53]]

The values of the matrix, indicates that there are no sample negative values incorrectly predicted as positives (i.e. False Positives (FP)); and also 53 or all sample positive values were correctly predicted as actual positives (i.e. True Positive (TP)); and no sample positive values were incorrectly or correctly predicted as negatives (i.e. TN= 0 and FN = 0). This therefore indicates that the KNN model experienced very negligible confusion during prediction and so its prediction accuracy is very good as also indicated by the f1-score. From the interpreted results briefly described in Figure 3.5, the predictions model and analytical descriptions of the development capacity of women globally and how such influences the creation of development opportunities for women within countries are important, innovative steps in the process of developing and optimizing effective systems, strategies, processes and resources for better management and creation of developmental opportunities for women. This will ensure better global connections, collaborations, and proper management and distribution of resources for an effective women and human development, in both poorer and richer countries; and also mitigate the unique challenges of gender inequality, limited education, limited basic resources and technology availability; and other socio-economic circumstances and laws that limit the creation of developmental opportunities for women globally.

The excellent prediction plot as shown in Figure 3.5 and the obtained values of the KNN analysis for women's development capacity ($W\psi$) also supports **Hypothesis 1(H1)**, that the global measure of the socio-economic potential of women or women's development capacity average internal flow of water can be measured and also predicted, which will also add to better management, deeper strategy development, design of a variety of systems, for an optimized development and opportunities creation system and laws for women to thrive. The results also indicates that the independent variables of *access to electricity as a percentage of population, gross tertiary education enrolment, proportion of youths and adults with information technology skills, legal frameworks for gender equality (GE*) achievement, women population* and *females with a mobile phone within a population* as earlier listed do also influence the amount of women's development capacity ($W\psi$) in line with

Hypothesis 2 (H2) and also Hypothesis 3 (H3). This can also mean that any of the independent variables can be modelled as a dependent variable and other UN SDG target metrics can be analyzed as independent variables, to determine the effect of variables on a particular dependent target metric variable. Such flexible analysis would further enhance insights to how well women's empowerment and opportunities creation can be better managed, measured, predicted, and optimized to aid greater women's empowerment, unlocking women potentials and great contribution of women to global development as needed. The KNN analysis has the advantages of being very easy to implement, faster and accepts new data seamlessly than other algorithms, because it requires no training before predictions; but it also has some cons, which are that it does not work well for categorical features due to distance calculations and larger datasets which will require a higher cost of more calculations during predictions (Swamynathan, 2017).

3.5 LIMITATIONS AND RECOMMENDATIONS OF STUDY

This research study was focused on the data-driven modern and innovative approach of certain factors that influence the creation and development of opportunities for women, with reference to selected UN SDG target metrics. Though a focused, detailed, and insightful analyses of the socio-economic factors affecting women's empowerment and the creation of developmental opportunities for women was carried out, there were also a few limitations within the scope of this study. In terms of data acquisition, the real life datasets used were from approved and dedicated data stores, repositories, and databanks, which in some cases were not fully complete with respect to certain regions or countries, thereby limiting further extensive regional or country based analyses. For example, in some specific scenarios for countries, up to date and/or yearly gender related data of women with respect to development and in line with the UN SDG targets, are not readily available, thereby providing a challenge in carrying out additional comparative analyses for regions and locations; though where appropriate acceptable estimates were also employed. Additionally, the data analysis did not involve sample surveys from respondents, which could further add more perspectives to the study, by employing mixed or data triangulation methods; though such comprehensive surveys could also be sometimes inconsistent across different countries; and carrying out such surveys for every individual could be impractical.

Furthermore, with respect to the analyses carried out, a non-exhaustive list of key variables was considered and analyzed in relation to the socio-economic factors influencing women's development capacity, women's empowerment and the management of the creation of opportunities for women, with reference to selected UN SDG target KPIs. However, some internal and/or indirect variables (which sometimes may not be easily measured, due to the conditional nature of such variables) were not part of this analysis, such as variables that relate to socio-political and socio-cultural biases, beliefs, and practices, which also influences developmental factors in certain countries, and can also provide additional insights on how such factors also influence women's empowerment opportunities and capabilities (Khan & Mazhar, 2017).

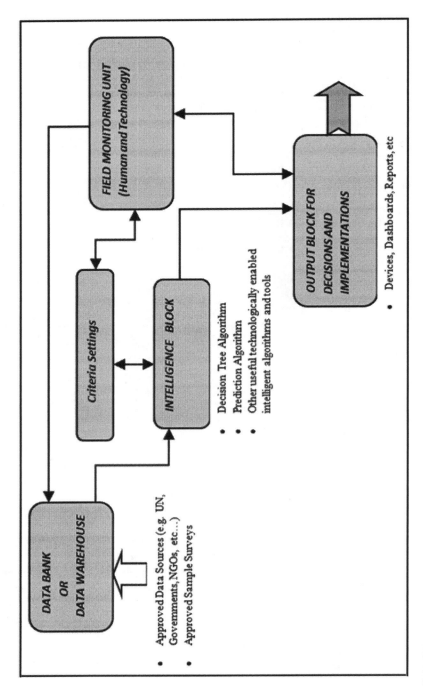

FIGURE 3.6 Proposed intelligent data-driven tool for women empowerment and development optimization. (Developed by author.)

Optimization of Development Opportunities Creation for Women 83

Thus, the data-driven research analysis was aimed at exploring the variables that more partly and directly influence women's empowerment and developmental opportunities for women across countries, with reference to the UN SDG targets in relation to socio-economic development. The already known modern machine learning engineering, data science and technological tools were employed in a new and innovative approach to address the challenges associated with availability of detailed insights to aid decision-making, strategy design, systems and process optimization for growth in development and quality of life.

The proposed system or tool to intelligently, efficiently and effectively manage and optimize the creation of developmental opportunities and development capacity of women across several countries of the world can be further investigated, with even the use of more algorithms and artificial intelligence systems developed, so as to increasingly 'unlock' the potentials of this system; with the required data also being available. This system promises efficient data and resource management, effective geographically diverse collaborations, intelligent monitoring, practically impactful realization of the UN SDG KPI targets, and other potentials which may not be fully known at this stage (see Figure 3.6).

3.6 FINDINGS

3.6.1 Analyses of UN SDG Women Targets and Socio-Economic Factors

3.7 CONCLUSIONS

More than ever in our modern world, women are needed to be equal partners in building a better, gender transformative agenda and sustainable developments in both public and private sectors of our societies and world at large. This theme is also an area of focus for the UN SDG programs and its related target KPIs for each country further emphasizes the need for increase in women empowerment for the ultimate good of humanity. Therefore, more precise, robust monitoring, innovative processes, tools and systems must continually be used in management, optimization, and creation of more developmental opportunities for women, to ensure a continuous increase in the development capacity and the socio and political potential of women to meet objectives of the UN SDG targets and developmental needs of societies. This research's main objective was to investigate, analyze, manage, and optimize some relevant factors and variables that influence the availability of socio-economic development opportunities and empowerment of women globally, with reference to specific target KPIs (key performance indicators) of the UN SDG (United Nations Sustainable Development Goals) using modern approaches to and within data analytics and machine learning engineering.

Datasets from trustworthy and recognized data stores, portals, and sources were used for this study's research work, analyses, demonstrations, and investigations, and as a result, it can be safely assumed that the datasets are reliable and credible. This analytically and mathematically data-driven research work investigated, analyzed and

TABLE 3.1
Socio-economic factors and UN SDG variables (infrastructure) affecting women's empowerment

Country	Wψ (X 100)	Women population 2022 (%)	GDP per capita (Year 2022) (USD) (per 500)	HDI_ (Between Year 2023-2024) (X 100)	Females with mobile phones (Av. Between Year 2016 to 2019) (%)	Population having access to electricity (Year 2016) (%)
Argentina	96.09720345	50.49338702	25.868	84.9	78**	100
Australia	135.451702	50.34518117	114.894	94.6	95**	100
Brazil	88.64130032	50.89345165	19.276	76	85.79	100
Canada	119.1633912	50.30386047	96.274	93.5	96**	100
China	86.88344398	48.96903717	21.42	78.8	94.88	100
Colombia	89.73786576	50.69669745	13.764	75.8	74.18	99.00445557
Egypt	79.3800152	49.43279672	6.356	72.8	98.36	100
Haiti	55.20897966	50.46428156	1.672	55.2	28**	38.69016647
India	65.52246956	48.40570567	4.676	64.4	75**	84.52681732
Israel	107.0865293	50.12234036	91.916	91.5	93.61	100
Jamaica	74.09461921	50.39999972	11.244	70.6	97.18	98.20426941
Japan	106.1323415	51.40053352	67.644	92	91.3	100
Nigeria	54.88624084	49.46118584	4.874	54.8	32.44	59.3
Netherlands	127.1467746	50.29758004	111.212	94.6	82.43	100
New Zealand	126.7220365	50.43052575	92.556	93.9	99.39	100
Palestine	74.26712711	49.10389139	7.578	71.6	70.46	100
Saudi Arabia	92.40783392	42.37870384	44.712	87.5	95.39	100
South Africa	73.50288608	51.30052759	12.882	71.7	61.04	84.2
Turkey	100.7482015	49.89242188	21.35	85.5	90**	100
United Kingdom	117.2947248	50.58361903	85.83	94	96**	100
United States	134.6178264	50.48679896	133.356	92.7	97**	100

Notes:
GE*: Means gender equality achievement measure of legal frameworks and **: Means estimated values.
For UN SDG metadata: https://unstats.un.org/sdgs/metadata/?Text=&Goal=1&Target=
For demographic data: World Bank Data Portal: https://data.worldbank.org/indicator/NY.GDP.PCAP.CD

TABLE 3.2
Socio-economic factors and UN SDG variables (equality and education) affecting women's empowerment

Country	Wψ (× 100)	GDP per capita (Year 2022) (USD) (per 500)	HDI (Between Year 2023-2024) (× 100)	legal_for_GE*achievement (Av. Between Year 2018 to 2020) (%)	Tertiary gross enrolment ratio for both sexes Year 2012 (%)
Argentina	96.09720345	25.868	84.9	60	78.9728775
Australia	135.451702	114.894	94.6	100	85.41391754
Brazil	88.64130032	19.276	76	80	45.2390213
Canada	119.1633912	96.274	93.5	95**	58.8835907 ##
China	86.88344398	21.42	78.8	80**	27.18442917
Colombia	89.73786576	13.764	75.8	90	46.21575165
Egypt	79.3800152	6.356	72.8	70**	27.61831093
Haiti	55.20897966	1.672	55.2	40**	1.02658999
India	65.52246956	4.676	64.4	40	22.86137009 #
Israel	107.0865293	91.916	91.5	70	67.79532623
Jamaica	74.09461921	11.244	70.6	50	29.06176949
Japan	106.1323415	67.644	92	70	61.45817184
Nigeria	54.88624084	4.874	54.8	30	10.07094955 #
Netherlands	127.1467746	111.212	94.6	100	78.50106812
New Zealand	126.7220365	92.556	93.9	90	80.84335327
Palestine	74.26712711	7.578	71.6	40	48.17982101
Saudi Arabia	92.40783392	44.712	87.5	50**	48.56230927
South Africa	73.50288608	12.882	71.7	60**	18.99403954
Turkey	100.7482015	21.35	85.5	70**	69.30194855
United Kingdom	117.2947248	85.83	94	90	59.22314072
United States	134.6178264	133.356	92.7	95**	94.84227753

Notes:
GE*: Means Gender Equality achievement measure of legal frameworks
**: Means estimated values; #: Means latest data available at year 2011; ##: Means latest data available at year 2000 and ###: Means latest data available at year 1986.
For UN SDG metadata: https://unstats.un.org/sdgs/metadata/?Text=&Goal=1&Target=

demonstrated that the socio-economic or the derived development capacity of women ($W\psi$) in a region and/or country and the variables or factors affecting and influencing $W\psi$, can be measured and also accurately predicted. Furthermore, this study suggests that these independent variables (which are also influenced by underlying factors) holistically affect $W\psi$ and can be accurately accounted for, as needed information for effective and efficient strategy, systems, policies, technologies development, to aid effective and efficient management and creation of empowerment opportunities for women globally. This was also in alignment with the three hypotheses, which were postulated, tested, then accepted in this research work. Analyses thereby inferred and revealed that the GDP per capita of a country, amount of modern ICT technological infrastructure, for example mobile phone usage, women education and legal frameworks needed for gender equality, are paramount towards the rapid and effective development of women, and also influences the creation of more empowerment opportunities for women. Additionally, these influencing factors affect women's empowerment at various degrees across countries.

Furthermore, the new data-centric techniques and theories aided by the application of the known machine learning KNN and decision tree algorithms, provided a unique/modern approach to measure, analyze, investigate, deduce, predict and draw meaningful conclusion with respect to the optimization of women's development and empowerment. This is to provide a more intelligent, precise, efficient process management and effective strategy design with respect to women's empowerment, in relation to selected UN SDG targets needed for a more coordinated and distributed empowerment of women, and all round development of our societies in our current data driven world.

ACKNOWLEDGMENTS

I am currently a PhD student on a fully funded Presidential Scholarship from the Chinese Government. I would like to sincerely thank my PhD supervisor and the Chinese Government for their support during my studies. Furthermore, I also want to appreciate the reviewers of this work and the editorial board for the opportunity to contribute to an excellent book and project. I would like to thank Assist. Prof. Esra Sipahi Dongul for her helpful support and excellent collaborative opportunities provided during my research studies.

REFERENCES

Ackon, L. (2010, February 6). Ghana: Achieving MDG. *Public Agenda*. www.africa.com

Afrah, S. H., & Fabiha, S. T. (2017). Empowering women entrepreneurs through Information and Communication Technologies (ICTs): A case study of Bangladesh. *Journal of Management, 7*(2), 1–6.

Agarwal, U. (2016). *Data mining and data warehousing* (2nd ed.). New Delhi, India.

Akinola, A. O., & Tella, O. (2013). Cultural adherence and the Nexus between women empowerment and millennium development goals in Africa. *Global Advanced Research Journal of Arts and Humanities* (GARJAH), *2*(5), 103–110.

Anurag, S. (2017). Women and sustainable development goals: A conceptual brief. *UN Women's Day International Academic Conference, 2017.*
Balk, W. L. (1996). *Managerial reform and professional empowerment in the public service.* Westport, CT: Quorum Books.
Benerõa, L., & Roldan, M. (1987). *The crossroads of class and gender: Industrial homework subcontracting, and household dynamics in Mexico city.* Chicago, IL: University of Chicago Press.
Boender, C., Malhorta A., & Schuler, R. S. (2002). Measuring women's empowerment as a variable in international development, background. *Paper Prepared for the World Bank Workshop on Poverty and Gender: New Perspectives.*
Central Intelligence Agency. (2013). *Facts on women from the world Factbook.* Washington, DC: CIA.
Davidson, A. (2012). Enhancing women empowerment through information and communication technologies. *Journal of Knowledge Review, 26*(1), 40–49.
Davis, A. (1981). *Women, race, and class.* New York: Vintage Books.
Deere, C. D. (1990). *Household and class relations: Peasants and landlords in Northern Peru.* Berkeley, CA: University of California Press.
Duflo, E. (2011). Women's empowerment and economic development. *NBER Working Paper.*
Hussain, F. (2016). *Women and ICT frontier initiative: "Women's empowerment, SDGs and ICT (Module C1).*
Jehan, Q. (2000). Role of women in economic development of Pakistan (PhD dissertation). Quetta: University of Balochistan.
Khan, M., & Mazhar, S. (2017, July). Socio-cultural impediments & women empowerment. *International Journal for Innovative Research in Multidisciplinary Field, 3*(7).
Lutfunnahar. (2022, December). Women entrepreneurship development through ICT in developing countries: Problems and prospects. *Social Science Review [The Dhaka University Studies, Part-D], 39*(3). https://doi.org/10.3329/ssr.v39i3.67432
Malhotra, A., & Schuler, R. S. (2005). Women's empowerment as a variable in international development. *Measuring Empowerment: Crossdisciplinary Perspectives, 2005,* 71–88.
Malik, R. (2018, July–December). HDI and gender development index: Current status of women development in India. *Journal of Indian Economy, 5*(2), 30–43. https://doi.org/10.17492/pragati.v5i2.14374
Mason, O. K., & Smith, L. H. (2003). *Women's empowerment and social context: Results from five Asian countries.* Washington, DC: Gender and Development Group, World Bank.
Moser, C. O. N. (1989). Gender planning in third world: Meeting practical and strategic gender needs. *World Development, 17*(11).
Mohammad, S., & Razmi, J. (2012, November). The relationship between women's empowerment and HDI in Islamic countries. *International Journal of Business and Behavioural Sciences, 2*(11).
Nelson, R., & Phelps, E. (1966). Investment in humans, technological diffusion, and economic growth. *American Economic Review, 56,* 69–75.
Nikulin, D. (2016). The impact of ICTs on women's economic empowerment. www.researchgate.net/publication/312938657
Prasad, P. N., & Sreedevi, V. (2007). Economic empowerment of women through Information Technology: A case study from an Indian State. *Journal of International Women's Studies, 8*(4), 107–108.
Queisser, M. (2016). *Gender equality and the sustainable development goals.* Organisation for Economic Co-operation and Development (OECD). https://doi.org/10.1787/9789264264687-16-en

Seguino, S. (2000). Gender inequality and economic growth: A cross-country analysis. *World Development, 28*(7), 1211–1230.
Singhal, R. (2003). Women, gender and development: The evolution of theories and practice. *Psychology Developing Societies.*
So, A., et al. (2020). *The applied artificial intelligence workshop.* Packt.
Swamynathan, M. (2017). *Mastering machine learning with Python in six steps.* Apress.
United Nations (UN). (2005, September). *Gender equality and empowerment of women through ICT. "Women 2000 and Beyond".* New York, USA.
UNDP Report. (2016). *Women and SDGs 2016.* New York: UNDP.
UN Statistics. (2015). *The World's women.* New York: UN.
Wright, E. O. (1996). *Class counts: Comparative studies in class analysis.* Cambridge, MA: Cambridge University Press.

4 Woman-Led Innovations
Strategies, Barriers, and Solutions for Women Entrepreneurs in a Changing Landscape

Almula Umay Karamanlıoglu and Iper Incekara

4.1 INTRODUCTION

Leadership constitutes a critical attribute within organizations and is indispensable for instigating change and innovation (Holt & Vardaman, 2013). Defining entrepreneurship involves understanding how to conceptualize and analyze it (Gartner, 1993). Innovation is a significant economic development engine and a driving force for social progress, recovery, and sustainability (Garcia & Calantone, 2002). Reflection on the social construction of gender and economics, particularly in business economics, began later compared to other scientific disciplines (Bruni, Gherardi, & Poggio, 2004:258). Despite the global importance of entrepreneurship and innovation for economies and societies, it may be considered that research on the relationship between innovation and gender is limited in the literature (Brush, Eddleston, Edelman, Manolova, McAdam, & Rossi-Lamastra, 2022). Based on this, studies showing the roles of women entrepreneurs and women in leadership suggest that they may have fallen short in explaining the potential of women entrepreneurs. In this regard, addressing entrepreneurship and innovation issues from a gender perspective among women leaders could be an important step in developing new strategies, generating solutions, and filling gaps in the field.

Innovation theories and research studies demonstrate that diversity impacts the success of innovation and digital transformation (Zhan, Bendapudi & Hong, 2015). This indicates that diversity encompasses the challenges and opportunities for established managers in innovation processes, digital transformation, and the path to digital leadership (Gfrerer, Rademacher, & Dobler, 2021). Studies reveal that gender diversity in management teams influences firm performance (Campbell & Minguez-Vera, 2007). Perceptions of leadership among male and female executives may vary (Wille et al., 2018). Research indicates that there is a disparity in self-confidence between female entrepreneurs and their male counterparts (Coleman & Robb, 2012), stemming from various factors such as societal gender norms and competing in male-dominated environments which may create differences between men and women.

Developing corporate foresight methodologies emphasizes the importance of the female perspective as a crucial component to complete a long-standing missing element (Chau & Quire, 2020). Significant differences between women and men are widely acknowledged in senior positions (O'Neill, 2011). One reason is the lack of data on how women behave in senior positions, creating a self-perpetuating cycle due to initially fewer women. Studies show that society demands women to prove themselves more in employment compared to their male colleagues. Additionally, innovation theories suggest that gender diversity in management teams, along with cognitive diversity, provides essential assets for an effective innovation and digitalization process (Gfrerer, Rademacher & Dobler, 2021).

Women's entrepreneurship enriches and diversifies entrepreneurship; more and better entrepreneurship leads to more sustainable economic growth (Nissan, Carrasco & Castaño, 2011). However, women entrepreneurs are thought to be more concerned with risks associated with rapid growth and loss of control, preferring a more moderate and manageable growth rate for their firms (Coleman & Robb, 2012).

Based on this, the chapter may contribute to the literature as a qualitative study revealing strategies and solutions for addressing the challenges faced by women entrepreneurs in changing conditions. It also serves as a valuable resource for researchers, practitioners, and students by offering insights into supporting women entrepreneurs and enhancing their success.

4.2　LITERATURE REVIEW

4.2.1　Definition and Importance of Innovation

The term innovation derived from the Latin word "nova" signifies the application of new methods in societal and cultural domains. Innovation decisions are crucial strategic choices for every company today as innovation serves as a vital tool for entering new markets, increasing current market share and enhancing competitive advantage (Gunday et al., 2011). Innovation largely depends on firms' abilities to assimilate external knowledge, integrate with their proprietary knowledge, and develop new market offerings (Chesbrough, 2003). Definitions of innovation are often associated with concepts such as entrepreneurship, the development of new products and processes and the generation of new ideas.

Joseph Schumpeter, a prominent economist, emphasized the connection between innovation and entrepreneurship. According to Schumpeter, innovation may be defined as any profitable novelty created by entrepreneurs resulting from technological advancements. Schumpeter's impact on economics, especially through his concepts of innovation and entrepreneurship, stands out as highly influential (Elgar, 2007). According to Schumpeter (1934), innovations do not typically emerge suddenly; they always have roots in the pre-existing economic structure. Defining the concept of innovation and subsequently explaining its relationship with similar concepts will aid in understanding innovation. The term invention denotes the discovery of new methods or materials whereas innovation encompasses the efforts to commercialize these inventions (Hill and Rothaermel,

2003). Another related concept is creativity. Creativity is the process of generating new and valuable ideas in any field; innovation, on the other hand, may be seen as the process of effectively implementing these creative ideas within an organization. Thus, the creativity of individuals and teams forms the foundation of innovation (Amabile et al., 1996).

After providing a general definition of innovation, organizational innovation refers to the systematic management and implementation of these innovative ideas within an organization. Organizational innovation is defined as the continuous adoption of a new idea or behavior within an organization. It may be described as the creation or acceptance of a new idea within companies (Zammuto & O'Connor, 1992). Organizational innovations encompass changes aimed at enhancing the effectiveness, efficiency, and creativity of companies. Such innovations include elements such as introducing and implementing new strategies, developing information management systems (information search, sharing, coding, and storage), and establishing new administrative and control systems and processes.

However, specific characteristics are required for successfully managing and implementing this process. These characteristics are effectively managed through the application of knowledge, specific leadership qualities, diversity and various perspectives. Increasing diversity plays a crucial role in both fostering innovative ideas and managing organizational innovation. However, understanding and evaluating the steps to manage organizational innovation involving women necessitates first examining the impact of female leadership on innovation.

4.2.2 The Power and Importance of Women Leaders in Innovation and Entrepreneurship

The role and significance of women leaders are increasingly crucial in effectively managing innovation. A key discussion surrounding the roles of women in innovation processes generally pertains to their place in science (Le Loarne-Lemaire, Bertrand, Razgallah, Maalaoui, & Kallmuenzer, 2021).

Women leaders not only hold strategic importance within organizations but also demonstrate their presence in operational activities. In this regard, women leaders enable innovative steps and facilitate the occurrence of innovative changes across various business models and processes. Women's entrepreneurship enriches and diversifies entrepreneurship; more and better entrepreneurship leads to more sustainable economic growth (Nissan, Carrasco, & Castaño, 2011:125). The role of women leaders in innovation may be considered pivotal for organizations to achieve competitive advantage, enhance efficiency and achieve sustainable growth. Innovative-thinking women within organizations are believed to play a significant role in achieving long-term success and developing innovative strategies. Studies indicate that women leaders excel in collaborative behavior, vision development, strategic planning, and innovative strategy development. The contribution of women's entrepreneurship to economic development not only provides job opportunities and economic growth but also increases the diversity of entrepreneurial activities and enhances overall entrepreneurial quality (Verheul, Stel & Thurik, 2006).

4.2.3 EXPECTANCY THEORY AND WOMEN ENTREPRENEURS

Expectancy theory is a dominant theoretical framework for explaining human motivation (Manolova, Brush & Edelman, 2008). One assumption of expectancy theory is that individuals make decisions among alternative courses of action based on their perceptions (expectations) of how likely a specific behavior will lead to desired outcomes (Mathibe, 2008). Expectancy theory suggests that individuals are driven to act when they perceive that their efforts may result in successful performance leading to desirable outcomes or other valued rewards (Vroom, 1964; Kreitner & Kinicki, 2007). Individuals base their actions on the belief that exerting effort towards a task or goal they desire may result in achievement and positive outcomes for the organization.

In such scenarios, individuals are more motivated when they believe that their efforts and actions may lead to achieving their goals. Expectancy theory focuses on three fundamental relationships to explain motivation: first, the subjective probability that exerting effort or action will lead to a desired outcome or performance; second, the attractiveness or desirability of that outcome to the individual; and finally, the perceived relationship between achieving one outcome and other outcomes (Vroom, 1964). Studies related to entrepreneurship may be explained within the framework of expectancy theory (Manolova, Brush, & Edelman, 2008). Research indicates that entrepreneurs who believe in their skills and abilities are motivated to exert the necessary effort (Shaver et al., 2001). Anna et al. (2000) found that women working in traditional sectors (such as services and retail) have stronger career expectations in balancing work-security and home-life demands compared to women in non-traditional sectors like manufacturing. Empowering women economically may be considered essential for sustainable growth (Muhamad et al., 2017; Mishra & Kiran, 2012).

4.2.4 INNOVATION STRATEGIES FOR WOMEN ENTREPRENEURS

In recent years, one of the most exciting aspects of entrepreneurship research has been the rapid growth of businesses owned by women and their increasing economic impact (Coleman & Robb, 2012). Women entrepreneurship is recognized as a crucial driver of economic growth and sustainable development (Hendratmi, Agustina, Sukmaningrum, & Widayanti, 2022:1). Some of the world's most profitable companies attribute their growth to innovation which they perceive as their ability to transform and reinvent themselves as a way to capitalize on opportunities (Idris, 2008). Organizations need to harness businesses and discover new innovative ideas (Gfrerer, Rademacher, & Dobler, 2021). Creative abilities are often seen as essential traits of successful entrepreneurs (Johnson, 2001). In this regard, creativity stands as a crucial attribute in effective leaders, necessitating the generation of novel and compelling ideas (Mayer & Maree, 2018). It may be thought that women entrepreneurs have a different dynamism that supports generating innovative and sustainable solutions within the organization. Studies show that leaders with diverse cultural backgrounds consider creativity as another skill necessary for effective leadership (Mayer & Oosthuizen, 2020). Moreover, women entrepreneurs effectively manage various support organizations and play crucial roles in their operations. Women-owned companies have gained importance in achieving sustainable development goals especially in small enterprises (Hendratmi, Agustina, Sukmaningrum, & Widayanti, 2022). Several factors contribute to the increasing number of women establishing innovative and growth-oriented companies (Coleman & Robb, 2012). First, educational initiatives have contributed to an increase in the number of women seeking career and entrepreneurial opportunities in non-traditional and technology-based fields (Coleman & Robb, 2012). This is because working in technology and innovation-oriented organizations may make women feel stronger and more competent as entrepreneurs. Second, it is thought that the number of women entrepreneurs who may act as role models and mentors to those who follow them has increased (Coleman & Robb, 2012). In this regard, women entrepreneurs engaged in innovation may serve as role models based on their own experiences which may lead to an increase in leadership positions in this area. In addition, it may be considered that funding sources have begun to open up more slowly for women entrepreneurs in growth-oriented sectors such as biotechnology and technology (Coleman & Robb, 2012). In this sense, women entrepreneurs may gain opportunities to expand their businesses and generate more innovative ideas. Regardless of the issues and obstacles, it is considered essential for women to develop ways to manage their businesses to increase their incomes and meet their families' demands (Muhamad et al., 2017). Women entrepreneurs undertake tasks such as exploring new business opportunities, managing risks, bringing innovations, coordinating, managing, and controlling the business effectively in all aspects (Gümüsay, 2014). Moreover, the use of internet technology provides entrepreneurs with better opportunities to develop business strategies and thus create a good business reputation (Muhamad et al., 2017).

4.2.5 Barriers for Women Entrepreneurs

Women transform modern business through their prominent role in the rapidly growing service sector, where they pioneer enterprises that cater to both clients and employees with innovative systems and flexible schedules (Winn, 2005). The rising prevalence of women entrepreneurship particularly in developing nations has contributed to overall household welfare and consumption (Minniti and Naudé, 2010). Despite approaching entrepreneurship with objectivity and open-mindedness, many women still encounter barriers specific to their gender (Winn, 2005).

Women entrepreneurs face many challenges in the organization. These include lack of access to education, experience, and training opportunities; lack of spatial mobility and family support; absence of institutional support; deficiency in entrepreneurial management; and difficulties in acquiring financial resources (Raghuvanshi, Agrawal, & Ghosh, 2017; Mathew, 2010; Roomi & Parrott, 2008). It may be considered that women entrepreneurs have limited access to education in management or finance topics. As a result, they may feel inadequate in business management and strategic decision-making processes. Their access to partnerships with large corporations or benefits from government support may also be limited. Additionally, women entrepreneurs often face significant challenges when trying to secure funding or obtain family support (Winn, 2005). Women entrepreneurs are often more frequently rejected by financial institutions when seeking loans or investments. For example, when a female entrepreneur requests a loan from a bank to expand her business, she may encounter additional hurdles such as requiring more documentation and guarantees compared to her male counterparts. Buttner and Rosen (1988) found that bank employees perceive women as less entrepreneurial than men and evaluate them as having a lower propensity for taking risks. Slovic (2000) concluded that sociopolitical factors play a crucial role in shaping gender disparities in risk perception, suggesting that these differences may stem from issues related to power dynamics and women's limited access to them.

4.3 RESEARCH METHOD

The purpose of this section is to identify the challenges faced by women entrepreneurs in evolving conditions and delineate to strategies that assist in overcoming these challenges and achieving success. This section addresses the following research questions accordingly.

1. How might women's entrepreneurship evolve in the future and what targeted strategies may be implemented to empower and support women entrepreneurs?
2. What are the significant challenges faced by women entrepreneurs and which strategies may be employed to overcome these challenges?

The research design of this study is phenomenology. The reason for this choice is to reveal women entrepreneurs' perceptions of specific issues and problems and the meanings they attribute to them. Thus, it is aimed to understand these phenomena by thoroughly examining participants' subjective experiences and comments (Creswell, 2013). Phenomenological research involves a systematic approach to gain in-depth understanding. In this regard, research questions aimed at understanding the phenomenon were formulated first. Subsequently, appropriate participants were selected, and the data collection process was initiated. The collected data were then analyzed in depth to understand participants' experiences; they were coded, themes were identified, and focus was placed on the uniqueness of the phenomenon. Finally, findings were interpreted, discussed, theoretical implications were drawn, and the research results were reported in writing. Data were obtained through the interview technique. The sample consists of 20 female executives who have established their businesses in various sectors in Ankara province determined within the scope of qualitative research. In this research, the sample was selected using convenience sampling with the participation of 20 female executives. In this regard, a sampling strategy was chosen based on participants' accessibility and their relevance to our research focus.

QDA Miner 6 software was used for data analysis in the research employing content analysis. QDA Miner is an easy-to-use software designed for qualitative data analysis, facilitating the organization, coding, annotation, retrieval, and analysis of document and image collections (provalisresearch). QDA Miner may be considered as a significant tool for conducting in-depth analysis of data used in qualitative research. Content analysis as a research method is a systematic and objective way of describing and measuring phenomena (Krippendorff, 1980). The process of content analysis culminates in organizing the data, applying coding to concepts, identifying themes, arranging codes based on themes and data, analyzing themes in relation to research questions, interpreting codes and themes, analyzing qualitative data, interpreting findings, and ultimately, reporting the results. In this way, themes were derived through systematic observation and analysis of data collected from participants supplemented by a review of the literature. Codes were generated based on participants' responses. The reliability and validity of the study were rigorously addressed and based on the opinions of experts in the field the evaluation process was conducted with a literature review being a crucial step in determining themes and codes. The research findings were communicated via email based on expert opinions and were confirmed by the feedback received.

4.4 FINDINGS

4.4.1 Strategies for Women Entrepreneurs

Table 4.1 reveals the participants' responses regarding the strategies. According to the results, the highest percentage rate at 22.6% is attributed to the code increasing access to investment funds. This is followed by collaboration and networking at 19.4% and improving digital skills and supporting trade agreements both at 12.9%. Codes such as increasing gender equality policies and training and mentoring programs stand out with rates of 9.7% while accessing financial technical and operational resources, and incentive reward and recognition programs are noted with a rate of 6.5%.

4.4.2 Barriers to Women Entrepreneurs

Table 4.2 reveals the participants' responses regarding the barriers. According to the results, the code with the highest percentage distribution is home and work-life

TABLE 4.1
Strategies for women entrepreneurs

Theme	Code	% Codes
Strategies for woman entrepreneurs	Increasing access to investment funds	22.6
Strategies for woman entrepreneurs	Collaboration and networking	19.,4
Strategies for woman entrepreneurs	Improving digital skills	12.9
Strategies for woman entrepreneurs	Supporting trade agreements	12.9
Strategies for woman entrepreneurs	Increasing gender equality policies	9.7
Strategies for woman entrepreneurs	Training and mentoring programs	9.7
Strategies for woman entrepreneurs	Accessing to financial, technical and operational resources	6.5
Strategies for woman entrepreneurs	Incentive reward and recognition programs	6.5

TABLE 4.2
Barriers to women entrepreneurs

Theme	Code	% Codes
Barriers to women entrepreneurs	Home and work life balance	23.1
Barriers to women entrepreneurs	Gender stereotypes and discrimination	19.2
Barriers to women entrepreneurs	Social expectations and roles	15.4
Barriers to women entrepreneurs	Lack of female network	11.5
Barriers to women entrepreneurs	Lack of female mentors	11.5
Barriers to women entrepreneurs	Lack to capital and asset accumulation	11.5
Barriers to women entrepreneurs	Difficulties in accessing finance	7.7

TABLE 4.3
Solutions for women entrepreneurs

Theme	Code	% Codes
Solution for women entrepreneurs	Networking events and supporting groups	19.1
Solution for women entrepreneurs	Financial support programs	17
Solution for women entrepreneurs	Tax breaks and incentives	14.9
Solution for women entrepreneurs	Awareness campaigns for gender equality	12.8
Solution for women entrepreneurs	Increasing training and mentoring opportunities	10.6
Solution for women entrepreneurs	Export incentives and support programs	10.6
Solution for women entrepreneurs	Cooperation between governments and non-governmental organizations	10.6
Solution for women entrepreneurs	Training programs	4.3

balance with a rate of 23.1%. Afterward, there is a rate of 19.2% for gender stereotypes and discrimination and a rate of 15.4% for social expectations and roles. The category with an 11.5% rate encompasses lack of female network, lack of female mentors and lack of capital and asset accumulation. The code with the least distribution is difficulties in accessing finance which stands at 7.7%.

4.4.3 SOLUTIONS FOR WOMEN ENTREPRENEURS

Table 4.3 reveals the participants' responses regarding the solutions. According to the results, the highest percentage rate is observed to be 19.1% for the code networking events and support groups. Following this, financial support programs are at 17% while tax breaks and incentives follow at 14.9%. Awareness campaigns for gender equality rank fourth with a rate of 12.8%, while codes such as increasing training and mentoring opportunities, export incentives and support programs and cooperation between governments and non-governmental organizations each account for 10.6%. Lastly, the code with the lowest percentage rate is training programs at 4.3%.

4.4 CONCLUSION

Women are actively engaged in nearly every sector of the economy, including agriculture, manufacturing, education, healthcare, tourism, and various other fields (Muhamad et al., 2017). Women entrepreneurs engage in entrepreneurship across all fields, take risks, and integrate resources with new methods to leverage opportunities in their surroundings (Hendratmi, Agustina, Sukmaningrum, & Widayanti, 2022). In this context, women entrepreneurs not only contribute to economic growth but also promote gender equality and diversity in the business world. Building on this premise, this study emphasizes critical strategies, barriers, and solutions to support the innovation potential of women entrepreneurs.

The findings underscore the importance of several key strategies in empowering women entrepreneurs. First, enhancing access to investment funds emerges as a significant strategy, aligned with literature emphasizing the necessity of financial resources for entrepreneurial success (Brush et al., 2009). Second, elements like collaboration and networking encourage fundamental connections and support systems that may enhance business growth and resilience (McAdam & Marlow, 2013). Third, developing digital skills represents another crucial strategy reflecting the increasing importance of technology in modern business operations. This aligns with research emphasizing the need for digital literacy to effectively compete in today's market. Consistent with studies highlighting the benefits of international trade for small businesses, supporting trade agreements is equally important as it can open new markets and opportunities for women entrepreneurs (Kelley et al., 2011). Moreover, the study's findings include other notable strategies such as increasing gender equality policies that create a more supportive and equitable work environment (Jennings & Brush, 2013) and promote critical guidance and skill development through education and mentorship programs (Carter & Marlow, 2007). Additionally, implementing incentive awards and recognition programs alongside access to financial, technical, and operational resources have been identified as a crucial strategy that may significantly help women entrepreneurs overcome common challenges (Henry et al., 2015). Based on all these findings, strategic initiatives focusing on investment access, networking, digital literacy, trade support, gender equality, and mentorship are seen as capable of reducing barriers and promoting gender equality in entrepreneurship. This suggests that these strategic initiatives may facilitate women entrepreneurs in overcoming the challenges they face in investment access, networking, digital literacy, trade support, gender equality, and mentorship, thereby promoting gender equality in entrepreneurship.

Furthermore, another significant finding of the study has revealed various critical barriers that women entrepreneurs encounter in the innovation processes. These include challenges in maintaining a balance between home and work life which align with findings emphasizing the disproportionate domestic responsibilities assigned to women in the literature (Brush et al., 2009). Gender stereotypes and discrimination also emerge as significant barriers, consistent with previous studies documenting how societal biases and discriminatory practices hinder women's entrepreneurial efforts (Marlow & Patton, 2005). Traditional gender roles often restrict professional opportunities, making the social expectations and roles further complicate women's entrepreneurial journeys (Welter, 2011). Additionally, the lack of women networks, mentors and capital accumulation is a recurring theme in the literature, underscoring the importance of access to supportive networks and resources for women entrepreneurs (Aldrich & Cliff, 2003). Finally, difficulties in accessing finance appear to be a notable barrier for women entrepreneurs. Women entrepreneurs frequently face challenges in securing funding compared to their male counterparts (Carter et al., 2007).

The study highlights several key solution proposals identified by participants to support the innovation processes of women entrepreneurs. One prominent solution is the promotion of networking activities and support groups that emphasize the importance of social capital and peer support in entrepreneurial success which aligns with existing literature (Brush et al., 2009; Mijid, 2014). Financial support programs also emerge as a crucial factor consistent with research indicating that access to financial resources is a significant barrier for women entrepreneurs (Carter et al., 2007; Coleman, 2000). Tax incentives and subsidies promote entrepreneurship among women reflecting the role of fiscal policies in encouraging entrepreneurial activities (Fairlie & Robb, 2009). Awareness campaigns promoting gender equality underscore the critical role of societal attitudes and perceptions in creating a conducive environment for women entrepreneurs (Klyver & Grant, 2010). Additionally, the findings support increased emphasis on education and mentorship opportunities which enhance entrepreneurial skills and confidence (Fielden & Hunt, 2011; Kickul et al., 2009). Recognizing the importance of export incentives and support programs suggests that facilitating access to international markets can stimulate growth for women-owned businesses (Orser et al., 2010). Collaboration between governments and civil society organizations also highlights the importance of cooperative efforts in providing comprehensive support to women entrepreneurs (Brush et al., 2009).

This section emphasizes the importance of access to policies, resources and networks for women entrepreneurs' innovation processes. These elements may help maximize their potential and sustainably grow their businesses. Moreover, they may contribute to maximizing the economic and societal impacts of women entrepreneurs. In this regard, incentives and funding programs established by governments and local authorities may provide crucial support for women entrepreneurs to expand their businesses and develop innovative projects. Furthermore, women entrepreneurs may use these programs to enhance their business management skills, explore new markets and overcome challenges. It is crucial to establish networks that facilitate the exchange of experiences among women entrepreneurs. These networks may facilitate mutual learning and support among women entrepreneurs. Therefore, empowering and supporting women entrepreneurs is crucial not only for fostering economic growth but also for achieving gender equality and supporting societal development. With future steps, women entrepreneurs may fully realize their potential and make a greater impact in the business world.

REFERENCES

Aldrich, H. E., & Cliff, J. E. (2003). The pervasive effects of family on entrepreneurship: Toward a family embeddedness perspective. *Journal of Business Venturing, 18*, 573–596.

Amabile, T. M., Conti, R., Coon, H., Lazenby, J., & Herron, M. (1996). Assessing the work environment for creativity. *Academy of Management Journal, 39*(5), 1154–1184. doi: 10.5465/256995

Anna, A, Chandler, G, Jansen, E, & Mero, N. (2000). Women business owners in traditional and non-traditional industries. *Journal of Business Venturing, 15*(2), 279–303. doi: 10.1016/s0883-9026(98)00012-3

Brush, C., Eddleston, K., Edelman, L., Manolova, T., McAdam, M., & Rossi-Lamastra, C. (2022). Catalyzing change: Innovation in women's entrepreneurship. *Strategic Entrepreneurship Journal, 16*(2), 243–254. https://doras.dcu.ie/28173/3/SEJ%20editorial%20for%20Catalyzing%20Innovation-%20final%205-15-2022.pdf

Brush, C. G., de Bruin, A., & Welter, F. (2009). A gender-aware framework for women's entrepreneurship. *International Journal of Gender and Entrepreneurship, 1*(1), 8–24. doi: 10.1108/17566260910942318

Bruni, A., Gherardi, S., & Poggio, B. (2004). Entrepreneur-mentality, gender and the study of women entrepreneurs. *Journal of Organizational Change Management, 17*(3), 256–268. doi: 10.1108/09534810410538315

Buttner, E., & Rosen, B. (1988). Bank loan officers' perceptions of the characteristics of men, women, and successful entrepreneurs. *Journal of Business Venturing, 3*(3), 249–58. doi: 10.1016/0883-9026(88)90018-3

Carter, S., Shaw, E., Lam, W., & Wilson, F. (2007). Gender, entrepreneurship, and bank lending: The criteria and processes used by bank loan officers in assessing applications. *Entrepreneurship Theory and Practice, 31*(3), 427–444. doi: 10.1111/j.1540-6520.2007.00181.x

Carter, S., & Marlow, S. (2007). Female entrepreneurship: Theoretical perspectives and empirical evidence. In N. M. Carter et al. (Eds.), *Women Entrepreneurship and the Venture Capital Market*. Edward Elgar Publishing. www.taylorfrancis.com/chapters/edit/10.4324/9780203013533-3/female-entrepreneurship-sara-carter-susan-marlow

Coleman, S. (2000). Access to capital and terms of credit: A comparison of men-and women-owned small businesses. *Journal of Small Business Management, 38*(3), 37. www.proquest.com/openview/cff9e10a4c69084a4f08da6eea1d062c/1?pq-origsite=gscholar&cbl=49244

Chau, V. S., & Quire, C. (2020). Back to the future of women in technology: Insights from understanding the shortage of women in innovation sectors for managing corporate foresight. In *Corporate Foresight and Innovation Management* (pp. 123–140).

Campbell, K., & Mínguez-Vera, A. (2007). Gender diversity in the boardroom and firm financial performance. *Journal of Business Ethics, 83*(3), 435–451. https://link.springer.com/article/10.1007/s10551-007-9630-y

Chesbrough, H. (2003). Open *Innovation: The New Imperative for Creating and Profiting from Technology*. Boston, MA: Harvard Business School Press.

Coleman, S., & Robb, A. (2012). Unlocking innovation in women-owned firms: Strategies for educating the next generation of women entrepreneurs. *JWEE*, (1–2), 99–125. https://library.ien.bg.ac.rs/index.php/jwee/article/view/91

Creswell, J. W. (2013). *Qualitative Inquiry and Research Design: Choosing Among Five Approaches*. Thousand Oaks, CA: SAGE Publications.

Elgar, E. (2007). *Elgar Companion to Neo-Schumpeterian Economics*. Augsburg: Economic Books.

Fairlie, R. W., & Robb, A. M. (2009). Gender differences in business performance: Evidence from the characteristics of business owners survey. *Small Business Economics, 33*(4), 375–395. doi: 10.1007/s11187-009-9207-5

Fielden, S. L., & Hunt, C. M. (2011). Online coaching: An alternative source of social support for female entrepreneurs during venture creation. *International Small Business Journal, 29*(4), 345–359. doi: 10.1177/0266242610369881

Garcia, R. (2002). A critical look at technological innovation typology and innovativeness terminology: A literature review. *Journal of Product Innovation Management, 19*(2), 110–132. doi: 10.1016/s0737-6782(01)00132-1

Gartner, W. B. (1993). Words lead to deeds: Towards an organizational emergence vocabulary. Journal of Business Venturing, 8(3), 231–239, https://doi.org/10.1016/0883-9026(93)90029-5

Gfrerer, A. E., Rademacher, L., & Dobler, S. (2021). Digital needs diversity: Innovation and digital leadership from a female managers' perspective. In *Digitalization: Approaches, Case Studies, and Tools for Strategy, Transformation and Implementation* (pp. 335–349). Cham: Springer International Publishing.

Gümüsay, A. A. (2014). Entrepreneurship from an Islamic perspective. *Journal of Business Ethics, 130*(1), 199–208. doi: 10.1007/s10551-014-2223-7

Gunday, G., et al. (2011). Effects of innovation types on firm performance. *International Journal of Production Economics,* 133, 662–676. doi: 10.1016/j.ijpe.2011.05.014

Hendratmi, A., Agustina, T. S., Sukmaningrum, P. S., & Widayanti, M. A. (2022). Livelihood strategies of women entrepreneurs in Indonesia. *Heliyon, 8*(9). www.cell.com/heliyon/fulltext/S2405-8440(22)01808-4

Henry, C., Foss, L., Fayolle, A., Walker, E., & Duffy, S. (2015). Entrepreneurial leadership and gender: Exploring theory and practice in global contexts. Journal of Small Business Management, 53(3), 581–586.

Holt, D. T., & Vardaman, J. M. (2013). Toward a comprehensive understanding of readiness for change: The case for an expanded conceptualization. *Journal of Change Management, 13*(1), 9–18. doi: 10.1080/14697017.2013.768426

Hill, C. W., & Rothaermel, F. T. (2003). The performance of incumbent firms in the face of radical technological innovation. *Academy of Management Review, 28*(2), 257–274. doi: 10.2307/30040712

Idris, A. (2009). Cultivating innovation through female leadership: The Malaysian perspective. *Asian Social Science, 4*(6). doi: 10.5539/ass.v4n6p3

Jennings, J.E. and Brush, C.G. (2013). Research on women entrepreneurs: challenges to (and from) the broader entrepreneurship literature?. Academy of Management Annals, 7(1), 663–715.

Johnson, D. (2001). What is innovation and entrepreneurship? Lessons for larger organisations. *Industrial and Commercial Training, 33*(4), 135–140. www.emerald.com/insight/content/doi/10.1108/00197850110395245/full/html

Kelley, D. J., Brush, C.G., Greene, Patricia G., Litovsky, Y., BABSON COLLEGE; GERA – Global Entrepreneurship Research Association (2011). *Global Entrepreneurship Monitor (GEM): 2010 Women's Report.* Babson College; Center for Women's Business Research; The Center of Women's Leadership at Babson College.

Kickul, J., Gundry, L. K., Barbosa, S. D., & Whitcanack, L. (2009). Intuition versus analysis? Testing differential models of cognitive style on entrepreneurial self-efficacy and the new venture creation process. *Entrepreneurship Theory and Practice, 33*(2), 439–453. doi: 10.1111/j.1540-6520.2009.00298.x

Kreitner, R., & Kinicki, A. (2007). *Organizational Behaviour.* Boston: McGraw-Hill.

Klyver, K., & Grant, S. (2010). Gender differences in entrepreneurial networking and participation. *International Journal of Gender and Entrepreneurship*, *2*(3), 213–227. doi: 10.1108/17566261011079215

Krippendorff, K. (1980). *Content Analysis: An Introduction to its Methodology*. Newbury Park: Sage Publications.

Le Loarne-Lemaire, S., Bertrand, G., Razgallah, M., Maalaoui, A., & Kallmuenzer, A. (2021). Women in innovation processes as a solution to climate change: A systematic literature review and an agenda for future research. *Technological Forecasting and Social Change*, 164. doi: 10.1016/j.techfore.2020.120440

Manolova, T. S., Brush, C. G., & Edelman, L. F. (2008). What do women entrepreneurs want? *Strategic Change*, *17*(3–4), 69–82. doi: 10.1002/jsc.817

Marlow, S., & Patton, D. (2005). All credit to men? Entrepreneurship, Finance, and Gender. Entrepreneurship Theory and Practice, 29(6), 717–735. https://doi.org/10.1111/j.1540-6520.2005.00105.x

Mathew, V. (2010). Women entrepreneurship in Middle East: Understanding barriers and use of ICT for entrepreneurship development. *International Entrepreneurship and Management Journal*, *6*(2), 163–181. doi: 10.1007/s11365-010-0144-1

Mathibe, I. (2008). Expectancy theory and its implications for employee motivation. Academic Leadership: *The Online Journal*, *6*(3), 8. doi: 10.58809/CUMG4502

Mayer, C.-H., Tonelli, L., Oosthuizen, R. M., & Surtee, S. (2018). "You have to keep your head on your shoulders": A systems psychodynamic perspective on women leaders. *SA Journal of Industrial Psychology*, *44*. doi: 10.4102/sajip.v44i0.1424

Mayer, C. H., & Oosthuizen, R. M. (2020). Concepts of creative leadership of women leaders in 21st century. *Creativity Studies*, *13*(1), 21–40. https://journals.vilniustech.lt/index.php/CS/article/view/10267

McAdam, M., & Marlow, S. (2013). Gender and entrepreneurship: Advancing debate and challenging myths; exploring the mystery of the under-performing female entrepreneur. International Journal of Entrepreneurial Behavior & Research, 19(1), 114–124.

Mishra, G. D. U. (2012). Rural women entrepreneurs: Concerns & importance. *International Journal of Science and Research*, *3*(9), 93–98. https://citeseerx.ist.psu.edu/document?repid=rep1&type=pdf&doi=304eb984cf302a1539fdc39b915a34b164e4ad2e

Minniti, M., & Naudé, W. (2010). What do we know about the patterns and determinants of female entrepreneurship across countries? *The European Journal of Development Research*, *22*(3), 277–293. doi: 10.1057/ejdr.2010.17

Muhamad, S., Ali, N., Jalil, A. B. D. U. L., Man, M., & Kamarudin, S. (2017). A sustainable e-business model for rural women: A case study. *Journal of Sustainability Science Management*, *3*, 130–140. https://jssm.umt.edu.my/wp-content/uploads/sites/51/2017/01/16-NRGS.suriyani.PV_.pdf

Mijid, N. (2014). Why are female small business owners in the United States less likely to apply for bank loans than their male counterparts? *Journal of Small Business & Entrepreneurship*, *27*(2), 229–249. doi: 10.1080/08276331.2015.1012937

Nissan, E., Carrasco, I., & Castaño, M.-S. (2011). Women entrepreneurship, innovation, and internationalization. *Women's Entrepreneurship and Economics*, 125–142. doi: 10.1007/978-1-4614-1293-9-9

O'Neil, D. A., Hopkins, M. M., & Sullivan, S. E. (2011). Do women's networks help advance women's careers? *Career Development International*, *16*(7), 733–754. doi: 10.1108/13620431111187317

Orser, B., Riding, A., & Manley, K. (2006). Women entrepreneurs and financial capital. *Entrepreneurship Theory and Practice*, *30*(5), 643–665. doi: 10.1111/j.1540-6520.2006.00140.x

Orser, B., Spence, M., Riding, A., & Carrington, C. A. (2010). Gender and export propensity. *Entrepreneurship Theory and Practice*, 34(5), 933–958. https://doi.org/10.1111/j.1540-6520.2009.00347.x

Raghuvanshi, J., Agrawal, R., & Ghosh, P. K. (2017). Analysis of barriers to women entrepreneurship: The DEMATEL approach. *The Journal of Entrepreneurship*, 26(2), 220–238. doi: 10.1177/0971355717708848

Roomi, M. A., & Parrott, G. (2008). Barriers to development and progression of women entrepreneurs in Pakistan. *The Journal of Entrepreneurship*, 17(1), 59–72. doi: 10.1177/097135570701700105

Schumpeter, J. A. (1934). *The Theory of Economic Development*. London: Oxford University Press.

Shaver, K. G., Gartner, W. B., Crosby, E., Bakalarova, K., & Gatewood, E. J. (2001). Attributions about entrepreneurship: A framework and process for analyzing reasons for starting a business. *Entrepreneurship Theory and Practice*, 26(2), 5–32. doi: 10.1177/104225870102600201

Slovic, P. (2000). *The Perception of Risk*. London: Earthscan Publications Ltd.

Verheul, I., Stel, A. V., & Thurik, R. (2006). Explaining female and male entrepreneurship at the country level. *Entrepreneurship and Regional Development*, 18(2), 151–183. doi: 10.1080/08985620500532053

Vroom, V. H. (1964). *Work and Motivation*. New York, NY: John Wiley & Sons.

Welter, F. (2011). Contextualizing entrepreneurship—Conceptual challenges and ways forward. *Entrepreneurship Theory and Practice*, 35, 165–184.

Wille, B., Wiernik, B. M., Vergauwe, J., Vrijdags, A., & Trbovic, N. (2018). Personality characteristics of male and female executives: Distinct pathways to success? *Journal of Vocational Behavior*, 106, 220–235. doi: 10.1016/j.jvb.2018.02.005

Winn, J. (2005). Women entrepreneurs: Can we remove the barriers? *The International Entrepreneurship and Management Journal*, 1(3), 381–397. doi: 10.1007/s11365-005-2602-8

Zhan, S., Bendapudi, N., & Hong, Y. (2015). Re-examining diversity as a double-edged sword for innovation process. *Journal of Organizational Behavior*, 36(7), 1026–1049. doi: 10.1002/job.2027

Zammuto, R., & O'Connor, E. (1992). Gaining advanced manufacturing technologies benefits: The role of organizational design and culture. *Academy of Management Review*, 17, 701–728. https://journals.aom.org/doi/abs/10.5465/amr.1992.4279062

ONLINE REFERENCE

https://provalisresearch.com/products/qualitative-data-analysis-software/

5 The Role of Multi-Stakeholder Partnerships in Promoting Women's Leadership in Southeast Asia
Challenges, Opportunities, and Trends

Putri Hergianasari, Michael Koks, and Rizki Amalia Yanuartha

5.1 INTRODUCTION

Based on the Council of Foreign Relations, Southeast Asia, a region comprising 11 countries including Brunei, Cambodia, Indonesia, Laos, Malaysia, Myanmar, the Philippines, Singapore, Thailand, Vietnam, and East Timor is known for its rich cultural diversity and varying degrees of economic development (CFR.org, 2023). This diversity extends to the status and representation of women in leadership roles across the region. The position of women in leadership in Southeast Asia reflects a complex interplay of historical, cultural, economic, and political factors that vary significantly from one country to another (Choi, 2019). These barriers are deeply rooted in cultural norms and stereotypes, legal and policy constraints, limited education and professional development opportunities, and insufficient support networks. However, notable examples of progress and success stories highlight the potential for further advancement. This background will explore the position of women in leadership in various Southeast Asian countries, explore possible communalities and root causes, examining the challenges, achievements, and prospects.

Historically, Southeast Asia has been characterized by patriarchal societies where leadership roles have predominantly been occupied by men (Ong, 1989). Cultural norms and traditions have often relegated women to emphasize their duties within the family and household. However, there have been significant shifts over the past century, influenced by modernization, education, and international movements advocating for gender opportunities. Whether this is a good development or actually a hurtful trend for various populations and the family structure is not investigated in this chapter. Traditional norms may sometimes be preferable over progress. This question may be addressed in a separate publication.

Countries like the Philippines and Indonesia have witnessed ground-breaking changes with women ascending to the highest political offices. For instance, the Philippines elected its first female president, Corazon Aquino, in 1986, followed by Gloria Macapagal Arroyo in 2001 (Beneria & Sen, 1981). In Indonesia, Megawati Sukarnoputri served as president from 2001 to 2004. These milestones were not only significant but also highlighted the ongoing struggle for broader representation of women in leadership (Kyveloukokkaliari & Nurhaeni, 2017).

Today, the representation of women in leadership varies across sectors and countries in Southeast Asia (Apriani & Zulfiani, 2020). While some countries have made notable strides, others still face significant challenges. This section provides an overview of the current status of women in leadership in each Southeast Asian country (Richter, 1990). In Brunei, women have limited representation in political leadership, largely due to the country's absolute monarchy and conservative Islamic values (Skrabakova, 2017). Although women participate in the workforce and have underrepresented to education, their presence in high-level leadership positions remains minimal (Tailassane, 2019). Efforts to promote gender opportunities are still nascent, and cultural norms continue to pose significant barriers (USAID, 2023).

Cambodia has seen gradual improvements in women's leadership, particularly in civil society and grassroots movements (USAID, 2016). However, women's representation in political and corporate leadership remains low. The country's turbulent history, including the Khmer Rouge era, significantly disrupted progress in gender opportunities (Qian, 2016). Current efforts focus on increasing women's participation in politics and decision-making processes. Indonesia has made notable progress in promoting women's leadership, especially in politics (Davies, 2005). The country's decentralization process has created opportunities for women to assume leadership roles at the local level (Siahaan, 2003). Indonesia also boasts a strong women's movement that has been instrumental in advocating for gender opportunities. Despite these advancements, women still face significant challenges in the corporate sector and higher echelons of political power because of systemic and cultural barriers (Robinson & Bessell, 2002).

In Laos, women's leadership is gradually increasing, particularly in the public sector. The Lao Women's Union has been a key organization in promoting women's rights and leadership (Lao Women's Union, 2009). However, limited access to education for women in rural areas continue to hinder their progress. The government has shown commitment to gender equality for access to opportunities, but such implementation remains a challenge. In many rural areas in Laos, women's educational participation remains very low (The World Bank, 2021). The main causes are limited infrastructure and transportation. Girls often have to walk long distances to the nearest school, leading many to drop out (UNICEF, 2020). Additionally, household chores and family responsibilities are often prioritized over education (UNICEF, 2021).

Malaysia presents a mixed picture regarding women's leadership. While women are well-represented in education and healthcare, their presence in political and corporate leadership remains limited (Ahmad, 2017). The country has implemented various policies to promote gender opportunities, including gender quotas in the public sector (Elias, 2011).

Myanmar has seen a slow but steady increase in women's participation in leadership, particularly in politics. Aung San Suu Kyi, the country's most prominent female leader, has been a symbol of democracy and gender opportunities (Thompson, 2002). Despite her leadership, women's representation in parliament and other leadership positions remains low. The ongoing political instability and ethnic conflicts further complicate efforts to promote women's leadership. The Philippines is often cited as a success story in terms of women's leadership in Southeast Asia (Sinpeng & Savirani, 2022). The country has a high representation of women in politics, business, and civil society. The Philippines ranks high on global gender opportunities indices, reflecting its progressive stance on gender issues (Abad, 2017). However, challenges remain, particularly in ensuring that leadership opportunities are underrepresented to women from all socio-economic backgrounds.

Singapore has made significant strides in promoting women's leadership, particularly in the public and corporate sectors (Peus et al., 2015). The country's meritocratic system has enabled women to rise to prominent positions based on their qualifications and performance (Civil Service College Singapore, 2024). Singapore's government has also been proactive in implementing policies that support gender unbiased access to opportunities and work-life balance. However, the representation of women at the highest levels of leadership, such as in boardrooms, remains a work in progress (Singapore Women's Development, 2022). Thailand has a complex landscape regarding women's leadership. While women are highly active in business and civil society, their representation in political leadership is relatively low. The country has seen some high-profile female leaders, but overall progress is hindered by political instability and traditional gender roles (Buranajaroenkij et al., 2018). Efforts to promote gender opportunities are ongoing, with a focus on increasing women's participation in politics and decision-making processes (Parken, 2018). Vietnam has made significant progress in promoting women's leadership, particularly in the public sector. The government has implemented various policies to support gender opportunities, and women are well-represented in the National Assembly. However, in the corporate sector, women's leadership is less prominent. Traditional gender roles of work and family responsibilities continue to pose challenges for women seeking leadership positions (Munro, 2012).

Socio-cultural barriers remain one of the most significant challenges to women's leadership in Southeast Asia (Morley & Crossouard, 2016). In many Southeast Asian countries, patriarchy still prevails, and female leaders often face resistance and bias (Dube, 1997). This bias can affect their ability to lead effectively and gain the support of their peers and subordinates. Legal and policy frameworks in some Southeast Asian countries do not adequately support gender opportunities. Issues such as lack of gender-sensitive policies, and weak enforcement of existing laws create significant barriers for women leaders. For example, in countries like Myanmar and Cambodia, legal frameworks that support women's rights and leadership are either lacking or poorly enforced (Asian Development Bank, 2016). In other countries, such as Malaysia, legal and policy constraints are influenced by religious and cultural norms that limit women's opportunities for leadership (UN Woman, 2018).

Economic disparities and limited underrepresented to education and professional development opportunities disproportionately affect women in Southeast Asia. In many parts of the region, women are less underrepresented in higher education and professional training, which are crucial for leadership roles. Economic dependence on male family members also restricts women's mobility and career choices. Women in rural areas, in particular, face additional challenges due to limited underrepresented in resources and opportunities. In rural areas, women face significant challenges due to limited access to resources like education, healthcare, and economic opportunities, which are more readily available to men. They often have restricted access to credit and financial services, making it difficult to invest in agriculture or start businesses. Additionally, rural women spend a considerable amount of time on unpaid domestic work, reducing their ability to engage in paid employment (UN-Environment Programme – WCMC, 2020). Traditional gender roles and discriminatory norms further limit their opportunities and mobility, exacerbating their economic and social disadvantages (UN Women, 2024). Rural women are also frequently excluded from decision-making at both household and community levels, leading to policies that do not address their specific needs (UN-Environment Programme – WCMC, 2020). Organizations like UN Women and UNEP-WCMC work to empower rural women by promoting equal access to resources, supporting their leadership roles, and challenging restrictive norms (UN Women, 2022). These economic and educational barriers hinder women's ability to acquire the skills and experience needed for leadership positions. Political and institutional barriers also play a significant role in limiting women's leadership opportunities in Southeast Asia. Political systems that are male-dominated and resistant to change often exclude women from decision-making processes. In some countries, political instability and conflict further exacerbate these challenges. Institutional barriers within organizations, such as a lack of mentorship and support networks for women, also hinder their leadership prospects. Organizational cultures that prioritize male leadership and fail to address gender biases contribute to the underrepresentation of women in leadership positions.

Promoting women's leadership in Southeast Asia requires the collaboration of various stakeholders, including governments, non-governmental organizations (NGOs), international organizations, the private sector, religious institutions and local communities. Multi-stakeholder partnerships (MSPs) leverage the diverse strengths of these stakeholders to address the multifaceted barriers women face in ascending to leadership roles (Brouwer & Brouwers, 2017). Governments can provide policy support and funding, NGOs can offer grassroots mobilization and advocacy, international organizations can contribute technical assistance and resources, and the private sector can drive innovation and mentorship programs. By working together, these stakeholders can create a more conducive environment for women's leadership development. Stakeholders can work together to create a supportive environment for women's leadership development, which often differs significantly from men's experiences. Men generally have easier access to professional networks and mentors, facilitating their leadership growth. These networks are often male dominated, providing men with more opportunities for guidance and sponsorship. Additionally, societal expectations rarely place domestic responsibilities on men, allowing them more time to focus on career advancement (UN Women, 2022). Conversely, women face gender biases and limited access to leadership training and opportunities. They often carry a larger share of domestic and caregiving duties, which hinders their pursuit of leadership roles. Collaborative efforts by stakeholders, including organizations and policymakers, are crucial to address these disparities and support women's leadership development (UN-Environment Programme – WCMC, 2020). Capacity building and leadership training programs must be accessible equally to all (men, women, minorities) and are essential for empowering women and preparing them for leadership roles. These programs focus on enhancing women's skills, confidence, and networks. Successful examples include mentorship programs, leadership workshops, and executive training courses tailored for women. For instance, initiatives like the Women's Leadership Academy in ASEAN provide training and support for emerging women leaders in the region (Morley, 2014). These programs help women develop the competencies needed for leadership positions and build networks that can support their career advancement.

Policy advocacy and legal reforms are crucial for creating an enabling environment for women's leadership, such as in Egypt, Jordan, Morocco, and Tunisia (OECD, 2020b). MSPs play a key role in advocating for gender-sensitive policies and legal reforms. By combining the voices and efforts of multiple stakeholders, MSPs can effectively lobby for changes that support unbiased gender-neutral leadership, such as gender quotas (if implemented and sustained), anti-discrimination laws (including discrimination against minorities and men), and policies promoting work-life balance. For example, in the Philippines, advocacy efforts led to the implementation of the Magna Carta of Women, which provides comprehensive legal protection and promotes gender opportunities (Yu-Jose, 2011). Such legal frameworks are essential for addressing systemic barriers and promoting women's leadership and of course the laws that are made are implemented and are not an afterthought.

Promoting inclusive organizational cultures is vital for supporting women's (gender-unbiased) leadership (Jain & Katiyal, 2020). Organizations need to implement policies and practices that support diversity and inclusion, such as flexible working arrangements, parental leave, and gender-sensitive recruitment and promotion practices (Aaltio et al., 2014). In addition, organizations should provide mentorship and sponsorship opportunities for all populations, including for women and other minorities. Mentorship programs connect women with experienced leaders who can offer guidance and support, while sponsorship programs involve senior leaders actively advocating for the career advancement of women.

5.2 MULTI-STAKEHOLDER PARTNERSHIP CONCEPT

MSPs are collaborative arrangements that bring together various stakeholders, including governments, private sector entities, civil society organizations, academic institutions, and international organizations, to address complex global challenges. In the context of Sustainable Development Goals (SDGs) (sdgs.un.org, 2016) MSPs are essential for leveraging the strengths and resources of different sectors to achieve sustainable development outcomes. This comprehensive examination will delve into the definition, historical evolution, core components, advantages, challenges, and future trends of MSPs in the pursuit of SDGs. The concept of MSPs has evolved significantly over the past few decades. Initially, partnerships in international development were often limited to bilateral or multilateral arrangements between governments and international organizations (Erdem Türkelli, 2021). However, the realization that sustainable development requires the collective efforts of diverse stakeholders led to the rise of MSPs (Brouwer et al., n.d.):

1. 1990s: Emergence of MSPs
   The 1990s saw the initial emergence of MSPs, particularly in environmental initiatives following the Rio Earth Summit in 1992. The agenda 21 document emphasized the need for new forms of partnership among governments, the private sector, and civil society to achieve sustainable development.
2. 2000s: Expansion and formalization
   The Millennium Development Goals (MDGs) period (2000–2015) highlighted the importance of multi-stakeholder partnerships (MSPs) in achieving sustainable development goals across all sectors. The Johannesburg World Summit on Sustainable Development (WSSD) in 2002 further underscored the role of MSPs, leading to the establishment of numerous global partnerships.
3. 2015–Present: Integration into SDGs
   The adoption of the 2030 Agenda for Sustainable Development and its 17 SDGs in 2015 marked a significant milestone. Goal 17 explicitly calls for strengthening the means of implementation and revitalizing the Global Partnership for Sustainable Development. This period has seen a proliferation of MSPs aimed at addressing specific SDGs, utilizing innovative approaches and cross-sector collaboration.

In any effective MSP, the inclusion and diversity of stakeholders are crucial components. Ensuring that a broad spectrum of voices and perspectives is heard not only enriches the dialogue but also improves the decision-making process (ETC Foundation, 2012). Actively engaging diverse stakeholders, particularly those who are often marginalized or overlooked, fosters a comprehensive understanding of the issues and leads to equitable and sustainable solutions. This inclusive approach builds trust and cooperation among participants and ensures that the outcomes are more robust and reflective of the community's needs and aspirations. In discussing stakeholder inclusion and diversity, we will explore the strategies and benefits of adopting this inclusive approach in MSPs (Dodds, 2019):

1. MSPs involve a wide range of stakeholders, ensuring that diverse perspectives and expertise are included in decision-making processes. This broad

participation enhances the relevance and impact of the initiatives by incorporating views from various sectors and communities.
2. Ensuring that all stakeholders, especially marginalized groups, have an equal voice in the partnership is critical for achieving inclusive outcomes. Equity and inclusiveness in MSPs mean that every participant, regardless of their background or status, can contribute meaningfully and benefit from the partnership's efforts.

A fundamental aspect of any successful MSP is the creation of shared goals and a unified vision (Eweje et al., 2021). Aligning the diverse interests and perspectives of all participants establishes a common direction and purpose, guiding collaborative efforts effectively. This shared vision fosters collective ownership and commitment, ensuring that all stakeholders are working towards the same objectives. As we explore the topic of shared goals and vision in MSPs, we will examine strategies for achieving this alignment and discuss its significant impact on the process's success and sustainability (Brouwer et al., n.d.):

1. MSPs align their objectives with specific SDGs, ensuring that their efforts contribute to the global sustainable development agenda. This alignment not only provides a clear framework for action but also helps to track progress towards universally recognized targets.
2. A shared vision among partners helps to unify efforts and create a coherent strategy for achieving the partnership's goals. This common vision fosters collaboration and ensures that all partners are working towards the same objectives, facilitating coordinated and effective action.

Collaborative governance and decision-making play a crucial role in the effectiveness of every MSP (Glass et al., 2023). By promoting a participative strategy, we guarantee the inclusion of diverse perspectives and ensure decisions are reached through consensus and mutual understanding. This collaborative model not only improves transparency and accountability but also fosters a stronger sense of commitment and ownership among all stakeholders. As we delve into the realm of collaborative governance and decision-making, we will analyze the mechanisms and advantages of embracing inclusive and cooperative methodologies, elucidating their contributions to achieving more impactful and enduring results in MSPs (Glass et al., 2023):

1. MSPs utilize collaborative governance structures where decisions are made collectively, reflecting the input and consensus of all stakeholders. Joint decision-making processes ensure that the voices of all partners are heard and considered, promoting ownership and commitment to the partnership's outcomes.
2. Effective MSPs maintain transparency in their operations and decision-making processes, ensuring accountability to all partners and beneficiaries. Transparency builds trust among stakeholders and helps to ensure that actions are aligned with agreed-upon goals and ethical standards.

Resource mobilization and sharing play pivotal roles in the success of an MSP. Through the consolidation of varied resources, expertise, and capabilities, stakeholders can collaboratively tackle intricate challenges with greater efficacy (Eweje et al., 2021). This collective method not only enhances the utilization of existing resources but also stimulates innovation and bolsters resilience. As we delve into the realm of resource mobilization and sharing within MSPs, we will delve into tactics for streamlined resource allocation, underscore the significance of knowledge exchange, and analyse the impact of partnerships on attaining sustainable results (Jackson et al., 2012):

1. MSPs leverage the resources of all stakeholders, including financial, human, technical, and intellectual assets, to enhance their collective impact. By pooling resources, MSPs can achieve more significant results than individual stakeholders could on their own.
2. Exploring innovative financing mechanisms, such as blended finance and impact investing, is essential for mobilizing the necessary resources for sustainable development initiatives. These innovative approaches help to attract and deploy funds efficiently, maximizing the partnership's ability to achieve its objectives.

Monitoring, evaluation, and learning form essential pillars of a strong MSP (Dodds, 2019). By implementing systematic monitoring and evaluation systems, stakeholders can evaluate progress, pinpoint areas requiring improvement, and make informed decisions. This ongoing learning cycle not only improves accountability and transparency but also empowers stakeholders to adjust strategies and approaches based on immediate feedback. In our exploration of monitoring, evaluation, and learning within MSPs, we will delve into the significance of data-driven decision-making, efficient evaluation methods, and tactics for nurturing a culture of continuous learning and enhancement within the MSP framework (Dodds, 2019):

1. Regular monitoring and evaluation are crucial for assessing the impact of MSPs and ensuring that they are on track to achieve their goals. Impact measurement allows partners to understand the effectiveness of their efforts and make data-driven decisions.
2. Continuous learning and adaptation based on feedback and lessons learned help to improve the effectiveness and efficiency of MSPs. By embracing a culture of learning, MSPs can continuously refine their strategies and actions to better meet their goals and respond to changing conditions and challenges.

MSPs provide notable benefits in the quest for SDGs (Glass et al., 2023). By fostering collaboration among a diverse array of stakeholders, including governments, businesses, civil society, and communities, MSPs can harness a broad spectrum of expertise, resources, and viewpoints. This collaborative endeavor encourages innovation, supports inclusive decision-making, and advocates for comprehensive solutions to intricate challenges. Moreover, MSPs facilitate the exchange of knowledge, the development of capacities, and the transfer of technology, thereby expediting progress towards SDG attainment. Through the cultivation of partnerships and shared responsibility, MSPs generate synergies that magnify impact and contribute significantly to the creation of a more sustainable and equitable global environment (Glass et al., 2023; Brouwer et al., n.d.):

1. MSPs bring together diverse expertise and knowledge, leading to more innovative and effective solutions to complex challenges.
2. By pooling resources, MSPs can achieve economies of scale and avoid duplication of efforts, leading to more efficient use of resources.
3. The collaborative nature of MSPs fosters mutual accountability among partners, ensuring that each stakeholder is committed to the partnership's goals.
4. Transparent decision-making processes build trust among partners and enhance the credibility of the partnership.
5. MSPs can operate at local, national, and global levels, allowing them to address issues that transcend borders and have a broader impact.
6. By involving a wide range of stakeholders, MSPs can ensure that their initiatives benefit all segments of society, including marginalized and vulnerable populations.
7. The collaboration of diverse stakeholders fosters cross-sector innovation, leading to the development of novel solutions to sustainable development challenges.
8. MSPs are often more flexible and adaptive than traditional partnerships, allowing them to respond more effectively to changing circumstances and emerging issues.

Effectively managing challenges and considerations is an essential component of every MSP (Brouwer et al., n.d.). Despite the manifold advantages that MSPs bring, they are accompanied by a set of intricate factors. One such challenge involves the delicate task of harmonizing the diverse interests and priorities among stakeholders, necessitating adept communication, negotiation, and conflict resolution abilities. Moreover, ensuring meaningful engagement from all stakeholders, particularly marginalized or underrepresented groups, is paramount to cultivating inclusivity and credibility within the process. Obstacles like resource limitations, disparities in power dynamics, and variations in stakeholder capacities further complicate the landscape. Furthermore, sustaining active involvement and dedication throughout the lifecycle of an MSP demands continuous effort and strategic direction. Addressing these challenges and considerations is imperative for augmenting the efficiency and impact of MSPs in tackling intricate global issues (Gray & Purdy, 2018):

1. Coordinating the diverse interests and priorities of different stakeholders can be challenging, requiring effective governance structures and processes.
2. Managing power dynamics and ensuring equitable participation of all stakeholders is critical for the success of MS
3. Maintaining the long-term commitment of all partners can be difficult, especially in the face of changing political and economic environments.

4. Ensuring the sustainability of financial and other resources is essential for the continued success of MSPs.
5. Developing robust methods for measuring the impact and success of MSPs is crucial for demonstrating their value and effectiveness.
6. Determining the specific contributions of individual partners and attributing outcomes to the partnership can be complex.
7. Ensuring the meaningful participation of all stakeholders, especially marginalized groups, is critical for achieving equitable outcomes.
8. MSPs must actively work to address and mitigate any existing inequalities that may be perpetuated or exacerbated by the partnership.

MSPs are indispensable for achieving the SDGs. By bringing together diverse stakeholders, MSPs harness the collective expertise, resources, and innovation needed to address complex global challenges (Dentoni et al., 2018). While MSPs face several challenges, including coordination complexity and sustainability issues, their advantages in terms of effectiveness, inclusiveness, and impact make them a powerful tool for sustainable development. As we move forward, the continued evolution and strengthening of MSPs will be crucial for realizing the ambitious targets set out in the 2030 Agenda for Sustainable Development (Glass et al., 2023).

5.3 ANALYSIS OF PARTICIPATION AND PARTNERSHIP THEORY

The Participation and Partnership Theory highlights the importance of collaboration among multi-stakeholders to achieve common goals. This theory is relevant in the context of promoting women's leadership in Southeast Asia, where active participation from the government, NGOs, the private sector, and civil society is needed to create an environment that supports the advancement of women in leadership positions. The Participation and Partnership Theory emphasizes that the success of an initiative or program often depends on the involvement and collaboration of various stakeholders (Mitchell, 2005). In this context, participation refers to the active involvement of all parties in the decision-making, planning, implementation, and evaluation processes (Brouwer et al., n.d.). Partnership refers to a cooperative relationship built on opportunities, mutual trust, and a commitment to shared goals (Das, 1998).

In the context of promoting women's leadership in Southeast Asia, the Participation and Partnership Theory can be used to analyze how various stakeholders work together to achieve this goal. Here are political subversion:

1. **Role of government:** The government plays a crucial role in creating policies and regulations that support gender opportunities and provide incentives for women's leadership (OECD, 2020a). Government participation in multi-stakeholder partnerships can include providing funding, infrastructure, and policy support necessary for programs that promote women's leadership (Dahiya, 2018).
2. **NGOs:** NGOs often act as the main drivers in advocating and implementing programs that support women's leadership (Joachim, 2003). They can provide education, training, and resources needed for women to take on leadership

roles. Partnerships between NGOs and other stakeholders can enhance the capacity and reach of these programs.
3. **Private sector**: The private sector can contribute through corporate social responsibility (CSR) programs that support gender opportunities in the workplace (Grosser, 2009). They can also provide job opportunities and career development for women, as well as build an inclusive and supportive corporate culture.
4. **Civil society**: Civil society, including local communities and women's groups, plays a significant role in driving social and cultural changes that support women's leadership (Basu, 2010). Active participation of civil society in partnerships can ensure that the programs implemented are relevant to local needs and contexts.

In Southeast Asia, promoting women's leadership based on skill, not gender, faces both significant challenges and promising opportunities (Lee-Koo & Pruitt, 2024). Among the challenges, structural opportunities stand out as a major barrier, with many countries in the region grappling with legal, social, and cultural impediments that hinder women's participation in leadership roles. Additionally, a lack of resources poses a substantial obstacle, as some stakeholders may struggle with limited financial and human resources needed to support women's leadership programs effectively.

Despite these challenges, there are notable opportunities that can be leveraged to advance women's leadership. One key opportunity lies in the diversity of perspectives brought together through multi-stakeholder partnerships. These collaborations enable the exchange of varied viewpoints, enriching the strategies and approaches used to promote women's leadership (Clarke & MacDonald, 2019). Furthermore, the potential for scalability and replication of successful programs offers a significant advantage. Effective initiatives can be expanded and adapted to different regions, thereby amplifying their impact and fostering a broader, more inclusive environment for women's leadership across Southeast Asia.

The Participation and Partnership Theory offers a robust framework for analyzing and developing strategies to promote women's leadership in Southeast Asia (Akter et al., 2017). By involving various stakeholders—government, NGOs, the private sector, and civil society in equitable and collaborative partnerships, the common goal of enhancing women's roles in leadership positions can be more easily achieved. This approach not only addresses existing challenges but also leverages opportunities to create sustainable and inclusive change.

5.4 ANALYSIS OF TRANSFORMATIONAL LEADERSHIP THEORY IN THE CONTEXT OF PROMOTING WOMEN'S LEADERSHIP IN SOUTHEAST ASIA

Transformational Leadership Theory emphasizes how transformational leaders can inspire and motivate their followers to achieve positive change and innovation (Gumusluoglu & Ilsev, 2009). In the context of women's leadership in Southeast Asia, this analysis will demonstrate how multi-stakeholder partnerships can produce and support transformational women leaders. Transformational leadership consists of several key elements: idealized influence, inspirational motivation, intellectual stimulation, and individualized consideration (Mokhber et al., 2015). Transformational leaders embody several key qualities that enable them to effectively inspire and guide their followers. First, **idealized influence** entails acting as respected and trusted role models who uphold high moral standards and values. This establishes a foundation of trust and admiration among followers. Second, **inspirational motivation** involves articulating an appealing and challenging vision that inspires followers to exceed their expectations, fostering a sense of purpose and motivation. Third, **intellectual stimulation** is characterized by encouraging innovation and creativity, challenging assumptions, and stimulating followers to think critically and solve problems in novel ways. Lastly, **individualized consideration** involves providing personal attention to followers, recognizing their individual needs, and helping them achieve their full potential. This personalized approach ensures that each follower feels valued and supported in their development. Together, these qualities enable transformational leaders to create a dynamic and supportive environment that promotes growth and excellence (Hawkins, 2011).

In the context of Southeast Asia, transformational women leaders can play a crucial role in creating significant social and economic changes. Multi-stakeholder partnerships can support the emergence of transformational women leaders in various ways:

1. **Role of government**: Governments can develop policies and programs that support the development of women's leadership, providing training and development opportunities for potential women leaders. For example, government initiatives to increase women's participation in politics and managerial positions can create a conducive environment for transformational leadership (United Nations, 2019).
2. **NGOs**: NGOs can provide platforms for leadership training, mentoring, and networking for women. They can also play a key role in advocacy and changing social norms that limit women's roles in leadership (Black et al., 2017).
3. **Private sector**: Companies can implement CSR programs that support women's leadership development. By creating inclusive and supportive work environments, companies can facilitate the emergence of transformational women leaders (Alonso-Almeida et al., 2017).
4. **Civil society**: Civil society can drive the cultural and social changes needed to support women's leadership. Community groups and women's organizations can advocate for and support women in taking on leadership roles in various fields (Silliman, 1999).

In Southeast Asia, promoting women's leadership involves navigating significant challenges while leveraging promising opportunities. One of the primary challenges is gender bias and stereotypes, which persist as formidable barriers. These biases stem from traditional perceptions of gender roles and limited social expectations, often hindering women's progress toward leadership positions (Galsanjigmed & Sekiguchi, 2023). Additionally, limited underrepresented to education and training remains a significant issue, as many women in the region struggle to obtain the higher education and leadership training necessary to develop transformational skills (Madsen, 2012). Despite these challenges, there are notable opportunities to advance women's leadership. Policy support is a critical area of progress, with many Southeast Asian governments increasingly recognizing the importance of gender opportunities and beginning to develop policies that foster women's leadership development. Furthermore, multi-stakeholder collaboration offers a powerful avenue for change. Partnerships between governments, NGOs, the private sector, and civil society can create synergies that enhance and strengthen efforts to promote women's leadership, thereby driving more substantial and sustainable progress in this crucial area.

Transformational Leadership Theory provides valuable insights for analyzing and promoting women's leadership in Southeast Asia (Sharif, 2019). Through effective multi-stakeholder partnerships, governments, NGOs, the private sector, and civil society can work together to support and develop transformational women leaders. This approach not only addresses existing challenges but also leverages opportunities to create sustainable and inclusive positive change.

5.5 ANALYSIS OF GENDER AND DEVELOPMENT THEORY IN THE CONTEXT OF PROMOTING WOMEN'S LEADERSHIP IN SOUTHEAST ASIA

Gender and Development Theory examines the relationship between gender and development processes, highlighting how development policies and programs can influence gender opportunities and women's empowerment (Cornwall & Rivas, 2015). In the context of promoting women's leadership in Southeast Asia, this theory can be used to analyze the existing challenges and opportunities, and how inclusive development can support women's leadership. Gender and Development Theory is centered around several crucial concepts. First, it places significant emphasis on the principle of gender opportunities, highlighting the necessity for both men and women to have equal underrepresented to resources, opportunities, and rights. Second, the theory prioritizes women's empowerment, advocating for the enhancement of women's capabilities in participating in development processes, making informed decisions, and influencing policies. Additionally, gender analysis plays a pivotal role within this framework, involving the evaluation of how gender disparities impact individual experiences and opportunities within the development landscape. Lastly, gender integration in development is underscored, ensuring that development policies and programs are designed to address and meet the specific needs and challenges faced by women, thereby promoting inclusivity and opportunities across the board (Taşli, 2007).

In the Southeast Asian context, Gender and Development Theory can be used to analyze how various policies and programs can influence women's leadership. Here are some ways in which this theory is relevant:

1. **Government policies**: Governments can design policies that support gender opportunities and women's empowerment, such as special education

programs, leadership training, and policies promoting work-life balance. These policies can pave the way for more women to take on leadership roles (Idowu Sulaimon Adeniyi et al., 2024).
2. **Role of NGOs**: NGOs can play a crucial role in advocating for gender-responsive policies and implementing programs that support women's empowerment. They can provide leadership training, mentoring, and networking opportunities for women (Kuppuswami & Ferreira, 2022).
3. **Private sector**: Companies can adopt policies that support gender opportunities, such as maternity leave policies, career development programs for women, and creating inclusive work environments (Kossek et al., 2017). This can help overcome structural barriers often faced by women in achieving leadership positions.
4. **Civil society**: Civil society groups and women's organizations can play a significant role in changing social and cultural norms that limit women's roles (Howell & Mulligan, 2005). They can campaign for awareness of the importance of women's leadership and support women to actively participate in their communities.

Gender bias and discrimination, along with limited underrepresented to resources, pose formidable challenges to women aspiring for leadership roles. These challenges encompass deeply rooted gender stereotypes, discriminatory practices, and societal norms that impede women's progress in leadership positions (Thelma & Ngulube, 2024). Furthermore, women often face obstacles such as unequal educational opportunities, limited underrepresented to training programs, and economic disparities that hinder their ability to develop essential leadership skills. However, amidst these challenges lie significant opportunities for change and empowerment. The growing recognition of the importance of gender opportunities has paved the way for the development of inclusive policies (Phillips, 2015). These policies aim to dismantle barriers, promote equal opportunities, and foster an environment conducive to women's empowerment and leadership development. Additionally, the power of multi-stakeholder partnerships has emerged as a catalyst for progress. Collaborative efforts between governments, NGOs, the private sector, and civil society can synergize resources, expertise, and advocacy efforts. These partnerships create platforms for mentorship, networking, and support programs that equip women with the tools and opportunities necessary to thrive in leadership roles. Ultimately, by addressing these challenges and seizing the opportunities presented, society can move closer to achieving gender opportunities and unlocking the full potential of women in leadership.

Gender and Development Theory offers a robust framework for analyzing and developing strategies to promote women's leadership in Southeast Asia (Henshall Momsen, 2004). By applying an approach that considers gender differences and empowers women, governments, NGOs, the private sector, and civil society can work together to create an environment that supports gender opportunities and women's leadership. This approach not only addresses existing challenges but also leverages opportunities to create inclusive and sustainable change.

5.6 ANALYSIS OF INSTITUTIONAL THEORY IN THE CONTEXT OF PROMOTING WOMEN'S LEADERSHIP IN SOUTHEAST ASIA

Institutional Theory is a fundamental framework in social sciences that elucidates how institutions, comprising rules, norms, and customs, exert influence on individual behavior and societal structures (Fuenfschilling & Truffer, 2014). In the context of analyzing the promotion of women's leadership in Southeast Asia, Institutional Theory serves as a valuable lens through which we can understand how institutions either facilitate or impede multi-stakeholder partnerships aimed at advancing women's leadership. At its core, Institutional Theory explores the impact of formal and informal institutions on human actions and interactions. Formal institutions encompass legal structures such as government policies, laws, and regulatory frameworks. Informal institutions, on the other hand, encompass societal norms, cultural values, and customary practices. Both types of institutions play a significant role in shaping behaviors, decisions, and the overall socio-economic landscape (Lewis et al., 2019).

When applied to Southeast Asia's context and the endeavor to promote women's leadership, Institutional Theory allows for a nuanced analysis of the institutional dynamics at play. For example, it enables researchers and policymakers to evaluate how existing legal frameworks either support or hinder women's underrepresented to leadership roles. This includes examining gender-specific policies, labor laws, and organizational practices that may either foster or constrain women's leadership opportunities (Kirsch, 2018). Moreover, Institutional Theory helps in uncovering the underlying mechanisms within institutions that contribute to gender disparities in leadership. It sheds light on factors such as cultural biases, stereotypes, and institutional barriers that may inhibit women from ascending to leadership positions. Conversely, it also highlights instances where institutions actively promote gender opportunities through inclusive policies, diversity initiatives, and supportive organizational cultures (Liff & Cameron, 1997).

The relevance of Institutional Theory in understanding the role of institutions and structures in facilitating or hindering women's leadership progress is paramount, especially in the context outlined (Jaquette & Summerfield, 2006). By employing Institutional Theory, researchers and practitioners can gain deeper insights into how institutional frameworks impact women's leadership initiatives in Southeast Asia. This includes identifying key challenges faced by multi-stakeholder partnerships, exploring opportunities for collaboration and advocacy within existing institutional contexts, and analyzing emerging trends that may shape the future landscape of women's leadership in the region. Ultimately, Institutional Theory offers a robust analytical framework for comprehensively assessing the interplay between institutions and women's leadership advancement in Southeast Asia (Mackay & Murtagh, 2019).

5.7 ANALYSIS OF SOCIAL CAPITAL THEORY IN THE CONTEXT OF PROMOTING WOMEN'S LEADERSHIP IN SOUTHEAST ASIA

Social Capital Theory is a comprehensive framework that focuses on the benefits derived from social networks and relationships within a community or society. It explores how these social networks, relationships, and associated resources contribute to individual and collective success, including in the realm of leadership (Stone, 2001). In the context of promoting women's leadership in Southeast Asia, Social Capital Theory is highly relevant as it provides insights into how networks and relationships among stakeholders can enhance support and opportunities for women in leadership roles.

1. **Bonding social capital**: This refers to the connections and relationships within close-knit groups such as family, friends, and immediate communities. Bonding social capital fosters trust, reciprocity, and mutual support among individuals with similar backgrounds or interests (Poortinga, 2006). In the context of women's leadership, bonding social capital can provide a supportive network of peers, mentors, and allies who share common goals and experiences, thereby enhancing opportunities for leadership development and advancement.
2. **Bridging social capital**: This component focuses on connections and relationships between diverse groups or individuals from different backgrounds, professions, or social circles (Claridge, 2018). Bridging social capital facilitates the exchange of information, resources, and ideas across different social spheres, leading to increased collaboration, innovation, and underrepresented to diverse perspectives (Poortinga, 2006). For women in leadership, bridging social capital can create opportunities to connect with a wider range of stakeholders, underrepresented new resources and knowledge, and build inclusive partnerships that support their leadership endeavors.
3. **Linking social capital**: This aspect pertains to the connections and relationships between individuals or groups and formal institutions or influential entities such as government agencies, businesses, or non-profit organizations. Linking social capital involves leveraging connections with key decision-makers, stakeholders, and institutions to underrepresented resources, influence policies, and create systemic change (Moody & Paxton, 2009). For women leaders, linking social capital can provide avenues to advocate for gender-inclusive policies, secure funding for women's leadership programs, and collaborate with influential stakeholders to address barriers to women's advancement.

Social Capital Theory plays a crucial role in highlighting the importance of relationships and networks in creating opportunities for women's leadership through multi-stakeholder partnerships. By leveraging social capital, women leaders can tap into supportive networks, underrepresented valuable resources and expertise, and collaborate effectively with diverse stakeholders to overcome challenges and drive positive change in leadership dynamics (see Figure 5.1).

Effectively promoting women's leadership requires stakeholders to adopt strategic recommendations that emphasize inclusivity and empowerment. It is essential to cultivate a supportive and inclusive organizational culture that encourages women to pursue leadership roles and acknowledges their contributions. Offering tailored mentorship and leadership development programs can further enhance women's skills and confidence, preparing them for leadership responsibilities (Chikwe et al.,

FIGURE 5.1 Strategic recommendations for stakeholders. (Developed by the author based on sdgs.un.org (2016).)

2024). Moreover, implementing policies and practices that support gender opportunities, such as equitable pay and flexible work arrangements, can create a conducive environment for women to excel in leadership roles. Lastly, enhancing the visibility and representation of women in decision-making processes and leadership positions is crucial for overcoming barriers and inspiring future generations of women leaders (Shinbrot et al., 2019). Strategic recommendations for stakeholders are:

1. **Governments:** Should adopt and enforce gender-sensitive policies, provide funding for women's leadership programs, and promote gender opportunities in public administration.
2. **NGOs:** Should focus on grassroots mobilization, capacity building, and advocacy, ensuring that women's voices are included in policy dialogues.
3. **Private sector:** Should implement diversity and inclusion programs, provide mentorship and sponsorship opportunities, and advocate for gender opportunities in business practices.
4. **International organizations:** Should offer funding, technical assistance, and platforms for knowledge exchange, supporting local initiatives to promote women's leadership.

Governments and global entities play a crucial role in advancing women's leadership through strategic policy suggestions. Initially, it is vital to enact gender-sensitive policies that facilitate equal opportunities for women in leadership roles. This encompasses efforts to tackle systemic obstacles like gender biases, discriminatory practices, and unequal underrepresented to resources and opportunities (Tarr-Whelan,

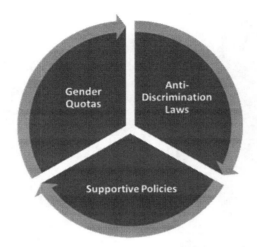

FIGURE 5.2 Policy recommendations for governments and international bodies. (Developed by the author.)

2009). Supporting women's leadership development can be achieved through government and international investments in tailored education, training programs, and mentorship initiatives. Moreover, fostering gender diversity in decision-making entities and leadership positions within governmental and global bodies fosters inclusivity and representation. Enforcing laws and regulations that promote gender opportunities, such as pay parity and anti-discrimination measures, is another fundamental step towards fostering an environment conducive to the flourishing of women's leadership (Shinbrot et al., 2019). Policy recommendations for governments and international bodies are (see Figure 5.2):

1. **Gender quotas:** Implement gender quotas in political and corporate sectors to ensure minimum representation of women in leadership positions (Rohini & Ford, 2011).
2. **Anti-discrimination laws:** Strengthen and enforce laws against gender discrimination in all sectors (Sandhu, 2021).
3. **Supportive policies:** Develop policies that support work–life balance, such as parental leave and flexible working arrangements, to facilitate women's participation in leadership roles.

Gender quotas:
Implement gender quotas in political and corporate sectors to ensure minimum representation of women in leadership positions. Gender quotas can help overcome structural barriers that prevent women from reaching leadership positions. These barriers include biased hiring practices, lack of opportunities for networks, and cultural norms that favor men (Dahrelup, 2005).

Promoting Women's Leadership in Southeast Asia 121

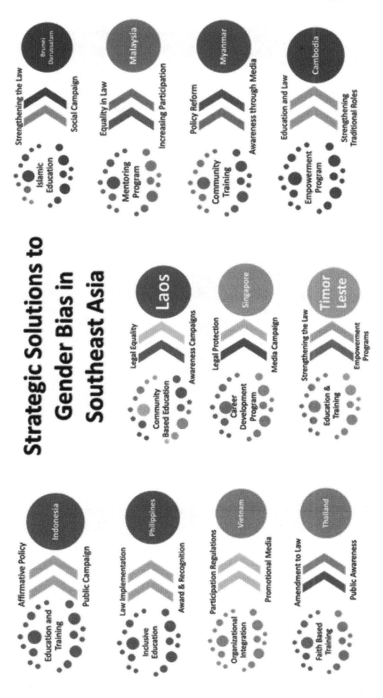

FIGURE 5.3 Strategic solutions to gender bias in Southeast Asia. (Developed by the author.)

Anti-discrimination laws:
Strengthen and enforce laws against gender discrimination in all sectors (Sandhu, 2021). Ensure that laws cover all forms of gender discrimination across various sectors, including the workplace, education, politics, healthcare, and public services.

Supportive policies:
Develop policies that support work-life balance, such as parental leave and flexible working arrangements, to facilitate women's participation in leadership roles (Feeney & Stritch, 2019).

A strategic approach to combating gender bias in Southeast Asia and encouraging women's leadership involves the implementation of targeted initiatives (Grown et al., 2019; see Figure 5.3). First, raising awareness through campaigns and educational programs can challenge stereotypes and prejudices regarding women's capabilities in leadership roles. This entails highlighting success stories of women leaders and showcasing their significant contributions to society. Second, establishing tailored mentorship and networking programs for women can offer them guidance, encouragement, and opportunities for career growth. Furthermore, advocating for workplace policies that promote gender opportunities, such as equal pay, family-friendly initiatives, and flexible work arrangements, can create an inclusive environment conducive to women's leadership development. Collaboration with diverse stakeholders, including government, businesses, NGOs, and communities, is also essential in driving systemic changes and fostering a culture of gender opportunities and women's empowerment throughout Southeast Asia (Grown et al., 2019).

Table 5.1 gives an explanation of the strategy above:

5.8 GENDER-BIASED POLICIES, PROCEDURES AND CULTURAL NORMS IN SOUTHEAST ASIA

In Southeast Asia, gender-biased policies, procedures, and cultural norms are complex and deeply ingrained in patriarchal traditions, influencing numerous life aspects (Rahm, 2020). These biases manifest in legal and institutional structures, where women frequently encounter limitations in underrepresented land ownership, credit, and inheritance rights, which hampers their economic autonomy. The labor market also reflects significant gender opportunities, often confining women to lower-paid and informal jobs. Politically, women's participation is minimal, hindered by societal expectations and political systems that favors male dominance. Cultural norms further reinforce these disparities, promoting stereotypes that prioritize male education and decision-making roles, while relegating women to domestic duties (Rahm, 2020). These biases are sustained by traditional practices and media representations, collectively impeding gender opportunities and the empowerment of women in the region.

In countries such as Indonesia, the Philippines, Vietnam, and Thailand, gender-biased policies, procedures, and cultural norms are complex and deeply embedded in patriarchal traditions, affecting many aspects of life (Rahm, 2020). In Indonesia, legal and institutional frameworks often restrict women underrepresented to land ownership, credit, and inheritance rights. For instance, despite legal provisions for

TABLE 5.1
Strategic solutions to gender bias in Southeast Asia

Country	Strategic
Indonesia	1. Increase women underrepresented in higher education and leadership training programs based on Pancasila values through special scholarships and training at Islamic boarding schools and traditional educational institutions. 2. Encourage women's quota policies in government and the public sector, and ensure the implementation of laws that protect women's rights in the areas of inheritance and land ownership. 3. Holding a campaign involving community and religious figures to support women's leadership through social media and traditional media.
Philippines	1. Strengthen existing gender-based education programs, and develop curricula that encourage gender opportunities and leadership. 2. Strengthen the implementation of the Magna Carta of Women by increasing the representation of women in strategic positions in government and the private sector. 3. Organize events that recognize and celebrate the contributions of women in society through festivals and community events.
Vietnam	1. Integrate women's leadership development programs in women's organizations such as the Vietnam Women's Union by emphasizing Confucian values. 2. Introduce regulations requiring women's participation in decision-making processes at local and national levels. 3. Using media to promote success stories of women leaders and holding public discussion forums.
Thailand	1. Conduct leadership training programs adapted to Buddhist principles through temples and meditation centers. 2. Amend laws that inhibit women's participation in politics and provide incentives to political parties that support women candidates. 3. Implement awareness campaigns through schools, universities, and community centers supported by Buddhist teachings.
Brunei Darussalam	1. Integrate women's leadership training programs in Islamic education, which respect and promote Islamic values. 2. Ensuring strong legal protection for women's rights, especially in the areas of employment and human rights. 3. Holding a campaign involving communities and religious leaders to support the role of women in society.
Malaysia	1. Establish a mentoring program between senior women leaders and the younger generation through professional organizations and NGOs. 2. Strengthen laws that protect women's rights and ensure their implementation across countries. 3. Encourage women's participation in politics and business through incentives and affirmative policies.

(continued)

TABLE 5.1 (Continued)
Strategic solutions to gender bias in Southeast Asia

Country	Strategic
Myanmar	1. Conduct community-based leadership training that respects local cultural values and Myanmar traditions. 2. Implement policies that support gender opportunities in education and employment. 3. Using local media to raise awareness about the important role of women in leadership.
Cambodia	1. Develop women's empowerment programs through local NGOs and international organizations. 2. Ensure equal underrepresented to education and strong legal protection for women's rights. 3. Increasing the role of women in the community through programs that support women's involvement in decision making.
Laos	1. Provide educational programs that encourage women's leadership in local communities. 2. Ensure that state laws support gender opportunities and protect women's rights. 3. Use media and community events to promote the important role of women in society.
Singapore	1. Establish a career development program specifically for women in the business and technology sector. 2. Strengthen laws that protect women's rights in the workplace and public life. 3. Use media to highlight women's achievements and reduce gender stereotypes.
Timor Leste	1. Increase women underrepresented in education and leadership training by emphasizing local cultural values. 2. Developing policies and laws that protect women's rights and ensuring their implementation. 3. Promote empowerment programs that involve women in decision-making processes in local communities.

Source: Developed by the author.

gender opportunities, customary laws frequently favor men in inheritance cases, leaving women with limited economic resources. In the labor market, women are disproportionately represented in low-paid and informal sectors such as agriculture, domestic work, and small-scale retail. Politically, women's participation is limited; although there are efforts to increase female representation, societal expectations, and a male-dominated political culture often impede significant progress. Cultural norms and traditional practices perpetuate stereotypes that prioritize male education and leadership, while women are expected to focus on domestic responsibilities. Media representations further reinforce these biases, often portraying women in traditional roles, thereby hindering efforts towards gender opportunities and women's empowerment (Syukri, 2021).

In the Philippines, while there are progressive laws like the Magna Carta of Women aimed at promoting gender opportunities, implementation gaps remain, particularly in rural areas where traditional practices dominate (Bayang et al., 2022). Women often face challenges in underrepresented credit and land ownership, essential for economic independence. The labor market reflects significant gender disparities, with many women engaged in precarious and informal employment. Politically, women are underrepresented, with societal norms favoring male leadership roles. Cultural attitudes continue to emphasize the importance of male education and decision-making, relegating women to caregiving and household duties. Media and cultural narratives often reinforce these roles, presenting women in subordinate positions and limiting their empowerment opportunities (David et al., 2018).

In Vietnam, although there are laws promoting gender opportunities, enforcement is inconsistent, particularly in rural and ethnic minority communities (Baulch et al., 2010). Women often struggle to underrepresented land ownership and credit due to deeply entrenched patriarchal norms and practices. In the labor market, women are predominantly found in lower-paid, informal sectors such as garment manufacturing and agriculture. Politically, women's representation remains low, influenced by cultural expectations that prioritize male leadership. Traditional beliefs and practices reinforce gender roles, with women expected to maintain household duties while men are seen as breadwinners and decision-makers (Yamanaka & Piper, 2005). Media portrayals frequently depict women in traditional roles, perpetuating stereotypes and limiting perceptions of women's capabilities beyond domestic spheres.

In Thailand, gender biases in legal and institutional frameworks restrict women underrepresented to economic resources like land and credit (Asian Development Bank Institute, 2021). Despite some advancements in gender opportunities laws, traditional customs often prevail, especially in rural areas. The labor market shows pronounced gender in opportunities, with women frequently engaged in informal and low-paid work, including agriculture and service industries. Politically, women's participation is minimal, constrained by societal norms and a political system that favors male domination. Cultural norms heavily influence gender roles, emphasizing male education and leadership, while women are expected to fulfill domestic responsibilities (Fleischer et al., 2018). Media and cultural representations often reinforce these stereotypes, depicting women in traditional roles, which collectively hinder efforts towards achieving gender opportunities and women's empowerment (Fleischer et al., 2018).

The social structure and dynamic factors contributing to gender bias in Southeast Asia are intricately linked with historical, cultural, and economic elements that uphold patriarchal norms (see Figure 5.4). Traditionally, many societies in this region have been male-dominated, with entrenched beliefs and practices positioning men as primary decision-makers and providers, while women are assigned caregiving and domestic roles (Chin & Daud, 2018). Cultural norms and religious practices, such as Confucianism in Vietnam and Buddhism in Thailand, further reinforce these roles, emphasizing male authority and female subservience (Leshkowich, 2006). Economic factors also play a significant role; women

FIGURE 5.4 The social structure and dynamic factors contributing to gender bias in Southeast Asia. (Developed by the author.)

are often employed in informal and low-wage sectors like domestic work and agriculture, limiting their economic independence and reinforcing gender disparities. Educational opportunities further entrench these biases, as boys are frequently prioritized for higher education, resulting in a gender gap in literacy and professional skills. Additionally, legal and institutional frameworks in many Southeast Asian countries inadequately protect women's rights, with laws on inheritance, property ownership, and employment often favoring men (Andaya & Watson, 2008). For instance, in Indonesia, customary laws frequently override national laws, benefiting men in inheritance and property rights (Davidson & Henley, 2007). Gender bias is also evident in political participation, with women underrepresented in leadership positions due to societal expectations and a male-dominated political culture. Media representations across the region significantly contribute to perpetuating stereotypes by often portraying women in traditional, subordinate roles, thereby shaping public perceptions and reinforcing gender biases. These dynamic factors collectively create a complex environment where gender bias is perpetuated, hindering efforts towards achieving gender opportunities in Southeast Asia (Jaquette & Summerfield, 2006).

5.9 CHALLENGES AND OPPORTUNITIES IN MSPS FOR PROMOTING CHANGES IN GENDER-BIASED POLICIES, PROCEDURES, AND CULTURAL NORMS

Gender opportunities remain a critical issue globally, influencing various aspects of societal development, including economic growth, health, education, and human rights (Soubbotina & Sheram, 2001). MSPs play a vital role in addressing gender biases and promoting opportunities by leveraging the strengths and resources of diverse stakeholders. However, the task of transforming gender-biased policies, procedures, and cultural norms is fraught with challenges and opportunities. This examination focuses on three significant areas: resource constraints and allocation, diverse stakeholder agendas and priorities, and cultural and religious considerations (see Table 5.2).

Promoting changes in gender-biased policies, procedures, and cultural norms through MSPs presents both significant challenges and opportunities. Addressing resource constraints and allocation issues, aligning diverse stakeholder agendas and priorities, and navigating cultural and religious considerations are critical to the success of these efforts. By leveraging innovative financing mechanisms, fostering inclusive governance structures, and adopting culturally sensitive approaches, MSPs

TABLE 5.2
Challenges and opportunities in MSPs

Challenges	Opportunities
1. Gender opportunities often competes with other pressing issues for funding, resulting in limited financial resources dedicated to gender-focused initiatives within MSPs.	1. Combining public, private, and philanthropic funding to create a more robust financial base for gender opportunities initiatives. This can help de-risk investments and attract more capital.
2. Many MSPs rely heavily on donor funding, which can be unpredictable and subject to shifting priorities, affecting the sustainability of gender opportunities programs.	2. Encouraging investments that generate social and gender-related impacts alongside financial returns can provide a sustainable funding source for gender-focused programs.
3. There is often a shortage of experts with specialized knowledge in gender issues, hindering the effective implementation of gender-focused programs.	3. Investing in training and capacity building to enhance the technical expertise of stakeholders in gender issues, ensuring that gender considerations are integrated into all aspects of MSP operations.

(continued)

TABLE 5.2 (Continued)
Challenges and opportunities in MSPs

Challenges	Opportunities
4. Limited resources for capacity building and training programs can restrict the ability of MSPs to develop the necessary skills and expertise among stakeholders.	4. Strengthening the capacity of local communities and grassroots organizations to advocate for gender opportunities can lead to more sustainable and contextually appropriate solutions.
5. MSPs often have to balance multiple priorities, leading to potential conflicts in resource allocation where gender opportunities initiatives may not always take precedence.	5. Encouraging stakeholders to share resources, knowledge, and best practices can optimize the use of available resources and enhance the overall impact of gender opportunities initiatives.
6. Resources within MSPs may be distributed inequitably, with larger, more influential stakeholders potentially receiving a disproportionate share, leaving gender-focused initiatives underfunded.	6. Utilizing digital tools and platforms to streamline operations, reduce costs, and improve the reach and effectiveness of gender-focused programs.
7. Different stakeholders often have divergent objectives and priorities, which can lead to conflicts and challenges in achieving consensus on gender opportunities initiatives.	7. Using skilled facilitators and mediators to navigate conflicts and build consensus among stakeholders can help align diverse agendas toward common gender opportunities goals.
8. Some stakeholders may prioritize short-term gains over long-term gender opportunities goals, affecting the commitment to sustained efforts.	8. Developing frameworks that clearly outline the roles, responsibilities, and contributions of each stakeholder can promote better alignment and cooperation.
9. Larger or more influential stakeholders may dominate decision-making processes, potentially sidelining gender opportunities concerns raised by smaller or less powerful partners.	9. Ensuring equitable representation of all stakeholders, particularly those advocating for gender opportunities, in governance structures can enhance the inclusivity and effectiveness of decision-making processes.
10. Ensuring that the voices of marginalized groups, including women and gender minorities, are heard and considered in MSPs can be challenging.	10. Adopting shared leadership models that distribute decision-making power more evenly among stakeholders can mitigate the dominance of powerful partners.
11. Aligning the gender opportunities goals of the MSP with the diverse missions and mandates of individual stakeholders can be difficult, leading to fragmented efforts.	11. Leveraging networks building alliances with other organizations and networks that share a commitment to gender opportunities can amplify the impact of MSPs and create synergies.
12. Deeply ingrained cultural norms and traditional gender roles can pose significant barriers to changing gender-biased policies and practices.	12. Engaging stakeholders at multiple levels, from local communities to international organizations, can ensure a more comprehensive approach to addressing gender biases.

TABLE 5.2 (Continued)
Challenges and opportunities in MSPs

Challenges	Opportunities
13. Efforts to promote gender opportunities may face resistance from communities and leaders who perceive such changes as threats to cultural identity and social stability.	13. Developing relationships with cultural and religious leaders who can act as champions for gender opportunities within their communities.
14. Different interpretations of religious texts and doctrines can influence attitudes towards gender roles and opportunities, sometimes reinforcing gender biases.	14. Facilitating inclusive dialogues that bring together stakeholders from different cultural and religious backgrounds to discuss and find common ground on gender opportunities issues.
15. The role of religious leaders and institutions in shaping public opinion and practices can either support or hinder efforts to promote gender opportunities.	15. Adapting gender opportunities initiatives to reflect local cultural contexts and values can enhance acceptance and effectiveness.
16. Gender issues and cultural norms vary widely across different regions and communities, requiring context-specific approaches to effectively address gender biases.	16. Supporting community-led solutions that are developed with the input and leadership of local stakeholders can ensure that initiatives are culturally appropriate and sustainable.
17. Balancing the need for gender opportunities with respect for cultural and religious traditions can be challenging, necessitating culturally sensitive strategies.	17. Providing cultural sensitivity training for MSP stakeholders to enhance their understanding of local contexts and improve their ability to engage effectively.
	18. Implementing education and awareness campaigns that highlight the benefits of gender opportunities and challenge harmful stereotypes and practices in a culturally respectful manner.
	19. Identifying and leveraging cultural and religious values that support gender opportunities, such as justice, respect, and compassion, to build support for gender-focused initiatives.
	20. Promoting positive role models and champions within cultural and religious communities who advocate for gender opportunities and serve as examples for others to follow.

Source: Developed by the author based on Brouwer et al. (n.d.) and Phillips (2015).

can effectively drive progress towards gender opportunities and contribute to the achievement of the SDGs. The collaborative efforts of diverse stakeholders, combined with strategic and context-specific strategies, will be essential in overcoming these challenges and realizing a more equitable and just world (Ayala-Orozco et al., 2018).

5.10 ANALYZING THE ROLE OF FAMILY IN ADVANCING GENDER OPPORTUNITIES

Families play a fundamental role in shaping gender norms and fostering educational opportunities. From early childhood, parents and caregivers influence children's perceptions of gender roles, encouraging behaviors and activities that align with societal expectations. By promoting an environment where both boys and girls are encouraged to pursue education and career aspirations, families lay the groundwork for equal opportunities. This support is particularly crucial for girls, as studies consistently show that parental encouragement and support significantly impact girls' educational attainment and career aspirations (Sewell & Shah, 1968). By challenging traditional stereotypes and providing equal support for educational pursuits, families contribute to breaking down barriers to gender equality. Family members serve as important role models and mentors, offering guidance and support for career aspirations. Through positive reinforcement and encouragement, parents and siblings can instill confidence in young women, empowering them to pursue their goals. Mentorship within the family provides invaluable insights and advice, helping women navigate career paths and overcome challenges. This support system not only fosters individual success but also reinforces the importance of women's participation in the workforce within the family unit. By championing women's career aspirations and providing mentorship opportunities, families contribute to advancing gender equality in the broader society (De Vries, 2015).

Equitable division of household responsibilities and advocacy for gender equality within the family are essential components of advancing gender opportunities. Families that promote shared caregiving and domestic duties challenge traditional gender roles, allowing women to balance their professional and personal lives effectively (Calasanti & Bowen, 2006). Additionally, families can advocate for gender equality within their communities, influencing societal attitudes and norms. By engaging in discussions, supporting initiatives, and leading by example, families contribute to creating a more inclusive and equitable society. Through collective efforts and shared responsibilities, families play a vital role in advancing gender opportunities and developmental trends, shaping a future where all individuals have equal access to opportunities and resources (Calasanti & Bowen, 2006).

5.11 ANALYZING THE ROLE OF MSPS IN ADVANCING GENDER OPPORTUNITIES AND DEVELOPMENTAL TRENDS

MSPs are critical in promoting gender opportunities and influencing developmental trends (Oburu & Yoshikawa, 2018). By engaging diverse stakeholders from different sectors, MSPs can address gender biases and create a more inclusive society. This section analyses the impact of MSPs on social, legal, and religious domains, explores the developmental trends that should be influenced by MSPs, and highlights the lessons learned and best practices. MSPs have played a significant role in advancing gender opportunities in social contexts by addressing norms, behaviors, and practices that perpetuate gender opportunities. MSPs have been effective in challenging traditional gender norms through awareness campaigns, educational programs, and community engagement initiatives (Kabeer, 2003). For example, the HeForShe campaign by UN Women has mobilized men and boys to advocate for gender opportunities, changing societal attitudes toward gender roles (Henry-White, 2015). By providing underrepresented to education, healthcare, and economic opportunities, MSPs have empowered women and girls, leading to improved social outcomes. Initiatives like the Global Partnership for Education (GPE) have increased girls' enrolment in schools, contributing to higher literacy rates and better health outcomes (Albright & Bundy, 2018). MSPs have been instrumental in addressing GBV by providing support services, legal aid, and advocacy (Gatimu, 2015). Partnerships such as the UN Trust Fund to End Violence against Women have funded projects that offer protection and support to survivors of violence (United Nations Trust Fund to End Violence against Women, 2013). MSPs have influenced legal reforms and policies that promote gender opportunities, ensuring that laws and regulations reflect and support the rights of women and girls. MSPs have contributed to the development and implementation of gender-sensitive laws and policies (Kabeer, 2003). For instance, the Equal Rights Trust collaborates with various stakeholders to advocate for legal reforms that eliminate discrimination against women. Through policy advocacy, MSPs have pushed for the adoption and enforcement of laws that protect women's rights. The Global Gender Opportunities Partnership, for example, has worked to ensure that national legislation aligns with international gender opportunities standards (Prügl & True, 2014). MSPs have improved underrepresented to justice for women by providing legal aid services, raising awareness of legal rights, and training law enforcement officials. This has led to increased reporting and prosecution of gender-based crimes. Religious beliefs and practices significantly influence gender norms and behaviors (Perales & Bouma, 2019). MSPs have engaged with religious leaders and communities to promote gender opportunities within religious contexts. MSPs have collaborated with religious leaders to reinterpret religious texts and doctrines in ways that support gender opportunities. For example, the Musawah movement engages Islamic scholars to advocate for gender justice in Muslim communities (Basarudin, 2009). Many faith-based organizations are part of MSPs, working to address gender biases within their communities. The Interfaith Partnership for Sustainable Development has brought together religious leaders from various faiths to promote gender opportunities and sustainable development. MSPs have utilized culturally sensitive approaches to address gender biases in religious communities, respecting religious traditions while promoting gender opportunities. This has led to increased acceptance and support for gender opportunities initiatives.

5.12 DEVELOPMENTAL TRENDS INFLUENCED BY MSPS

MSPs have the potential to influence several key developmental trends, driving progress towards gender opportunities and sustainable development. MSPs can promote inclusive economic growth by supporting women-owned businesses, providing vocational training, and advocating for equal pay and workplace policies that support gender opportunities. Initiatives like the Women's Empowerment Principles (WEPs) encourage companies to adopt practices that empower women economically (de Souza Mauro et al., 2019). MSPs can enhance women's financial inclusion by facilitating underrepresented to banking services, credit, and financial literacy programs. The Global Banking Alliance for Women (GBA) works with financial institutions to design products and services that meet the needs of women (Odebrecht, 2013). MSPs can ensure equal underrepresented to quality education for girls and women, addressing barriers such as school fees, safety concerns, and cultural attitudes. Programs like Malala Fund's initiatives have increased girls' enrolment and retention in schools (Sperling et al., 2016). Promoting STEM education for girls is crucial for their participation in high-growth sectors (Serrano et al., 2023). MSPs can support STEM programs, scholarships, and mentorship opportunities for girls and women. MSPs can improve maternal and child health outcomes by providing healthcare services, education, and support to pregnant women and mothers. Partnerships like Every Woman Every Child have mobilized resources and political support for maternal and child health (WHO, 2017). MSPs can advocate for reproductive rights and underrepresented to reproductive health services, ensuring that women can make informed choices about their bodies and health. Organizations like Planned Parenthood Global work with local partners to expand underrepresented to reproductive health services (Daniel & Riley, 2023). MSPs can advocate for policies and practices that increase women's representation in political and decision-making processes. Initiatives like the Women in Parliaments Global Forum work to enhance women's political leadership and participation (Bullough et al., 2012). MSPs can provide leadership training and mentorship programs for women, equipping them with the skills and confidence needed to take on leadership roles in various sectors.

Ensuring diverse representation of stakeholders, including women and marginalized groups, is crucial for the success of MSPs (Brouwer et al., n.d.). This diversity brings different perspectives and experiences, enriching the partnership and its outcomes. Establishing equitable decision-making processes that give all stakeholders a voice and influence is essential. This includes setting up governance structures that reflect the diversity of the partnership. Trust among stakeholders is fundamental for effective collaboration. MSPs should prioritize building trust through transparency, accountability, and consistent communication(Brouwer et al., n.d.). Adopting collaborative approaches that leverage the strengths and resources of all partners can enhance the impact of MSPs. This includes joint planning, implementation, and evaluation of initiatives. Understanding the local context, including cultural, social, and economic factors, is vital for the success of gender opportunities initiatives. MSPs should conduct thorough context analyses and engage local stakeholders to ensure relevance and acceptance. Implementing culturally sensitive strategies that respect local traditions while promoting gender opportunities can lead to more sustainable outcomes (Aina, 2011). This involves working with local leaders and communities to co-create solutions. Regular monitoring and evaluation of MSP initiatives are crucial for assessing their impact and effectiveness. This includes setting clear indicators and benchmarks for gender opportunities outcomes. MSPs should foster a culture of learning and adaptation, using insights from evaluations to improve strategies and approaches. Sharing lessons learned and best practices can enhance the effectiveness of other MSPs. MSPs should engage in policy advocacy to influence laws and regulations that promote gender opportunities (Brouwer et al., n.d.). This involves working with governments, policymakers, and other stakeholders to create an enabling environment for gender opportunities. Raising public awareness about gender opportunities issues through campaigns, media, and education can build support for MSP initiatives and drive societal change. MSPs play a critical role in advancing gender opportunities by addressing social, legal, and religious biases and influencing key developmental trends. By leveraging the strengths and resources of diverse stakeholders, MSPs can drive significant progress towards gender opportunities. However, this requires overcoming challenges related to resource constraints, diverse stakeholder agendas, and cultural and religious considerations. The lessons learned and best practices from successful MSPs highlight the importance of inclusive partnerships, collaborative approaches, cultural sensitivity, and effective advocacy in achieving gender opportunities. As MSPs continue to evolve, they hold great potential to create a more equitable and just world for all (Brouwer et al., n.d.).

5.13 CONCLUSION

Families play a fundamental role in shaping gender norms and fostering educational opportunities. From early childhood, parents and caregivers influence children's perceptions of gender roles, encouraging behaviors and activities that align with societal expectations. By promoting an environment where both boys and girls are encouraged to pursue education and career aspirations, families lay the groundwork for equal opportunities. This support is particularly crucial for girls, as studies consistently show that parental encouragement and support significantly impact girls' educational attainment and career aspirations. By challenging traditional stereotypes and providing equal support for educational pursuits, families contribute to breaking down barriers to gender equality. Family members serve as important role models and mentors, offering guidance and support for career aspirations. Through positive reinforcement and encouragement, parents and siblings can instill confidence in young women, empowering them to pursue their goals. Mentorship within the family provides invaluable insights and advice, helping women navigate career paths and overcome challenges. This support system not only fosters individual success but also reinforces the importance of women's participation in the workforce within the family unit. By championing women's career aspirations and providing mentorship opportunities, families contribute to advancing gender equality in the broader society.

Equitable division of household responsibilities and advocacy for gender equality within the family are essential components of advancing gender opportunities. Families that promote shared caregiving and domestic duties challenge traditional gender roles, allowing women to balance their professional and personal lives effectively. Additionally, families can advocate for gender equality within their communities, influencing societal attitudes and norms. By engaging in discussions, supporting initiatives, and leading by example, families contribute to creating a more inclusive and equitable society. Through collective efforts and shared responsibilities, families play a vital role in advancing gender opportunities and developmental trends, shaping a future where all individuals have equal access to opportunities and resources.

In Southeast Asia, the position of women in leadership is influenced by historical, cultural, economic, and political factors. While progress has been made, substantial barriers persist. Families in the region play a crucial role in shaping attitudes toward women's leadership by fostering supportive environments for education and career aspirations. By encouraging girls to pursue leadership roles and providing mentorship and guidance, families contribute to the development of future women leaders. Additionally, families can advocate for gender equality within their communities, challenging traditional norms and promoting inclusive leadership practices. As Southeast Asia strives for gender equality in leadership, the support and advocacy of families will be instrumental in driving meaningful change and creating a more equitable society for all.

REFERENCES

Aaltio, I., Salminen, H. M., & Koponen, S. (2014). Ageing employees and human resource management - Evidence of gender-sensitivity? *Equality, Diversity and Inclusion, 33*(2), 160–176. https://doi.org/10.1108/EDI-10-2011-0076

Abad, M. (2017). *Philippines improves in 2023 world gender equality ranking*. Philippine Institute for Development Studies.

Adeniyi, I. S., Al Hamad, N. M., Adewusi, O. E., Unachukwu, C. C., Osawaru, B., Onyebuchi, C. N., Omolawal, S. A., Aliu, A. O., & David, I. O. (2024). Gender equality in the workplace: A comparative review of USA and African Practices. *World Journal of Advanced Research and Reviews, 21*(2), 763–772. https://doi.org/10.30574/wjarr.2024.21.2.0491

Ahmad, S. B. (2017). *A qualitative inquiry into women principals' leadership in Malaysia*. University of Warwick. http://wrap.warwick.ac.uk/130010

Aina, O. I. (2011). *Promoting gender equality for sustainable development*. https://d1wqtxts1xzle7.cloudfront.net/32802897/PROF_AINA_-_EKITI_SUMMIT_-_LEAD_PAPER_-_NOV_1-libre.pdf?1391187022=&response-content-disposition=inline%3B+filename%3DPROMOTING_GENDER_EQUALITY_FOR_SUSTAINABL.pdf&Expires=1721379880&Signature=aL8cGcdVbKjuOcnVk5ujoE88nRcq~TkNgsJ4NYi2UOEIRzqvQnxw6Z5oQ~JXPfeQVR6WJuhDawWpRygWCLnIfczYNdGeQ~4g5VCGZucEIw3xQa8n6Zst0OQRBJuk-1uuRuT85e5ybkO4Ki~KisQboUxhEn~x-RJs-JcwWHQixvcmkXBfDEvQh2EOcamEicirE5Eh~vaiixO8Ya1pVnET1st3Htrqu9HBXqaEesOBU-DsV5TvBa8SxhIc2W4c0OFIJAMOfFCm~m5NzAdQ3sn-bZU3GTSQryKVZ~S0QuPyE7w0eIoxefVp~T3FonmZhoMGCmqbJDgLWbqkEiSRCFLEHA__&Key-Pair-Id=APKAJLOHF5GGSLRBV4ZA

Akter, S., Rutsaert, P., Luis, J., Htwe, N. M., San, S. S., Raharjo, B., & Pustika, A. (2017). Women's empowerment and gender equity in agriculture: A different perspective from Southeast Asia. *Food Policy, 69*, 270–279. https://doi.org/10.1016/j.foodpol.2017.05.003

Albright, A., & Bundy, D. A. P. (2018). The global partnership for education: Forging a stronger partnership between health and education sectors to achieve the Sustainable Development Goals. In *The Lancet Child and Adolescent Health* (Vol. 2, Issue 7, pp. 473–474). Elsevier B.V. https://doi.org/10.1016/S2352-4642(18)30146-9

Alonso-Almeida, M. del M., Perramon, J., & Bagur-Femenias, L. (2017). Leadership styles and corporate social responsibility management: Analysis from a gender perspective. *Business Ethics, 26*(2), 147–161. https://doi.org/10.1111/beer.12139

Andaya, & Watson, B. (2008). *The flaming womb: Repositioning women in early modern Southeast Asia*. University of Hawai'i Press.

Apriani, F., & Zulfiani, D. (2020). Women's leadership in Southeast Asia: Examining the authentic leadership implementation potency. *Policy & Governance Review*, *4*(2), 116–127. https://doi.org/10.30589/pgr.v4i2.275

Asian Development Bank. (2016, October). *Gender equality and women's rights in Myanmar: A situation analysis*. Asian Development Bank.

Asian Development Bank Institute. (2021). *Women's economic empowerment in Asia*. Asian Development Bank Institute.

Ayala-Orozco, B., Rosell, J. A., Merçon, J., Bueno, I., Alatorre-Frenk, G., Langle-Flores, A., & Lobato, A. (2018). Challenges and strategies in place-based multi-stakeholder collaboration for sustainability: Learning from experiences in the Global South. *Sustainability (Switzerland)*, *10*(9). https://doi.org/10.3390/su10093217

Basarudin, A. (2009). *Musawah movement: Seeking equality and justice in Muslim family law*. https://escholarship.org/uc/item/1gz83404

Basu, A. (2010). *Women's movements in the global era*. Westview Press.

Baulch, B., Nguyen, H. T. M., Phuong, P. T. T., & Pham, H. T. (2010). *Ethnic minority poverty in Vietnam* (vol. 169). Chronic Poverty Research Centre (CPRC).

Bayang, J. N., Cardenas, M. S., Melgar, J. D., R., M. A., Papa, N. V., Tadem, E. C., & Viajar, R. (2022). *Promoting gender equality and women empowerment among vulnerable groups in the Philippines continuing challenges and ways forward*. www.flickr.com/photos/worldremit/25829009933/in/photostream/

Beneria, L., & Sen, G. (1981). Accumulation, reproduction, and women's role in economic development: Boserup revisited. *Journal of Women in Culture and Society*, *7*(2), 279–298. http://www.journals.uchicago.edu/t-and-c

Black, A., Henty, P., Sutton, K., & Watson, J. (2017). *Women in humanitarian leadership*. Centre for Humanitarian Leadership.

Brouwer, H., & Brouwers, J. (2017). *The MSP tool guide*. Wageningen University & Research. www.mspguide.org/tools-and-methods

Brouwer, H., Woodhill, J., Hemmati, M., Verhoosel, K., & Vugt, S. van. (n.d.). *The MSP guide: How to design and facilitate multi-stakeholder partnerships*. Practical Action Publishing.

Bullough, A., Kroeck, K. G., Newburry, W., Kundu, S. K., & Lowe, K. B. (2012). Women's political leadership participation around the world: An institutional analysis. *Leadership Quarterly*, *23*(3), 398–411. https://doi.org/10.1016/j.leaqua.2011.09.010

Buranajaroenkij, D., Doneys, P., Kusakabe, K., & Doane, D. L. (2018). Expansion of women's political participation through social movements: The case of the Red and Yellow shirts in Thailand. *Journal of Asian and African Studies*, *53*(1), 34–48. https://doi.org/10.1177/0021909616654508

Calasanti, T., & Bowen, M. E. (2006). Spousal caregiving and crossing gender boundaries: Maintaining gendered identities. *Journal of Aging Studies*, *20*(3), 253–263. https://doi.org/10.1016/j.jaging.2005.08.001

CFR.org. (2023, September 18). *What is ASEAN?* www.Cfr.Org.

Chikwe, C. F., Eneh, N. E., & Akpuokwe, C. U. (2024). Navigating the double bind: Strategies for women leaders in overcoming stereotypes and leadership biases. *GSC Advanced Research and Reviews*, *18*(3), 159–172. https://doi.org/10.30574/gscarr.2024.18.3.0103

Chin, G. V. S., & Daud, K. M. (2018). *The Southeast Asian woman writes back: Gender, identity and nation in the literatures of Brunei Darussalam, Malaysia, Singapore, Indonesia and the Philippines: Vol. Asia in Transition 6*. Springer.

Choi, N. (2019). Women's political pathways in Southeast Asia. *International Feminist Journal of Politics*, *21*(2), 224–248. https://doi.org/10.1080/14616742.2018.1523683

Civil Service College Singapore. (2024). *Meritocracy: Time for an update?* Civil Service College Singapore.
Claridge, T. (2018). Functions of social capital – bonding, bridging, linking. *Social Capital Research*. https://doi.org/10.5281
Clarke, A., & MacDonald, A. (2019). Outcomes to partners in multi-stakeholder cross-sector partnerships: A resource-based view. *Business and Society, 58*(2), 298–332. https://doi.org/10.1177/0007650316660534
Cornwall, A., & Rivas, A. M. (2015). From 'gender equality and 'women's empowerment' to global justice: Reclaiming a transformative agenda for gender and development. *Third World Quarterly, 36*(2), 396–415. https://doi.org/10.1080/01436597.2015.1013341
Dahiya, B. (2018). *Building successful multi-stakeholder partnerships to implement the 2030 Agenda in Asia-Pacific*. www.researchgate.net/publication/327221104
Dahrelup, D. (2005). Increasing women's political representation: New trends in gender quotas. In Julie Ballington and Azza Karam (Eds.) *Women in parliament: Beyond numbers* (p. 265). International IDEA.
Daniel, C., & Riley, G. (2023). Building equitable partnerships and a social justice mindset through a donor-funded reproductive rights and health internship program. *Journal of Higher Education Outreach and Engagement, 27*(1), 157. https://openjournals.libs.uga.edu/jheoe/article/view/2722
Das, T. K. (1998). Between trust and control: Developing confidence in partner cooperation in alliances. *Academy of Management Review, 23*(3), 491–512. https://doi.org/10.5465/amr.1998.926623
David, C. C., Albert, J. R. G., & Vizmanos, J. F. V. (2018). *Sustainable development goal 5: How does the Philippines fare on gender equality? Standard-Nutzungsbedingungen* (PIDS Discussion Paper Series No. 2017-45). www.pids.gov.ph
Davidson, J. S., & Henley, D. (2007). *The revival of tradition in Indonesian politics*. Routledge. www.api.taylorfrancis.com
Davies, S. G. (2005). Women in politics in Indonesia in the decade post-Beijing. *ISSJ, 57*(194), 231–242.
de Souza Mauro, A. J., Guilhen Mazaro Araújo, G., & de Andrade Guerra, J. B. S. O. (2019). *Women's empowerment principles (WEPs)* (pp. 1–13). https://doi.org/10.1007/978-3-319-70060-1_15-1
De Vries, J. A. (2015). Champions of gender equality: Female and male executives as leaders of gender change. *Equality, Diversity and Inclusion, 34*(1), 21–36. https://doi.org/10.1108/EDI-05-2013-0031
Dentoni, D., Bitzer, V., & Schouten, G. (2018). Harnessing wicked problems in multi-stakeholder partnerships. *Journal of Business Ethics, 150*(2), 333–356. https://doi.org/10.1007/s10551-018-3858-6
Dodds, F. (2019). *Stakeholder democracy: Represented democracy in a time of fear*. Routledge.
Dube, L. (1997). *Women and kinship: Perspectives on gender in South and South-East Asia*. United Nations University Press.
Elias, J. (2011). The gender politics of economic competitiveness in Malaysia's transition to a knowledge economy. *Pacific Review, 24*(5), 529–552. https://doi.org/10.1080/09512748.2011.596564
Erdem Türkelli, G. (2021). Transnational multistakeholder partnerships as vessels to finance development: Navigating the accountability waters. *Global Policy, 12*(2), 177–189. https://doi.org/10.1111/1758-5899.12889
ETC Foundation. (2012). *Power dynamics in multi-stakeholder processes: A balancing act*. www.dlprog

Eweje, G., Sajjad, A., Nath, S. D., & Kobayashi, K. (2021). Multi-stakeholder partnerships: A catalyst to achieve sustainable development goals. *Marketing Intelligence and Planning*, *39*(2), 186–212. https://doi.org/10.1108/MIP-04-2020-0135

Feeney, M. K., & Stritch, J. M. (2019). Family-friendly policies, gender, and work–life balance in the public sector. *Review of Public Personnel Administration*, *39*(3), 422–448. https://doi.org/10.1177/0734371X17733789

Fleischer, L., Bogiatzis, A., Asada, H., & Koen, V. (2018). *Making growth more inclusive in Thailand* (Vol. 1469). https://doi.org/10.1787/263a78df-en

Fuenfschilling, L., & Truffer, B. (2014). The structuration of socio-technical regimes–Conceptual foundations from institutional theory. *Research Policy*, *43*(4), 772–791. https://doi.org/10.1016/j.respol.2013.10.010

Galsanjigmed, E., & Sekiguchi, T. (2023). Challenges women experience in leadership careers: An integrative review. *Merits*, *3*(2), 366–389. https://doi.org/10.3390/merits3020021

Gatimu, C. (2015). An analysis of Sexual and Gender Based Violence (SGBV) interventions. In M. Edith (Ed.), *South Sudan an analysis of sexual and gender based violence (SGBV) interventions in South Sudan*. International Peace Training Centre (IPSTC). www.ipstc.org

Glass, L. M., Newig, J., & Ruf, S. (2023). MSPs for the SDGs – Assessing the collaborative governance architecture of multi-stakeholder partnerships for implementing the Sustainable Development Goals. *Earth System Governance*, *17*. https://doi.org/10.1016/j.esg.2023.100182

Gray, B., & Purdy, J. (2018). *Collaborating for our future: Multistakeholder partnerships for solving complex problems*. Oxford University Press.

Grosser, K. (2009). Corporate social responsibility and gender equality: Women as stakeholders and the European Union sustainability strategy. *Business Ethics: A European Review*, *18*(3), 217–230. https://doi.org/10.1111/j.1467-8608.2009.01564.x

Grown, C., Gupta, G. R., Kes, A., & UN Millennium Project. Task Force on Education and Gender Equality. (2019). *Taking action: Achieving gender equality and empowering women*. Routledge. https://doi.org/10.4324/9781849773560

Gumusluoglu, L., & Ilsev, A. (2009). Transformational leadership, creativity, and organizational innovation. *Journal of Business Research*, *62*(4), 461–473. https://doi.org/10.1016/j.jbusres.2007.07.032

Hawkins, P. (2011). *Leadership team coaching: Developing collective transformational leadership*. Kogan Page.

Henry-White, J. M. (2015). *Gender equality?: A transnational feminist analysis of the UN HeForShe Campaign as a global "solidarity" movement for men*. University of Missouri. https://hdl.handle.net/10355/46586

Henshall Momsen, J. (2004). *Gender and development*. Routledge.

Howell, J., & Mulligan, D. (2005). *Gender and civil society*. Routledge Taylor & Francis Group.

Jackson, H. M., Sanjeev, A., Evaluator, K., Members, V. R., Iara, L. A., Coordinator, B., Delsman, A., Kamran, T. E., Walter, E., Ayana, H., Sam, H., Coordinator, H., Jung, D., Lin, U., Novosyolova, A., Park, Z., Serizawa, S., & Van, A. (2012). *Review of best practices for multi-stakeholder initiatives: Recommendations for GIFT*. Global Initiative for Fiscal Transparency (GIFT).

Jain, K., & Katiyal, D. (2020). Exploratory research on women leadership at workplace: In Different organizations of Indore. *UNNAYAN - Conference Special Issue* (pp. 311–323). IPS Academy. www.ipsacademy.org/unnayan/v13/25-Paper.pdf

Jaquette, J. S., & Summerfield, G. (2006). *Women and gender equity in development theory and practice institutions, resources, and mobilization*. Duke University Press.

Joachim, J. (2003). Framing issues and seizing opportunities: The UN, NGOs, and women's rights. *International Studies Quarterly*, *47*(2), 247–274. https://doi.org/10.1111/1468-2478.4702005

Kabeer, N. (2003). Gender mainstreaming in poverty eradication and the millennium development goals: A handbook for policy-makers and other stakeholders. In T. Johnson (Ed.), *Commonwealth Secretariat*. International Development Research Centre.

Kirsch, A. (2018). The gender composition of corporate boards: A review and research agenda. *Leadership Quarterly*, *29*(2), 346–364. https://doi.org/10.1016/j.leaqua.2017.06.001

Kossek, E. E., Su, R., & Wu, L. (2017). "Opting out" or "Pushed out"? Integrating perspectives on women's career equality for gender inclusion and interventions. *Journal of Management*, *43*(1), 228–254. https://doi.org/10.1177/0149206316671582

Kuppuswami, D., & Ferreira, F. (2022). Gender equality and women's empowerment capacity building of organisations and individuals. *Journal of Learning*, *9*(3), 394–419.

Kyveloukokkaliari, L. K., & Nurhaeni, I. D. A. (2017). Women leadership: A comparative study between Indonesia and Greece. *Jurnal Studi Pemerintahan*, *8*(4), 514–535. https://doi.org/10.18196/jgp.2017.0057.514-535

Lao Women's Union. (2009). *Lao women's union: Promotion and protection of Lao Women's Right*. Lao Women's Union.

Lee-Koo, K., & Pruitt, L. (2024). Prospects for intergenerational peace leadership: Reflections from Asia and the Pacific. *Cooperation and Conflict*. https://doi.org/10.1177/00108367241246535

Leshkowich, A. M. (2006). Woman, Buddhist, entrepreneur: Gender, moral values, and class anxiety in late socialist Vietnam. *Journal of Vietnamese Studies*, *1*(1–2), 277–313. https://doi.org/10.1525/vs.2006.1.1-2.277

Lewis, A. C., Cardy, R. L., & Huang, L. S. R. (2019). Institutional theory and HRM: A new look. *Human Resource Management Review*, *29*(3), 316–335. https://doi.org/10.1016/j.hrmr.2018.07.006

Liff, S., & Cameron, I. (1997). Changing equality cultures to move beyond "women's problems. *Gender, Work and Organization*, *4*(11), 35–46.

Mackay, F., & Murtagh, C. (2019). New institutions, new gender rules? A feminist institutionalist lens on women and power-sharing. *Feminists@law*, *9*(1). https://doi.org/https://doi.org/10.22024/UniKent/03/fal.745

Madsen, S. R. (2012). Women and leadership in higher education: Current realities, challenges, and future directions. *Advances in Developing Human Resources*, *14*(2), 131–139. https://doi.org/10.1177/1523422311436299

Mitchell, B. (2005). Participatory partnerships: Engaging and empowering to enhance environmental management and quality of life? *Social Indicators Research*, *71*(1), 123–144. https://doi.org/10.1007/s11205-004-8016-0

Mokhber, M., Khairuzzaman, W., Ismail, W., & Vakilbashi, A. (2015). Optimization of the inflationary inventory control the effect of transformational leadership and its components on organizational innovation. *Iranian Journal of Management Studies*, *8*(2), 221–241. http://ijms.ut.ac.ir/

Moody, J., & Paxton, P. (2009). Building bridges: Linking social capital and social networks to improve theory and research. *American Behavioral Scientist*, *52*(11), 1491–1506. https://doi.org/10.1177/0002764209331523

Morley, L. (2014). Lost leaders: Women in the global academy. *Higher Education Research and Development*, *33*(1), 114–128. https://doi.org/10.1080/07294360.2013.864611

Morley, L., & Crossouard, B. (2016). Women's leadership in the Asian Century: Does expansion mean inclusion? *Studies in Higher Education*, *41*(5), 801–814. https://doi.org/10.1080/03075079.2016.1147749

Munro, J. (2012). *Women's representation in leadership in Viet Nam*. UN Women.

Oburu, P. O., & Yoshikawa, H. (2018). *Roles of multiple stakeholder partnerships in addressing developmental and implementation challenges of sustainable development goals* (pp. 421–438). Springer. https://doi.org/10.1007/978-3-319-96592-5_24

Odebrecht, C. N. (2013). *Women, entrepreneurship and the opportunity to promote development and business*. www.brookings.edu/wp-content/uploads/2016/07/niethammer-policy-brief.pdf

OECD. (2020a). *Policies and practices to promote women in leadership roles in the private sector*. OECD.

OECD, I. L. O. and C. of A. W. for T. and R. (2020b). *Changing laws and breaking barriers for women's economic empowerment in Egypt, Jordan, Morocco and Tunisia*. OECD. https://doi.org/10.1787/ac780735-en

Ong, A. (1989). *Center, periphery and hierarchy: Gender in Southeast Asia*. Southeast Asia Program Publications.

Parken, A. (2018). *Putting equality at the heart of decision-making Gender Equality Review (GER) phase one: International policy and practice*. www.wcpp.org.uk

Perales, F., & Bouma, G. (2019). Religion, religiosity and patriarchal gender beliefs: Understanding the Australian experience. *Journal of Sociology*, *55*(2), 323–341. https://doi.org/10.1177/1440783318791755

Peus, C., Braun, S., & Knipfer, K. (2015). On becoming a leader in Asia and America: Empirical evidence from women managers. *Leadership Quarterly*, *26*(1), 55–67. https://doi.org/10.1016/j.leaqua.2014.08.004

Phillips, R. (2015). How 'Empowerment' may miss its mark: Gender equality policies and how they are understood in women's NGOs. *Voluntas*, *26*(4), 1122–1142. https://doi.org/10.1007/s11266-015-9586-y

Poortinga, W. (2006). Social relations or social capital? Individual and community health effects of bonding social capital. *Social Science and Medicine*, *63*(1), 255–270. https://doi.org/10.1016/j.socscimed.2005.11.039

Prügl, E., & True, J. (2014). Equality means business? Governing gender through transnational public-private partnerships. *Review of International Political Economy*, *21*(6), 1137–1169. https://doi.org/10.1080/09692290.2013.849277

Qian, M. (2016). *Women's leadership and corporate performance* (vol. 472). Available at: http://ssrn.com/abstract=2737833; https://ssrn.com/abstract=2737833Electroniccopy;

Rahm, L. (2020). Gender-biased sex selection in South Korea, India and Vietnam assessing the influence of public policy. In Nisha Dhanraj and Niranjan Sahoo (Eds.) *Demographic transformation and socio-economic development*. Springer. https://doi.org/10.1007/978-3-030-20234-7

Richter, L. K. (1990). Exploring theories of female leadership in South and Southeast Asia. *Source: Pacific Affairs*, *63*(4), 524–540.

Robinson, K., & Bessell, S. (2002). *Women in Indonesia*. Institute of Southeast Asian Studies.

Rohini, P., & Ford, D. (2011). GENDER QUOTAS AND FEMALE LEADERSHIP Gender quotas and female leadership: A review. *Background Paper for the World Development Report on Gender*. World Bank.

Sandhu, S. (2021). Women's rights to vote and laws against gender discrimination: The makeup for gender equality and women's empowerment. *UC Merced Undergraduate Research Journal*, *13*(1). https://doi.org/10.5070/m4131052985

sdgs.un.org. (2016). *THE 17 GOALS*. Sdgs.Un.Org. https://sdgs.un.org/goals

Serrano, D. R., Fraguas-Sánchez, A. I., González-Burgos, E., Martín, P., Llorente, C., & Lalatsa, A. (2023). Women as Industry 4.0. Entrepreneurs: Unlocking the potential of entrepreneurship in Higher Education in STEM-related fields. *Journal of Innovation and Entrepreneurship, 12*(1). https://doi.org/10.1186/s13731-023-00346-4

Sewell, W. H., & Shah, V. P. (1968). Social class, parental encouragement, and educational aspirations'. *American Journal of Sociology, 73*(5), 559–572. www.journals.uchicago.edu/t-and-c

Sharif, K. (2019). Transformational leadership behaviours of women in a socially dynamic environment. *International Journal of Organizational Analysis, 27*(4), 1191–1217. https://doi.org/10.1108/IJOA-12-2018-1611

Shinbrot, X. A., Wilkins, K., Gretzel, U., & Bowser, G. (2019). Unlocking women's sustainability leadership potential: Perceptions of contributions and challenges for women in sustainable development. *World Development, 119*, 120–132. https://doi.org/10.1016/j.worlddev.2019.03.009

Siahaan, A. Y. (2003). *The politics of gender and decentralization in Indonesia*. The Asia Foundation.

Silliman, J. (1999). Expanding civil society: Shrinking political spaces-the case of women's nongovernmental organizations. *Social Politics, 6*(1), 45–71. http://sp.oxfordjournals.org/

Singapore Women's Development. (2022). *Towards a fairer and more inclusive society*. Ministry of Social and Family Development (MSF).

Sinpeng, A., & Savirani, A. (2022). *Women's political leadership in the ASEAN region research report*. ASEAN Studies Centre.

Skrabakova, K. (2017). Islamist women as candidates in elections. *Brill, 57*, 329–359. https://doi.org/10.2307/26568529

Soubbotina, T. P., & Sheram, K. A. (20001). *Beyond economic growth: Meeting the challenges of global development*. World Bank.

Sperling, G. B., Winthrop, R., & Kwauk, C. (2016). *What works in girls: Education evidence for the World's best investment*. Brookings Institution Press.

Stone, W. (2001). *Measuring social capital towards a theoretically informed measurement framework for researching social capital in family and community life*. www.aifs.org.au/

Syukri, M. (2021). Gender equality in Indonesian new developmental state: The case of the new participatory village governance. *SMERU Working Paper*. Australian Institute of Family Studies (AIFS). www.smeru.or.id

Tailassane, R. (2019). Women' s rights and representation in Saudi Arabia, Iran, and Turkey: The patriarchal domination of religious interpretations. *International Relations Honors Papers*, 1–122. https://digitalcommons.ursinus.edu/int_hon

Tarr-Whelan, L. (2009). *Women lead the way: Your guide to stepping up to leadership and changing the world*. Berrett-Koehler Publishers.

Taşli, Kaan. (2007). *A conceptual framework for gender and development studies: From welfare to empowerment*. Südwind-Verl.

The World Bank. (2021). *Gender equality in education*. World Bank Group.

Thelma, C. C., & Ngulube, L. (2024). Women in leadership: Examining barriers to women's advancement in leadership positions. *Asian Journal of Advanced Research and Reports, 18*(6), 273–290. https://doi.org/10.9734/ajarr/2024/v18i6671

Thompson, M. R. (2002). Female leadership of democratic transitions in Asia. *Pacific Affairs, 75*(4), 535–555. www.terra.es/personal2/monolith/

UN Women. (2018, June). *The magazine for gender-responsive evaluation the women's political participation and leadership issue*. www.unwomen.org/en/about-us/

UN Women. (2022, October). *Three challenges for rural women amid a cost-of-living crisis*. UN Women.

UN Women. (2024). *Rural women*. UN Women.

UN-Environment Programme – WCMC. (2020, October). *Empowering rural women and girls as a solution to environmental sustainability and food security*. UN-Environment Programme - WCMC.

UNICEF. (2020). *Lao People's Democratic Republic - Update on the context and situation of children*. UNICEF.

UNICEF. (2021). *Government of Lao PDR - UNICEF country programme document 2022-2026*. UNICEF Lao PDR Home.

United Nations. (2019). *Pathways to influence: Promoting the role of women's transformative leadership to achieve the SDGs in Asia and the Pacific*. United Nations Publication.

United Nations Trust Fund to End Violence against Women. (2013). *UN trust fund to end violence against women annual report 2013*. www.unwomen.org/sites/default/files/Headquarters/Attachments/Sections/Library/Publications/2014/UNTF-AnnualReport2013-en%20pdf

USAID. (2016). *USAID Cambodia gender assessment*. UNICEF.

USAID. (2023). *2023 gender equality and women's empowerment policy*. USAID.

WHO. (2017). *2017 progress report on the every woman every child global strategy for women's, children's and adolescents' health progress in partnership*. WHO. http://apps.who.int/bookorders.

Yamanaka, K., & Piper, N. (2005). *Feminized migration in East and Southeast Asia policies, actions and empowerment*. UNRISD. https://hdl.handle.net/10419/148824

Yu-Jose, L. N. (2011). *Civil society organizations in the Philippines: A mapping and strategic assessment*. Civil Society Resource Institute (CSRI).

6 Unpacking the Effect of Family Support for Women's Entrepreneurial Success in Pakistan

Aemin Nasir and Shajara Ul-Durar

6.1 INTRODUCTION AND BACKGROUND

The term 'entrepreneurship' was first coined in the 17th century for male businessmen; however, women's entrepreneurship was accepted in the early 19th century due to the feminism phenomenon. Furthermore, the 21st century, a highly technologically equipped era, enabled females to become more prevalent in business. It has been observed that female business owners have higher discontinuation as compared to their male counterparts despite the increasing trend of women owned business (Brush & Brush, 2006). The trend of women's entrepreneurship has increased in recent decades as seen in Asian countries including Bangladesh, Nepal, Pakistan, India, Korea and Southeast Asia. Micro businesses or enterprises can be operated with low capital, limited skills, and technological know-how. The most striking barrier in female entrepreneurial activities is the need for more institutional support for females to become entrepreneurs, as limited opportunity is given to females to become businesspersons; the macro-environmental issues also cause limited female participation. Supportive governmental policies enable females to initiate business ventures and entrepreneurial activities. However, it is observed that females lack the understanding, knowledge, skills, and abilities to initiate and support business ventures (Tanusia et al., 2016).

Long-term objectives and goals can be attained by employing the full potential of businesses while utilizing the diverse workforce, specifically females' participation in business ventures. Female participation in initiating business ventures introduces new products and services that create new ideas for the future generation that address environmental issues (Mahajan & Bandyopadhyay, 2021). Furthermore, the participation of females in businesses contributes to sustainable development through innovative solutions, and innovation that creates a lot of jobs and career opportunities, increases socialization, reduces poverty, and plays a significant role in career success (Diaz-Sarachaga & Ariza-Montes, 2022; Mazhar et al., 2022; Shkabatur et al., 2022; Trivedi & Petkova, 2022). In the subcontinent, the significant population and persistent gender disparity make it challenging for women to participate in business and economic activities. It highlights the importance of support and equality in opportunities. The overall population of Pakistan is estimated to be approximately 222.4 million, which consists of 49% females and 51% male members, and the total workforce is declared as 71.76 million (male) and 15.34 million (females). Females were 22.8% of the total employed workforce in Pakistan, as reported in an economic survey. The service sector contributes almost 58% to the economy with the utilization of 37.2% labor force. Female employers are only 0.1% of Pakistan's entrepreneurship ratio. The overall ratio is 2.6%, of which 0.3% of females contribute economic strength, according to the Pakistan Labor Force Survey 2022 (Rizvi et al., 2023). This ratio is relatively low compared to any other developing or developed nation; it is required to consider the research study on such lower female participation.

Pakistani society is much more restrictive of women's empowerment as Pakistan is ranked 133 out of 160 countries on the Gender Inequality Index, which shows gender bias that restricts females from participating in business ventures and prevents them from participating in the economy. In the same region, women entrepreneurs were found to be engaged in self-employment; this depicts limited opportunities for females to participate in business ventures and business developments. Female business owners reported low rate of return and limited business growth opportunities, and it has been reported that the success rate could be higher compared to male members, ultimately leading to business closure. Pakistani females have entrepreneurial intentions for financial strength and to meet living expenses, but limited opportunities are available (Shaheen et al., 2022).

According to economic growth theory, one country can increase economic growth through effective business initiatives and ventures. Females in Pakistan face various restrictions in business activities, such as structural, cultural, and historical factors that prevent females from participating in business ventures. The development of entrepreneurship has vital and central importance for economic progress; however, females are limited in entrepreneurial activities as the female gender faces several problems. These factors relate to cultural issues, power point of view, abases of feminist approach, gender inequality, and structural issues. So, Pakistani females participate less compared to the USA, where 38% of organizations are owned by females, significantly contributing to economic growth (Shahid & Venturi, 2022).

The female population is significantly higher but the participation in business and workforce as an entrepreneurship is underrepresented. The participation of females in business ventures has increased compared to the previous decade; females' participation is considered an integral part of the country's economic system for its significance (Strawser et al., 2021). Entrepreneurial activities are important in job creation and economic growth; notably, women's presence has increased in labor significantly in developing countries, so female participation in these countries is found to be more relevant. Another crucial factor has been identified that despite economic contribution, females receive lower wages than the male workforce (Corrêa et al., 2022). In recent years, it has been noted that females' socio-economic status has increased; however, various issues and problems related to social equality, discrimination, violence, and equality against women still prevail. The research scholars highlighted the knowledge gap in studying entrepreneurship in developing nations, and the lack of views on entrepreneurial activities of females in general is evident (Foss et al., 2019; Corrêa et al., 2022). The comprehensive knowledge structure has not yet been incorporated in previous studies; empirical evidence has been lacking in prior literature.

This study is motivated to investigate the issues, challenges, and problems female entrepreneurs face, regardless of improved women's contribution to business development and growth, assisting in providing job opportunities and increasing wealth maximization and innovation while focusing on the diverse genders. Scholars have also suggested ensuring equal investment opportunities by reducing impeding factors of female entrepreneurial activities for equality (Neumeyer et al., 2019; Nair, 2020). It is also documented that women's entrepreneurship has yet to reach its full potential, and there are both aspects including qualitative. There is a dire need to recognize different factors, including implicit knowledge, the structure of work, and entrepreneurial activities while also fostering a more theoretical understanding of entrepreneurial activities (Strawser et al., 2021).

The economic crisis has a long-lasting impact anywhere in the world; therefore, countries initiate various business activities to pull the country from crisis-like situations while becoming competitive and inclusive. The government plays an interactive role in encouraging people to take advantage of potential growth to take initiatives for business ventures for wealth creation. This situation unlocks the economic power of females; a significant population can crucially influence economic revival development as well as sustainable growth. Women's entrepreneurial activities enable the countries to gain economic strength and can potentially increase economic growth (Sajjad et al., 2020). Interestingly, it has been reported that female entrepreneurs have outperformed male businesspersons in the previous decade with their skills, capabilities, and initiatives. The statistics presents that females are successful than men, as a 70% success rate has been determined for female entrepreneurs compared to 40% in the last seven years (Alsaad et al., 2023).

Female entrepreneurs elevate development in diverse sectors and become the rising stars of financial strength (Thaddeus et al., 2022). The World Economic Forum also labelled female businesspeople as 'the way forward' due to the level of struggle of females towards entrepreneurial activities to boost economic strength and for a rightful place in economic life. It has also been reported that female businesspeople are fewer in number compared to male members of society. Business initiatives create the opportunities for employment, an essential factor for reduction of poverty and increase in economic growth. It is essential to encourage empowered women to initiate business ventures for entrepreneurial activities that enhance economic stability (Zeidan & Bahrami, 2020). The entrepreneurial activities of females must be aligned with the opportunities available in the country to harvest the potential benefits. The professional approach should be considered for success in the entrepreneurial field. For this purpose, empowerment becomes crucial for women entrepreneurs who facilitate their role in the emergence of entrepreneurial activities (Alsaad et al., 2023).

This chapter addresses the crucial role of women's entrepreneurial success, influenced by women's innovation capability, risk-taking behavior, and psychological empowerment of females. Furthermore, family support is essential to female business initiatives in Pakistan. Thus, this chapter focused entrepreneurial success of women business owners who receive impact from diverse factors, with an essential factor of family support as a moderating variable between women's innovation capability and the success of female business owners, the association between risk-taking behavior to predict the entrepreneurial success of female business owners.

6.2 REVIEW OF THE LITERATURE

This section of the chapter entails previously published literature depicting the phenomenon of women's entrepreneurial success and its relationship with women's innovation capability, risk-taking behavior, and psychological empowerment.

6.3 WOMEN'S INNOVATION CAPABILITY AND WOMEN'S ENTREPRENEURIAL SUCCESS

It is not easy to distinguish women's efforts and continuous participation and contribution to entrepreneurial activities to enhance a nation's growth. The contribution of women in entrepreneurial activities has dramatically expanded over the past ten years; women's entrepreneurship happens when a female has taken part in an innovation business (Riandika & Mulyani, 2020). Innovation is vital in improving and enhancing women's entrepreneur's success. The entrepreneurial activities performed through innovation create quality and produce the value of entrepreneurship due to effective innovation implementation for value-maximization (Maziriri et al., 2024). The progress in products and services through innovation growth in sales can be enjoyed, inclination in profit can be harvested, and decision-making power can be exercised. These are crucial elements and essential strategies for every organization to gain a competitive advantage and overcome competitors. Innovation capability allows organizations to fulfil the requirements of consumers, analyze the market situation from the perspective of competition, integrate strengths, and avail opportunities (de Souza Barbosa et al., 2023).

Innovation capabilities enable the firms to create new ideas, products, or services for the organization, which enhances the performance of the firms and helps to increase the growth of the economy, create opportunities for individuals to decrease unemployment, and generate benefits for such innovative initiatives (Taleb et al., 2023). The literature has discussed that organizational management dynamic capabilities significantly impact the firm's performance; it helps to increase the firm's efficiency in terms of rivals and marketing to match the needs and consumer demands, employees, and shareholders. The consistency of innovation influence the procedure for various organizations to create new ideas for the stage of implementation (Kongrode et al., 2023). Innovation is defined based on technology, management, and the strategic approaches employed by the firms to facilitate its implementation. It is a critical process for firms to put essential efforts into innovation which influences the performance, product, and process innovation capabilities necessary for every organization's success. Innovation capabilities enable enterprises to create new innovative ideas, change the product process and design, develop management systems, and increase the chances of competing (Somwethee et al., 2023).

The definition of innovation capability is a process of generating and managerial opportunities to gain competitiveness for the firm; it encircles the problem-solving and decision-making process to achieve the goals of innovation through technology to enhance the sustainable performance (Shahzad et al., 2020; Maziriri et al., 2024). The researchers have given attention to the critical perspective of entrepreneurial leadership and success with the moderation effect of innovation and opportunity capitalization. Further, entrepreneurial opportunity and innovation capability significantly mediate the relationship between leadership approach and success (Taleb et al., 2023). Another article revealed entrepreneurial education moderates between 'need for achievement' and exogenous construct. Moreover, innovation conviction, career, and mindset significantly predict entrepreneurial success depends upon education and 'need for achievement'. The education tends to enhance the relationship strength among 'need for achievement' and entrepreneurship success (Maziriri et al., 2024). Thus, the following hypothesized statement is derived:

H1: Women's innovation capability influences success of female entrepreneurs in Pakistan

6.4 RISK-TAKING BEHAVIOR AND WES

This section addresses the risk-taking behavior of female entrepreneurs. The argument claims that entrepreneurs' traits are to take risk in investing, and female entrepreneurs tend to be risk-taking personalities that significantly reveal success in business ventures. Risk-taking is the pursuit of risky assets in business to achieve higher value in the long run and maximize shareholder wealth. It is the ability to take financial risk during business activities, services, and markets by exploiting the opportunities to enhance business performance. Risk-averse business people cannot effectively utilize the capabilities and exploit new business ventures in a dynamic market and fail to achieve business performance. Prior literature argued that females are found to be low-risk takers in business decisions as females are more inclined towards seeking security in business activities. The literature also posits that risk-taking generally is significantly positive to higher business performance, but risk-averse behavior hinders performance (Lim & Envick, 2013). The evidence of risk-taking behavior that leads to business success is still inconclusive; establishing that low-risk behavior of females hurts business success (Mozumdar et al., 2022).

The literature review shows that passion of an individual influence the career success and examined the moderation impact of risk-taking. Results of the study stated that passion success, risky initiatives, and risk-taking success have positive and significant impacts. The study showed that risk-taking approach mediate the relationship between endogenous construct of passion and exogenous construct of career success (Ratanavanich & Charoensukmongkol, 2023). A recent study examined the relationship between self-esteem and 'need for achievement' and between risk-taking opportunities and entrepreneurial intentions in South African universities. It has been highlighted that risk-taking behavior is considered a key determinant of entrepreneurial activity; it supports the student's self-esteem in achieving entrepreneurial goals (Steenkamp et al., 2024). Further, women's entrepreneurial success depends upon internally related elements, that entails confidence and risk-taking, and 'need for achievement'. Further, the factors also predict the success as of economic situation, and sociocultural factors. The study highlights the recommendations for SMEs and policymakers to encourage female entrepreneurs for long-term perspective and support (Khan et al., 2021).

Previous research has explored risk-taking behavior of constructors that expresses the role of the risk-taking approach of constructors. This entails personal factors such as attitude of employees having a tendency to take risks, expected outcomes, perception of risk, and feelings of anxiety or insecurity, all of which significantly promote the risk- related behavior. The study explored the indirect mediating role policies on behavior and the indirect significant impact of training on behavior mediated through attitude towards risky behavior and perception of worry unsafe (Man et al., 2021). Another research effort has been made to explore the correlation between mobile usage addiction, risk-taking, and self-control behavior. The moderation effect of adolescent sex has been explained, and the mediating effect of self-control is discussed as well. The study's findings demonstrated that the moderation and mediation model suggests that after controlling the demographic variables including self-control and risky behavior, the addiction towards technology or mobile phones usage positively predict the behavior and reduce the self-control (Dou et al., 2020).

Previously the researcher explored predictors that influence taking risk, and explain the phenomenon of emotional stability, and self-control in the context of college-level students in China. A positive correlation is depicted between chronotype, emotional stability, and self-control. However, there is a negative relation between self-control, emotional stability, and risk-taking behavior. Results of the study stated that chronotype significantly impacts behavior in the series of mediation; risky behavior is affected by two pathways: self-control and emotional stability (Zhang et al., 2022).

Risk-taking behavior is an important factor that predicts significant outcomes; they hypothetical statement is:

H2: Risk-taking behavior influences WES among Pakistani female entrepreneurs

6.5 PSYCHOLOGICAL EMPOWERMENT AND WES

Entrepreneurs are those individuals who have freedom of spirit; they are economic heroes, stuff of legends, create opportunities for others, identify the future needs and demands of the customers and adapt the products according to their needs, create new ideas, take risks, reshape the businesses and bring innovation in their business and develop the growth of the country (Hermayen et al., 2022). In various countries, women entrepreneurs face criticism, gender equality, and issues including financial problems, opportunities for growth and development in the market, approach for management, training skills, and problems with owning land to operate their businesses (Hermayen et al., 2022). In many organizations, women leave due to unequal pay, inadequate benefits, and lack of opportunities, leading them to pursue entrepreneurship. It has been clearly observed that women particularly mothers, often choose entrepreneurship for the flexibility it provides, empowering them to balance work and family responsibilities. The supportive family system plays a vital role in female entrepreneurs' lives; a family support system provides psychological, physical, and financial support. Psychological empowerment is a form of motivation that promotes the self-control and active participation in one's working role (Boudreaux et al., 2023).

Previous literature has explained psychological women's empowerment and the field of entrepreneurship has received significant attention. Moreover, the effect of social exclusion and social inequality is important to explain the phenomenon. It has been documented that the ability of individuals to select strategic choices, social inequality, and discrimination must be addressed. The empowerment process is vital in this context (Hibbs, 2022; Malik et al., 2021). It has been reported that economic strength improves due to entrepreneurial orientation and psychological empowerment of females who own, run, and control business ventures. Psychological empowerment was observed to be positively associated with entrepreneurial orientation. It has been reported that psychological empowerment and entrepreneurial orientation increase the productivity and performance o females SMEs businesses (Kadiyono & Fathoni Cahyono, 2023). Another research effort has been made to determine the predictors of innovative work behaviors; the study revealed that culture, and empowerment are striking influencers with mediating effects of psychological empowerment. It is reported that a significant link between entrepreneurial culture and psychological empowerment; psychological empowerment significantly influences innovative work behavior. The study revealed that psychological empowerment fully mediates the association between entrepreneurial culture and innovative work behavior. It has been found that entrepreneurial culture has no impact on the innovative work behavior of the employees in the organization; psychological empowerment has been found significant and supports innovative work behavior (Nguyen et al., 2023).

According to the United Nations report, gender inequality is one of the major concerns in Pakistan; in education and work, females face inequality that causes an increase in unemployment and reduces literacy. The United Nations stressed ensuring the empowerment of females and suggested the Ministry of Planning and Development take specific initiatives to empower females. Further, there is a dire need to develop plans for reducing inequality to contribute to the economic perspective. The gender-based violence is another challenge for Pakistan, and it is considered a crucial issue to deal with and eliminate all kinds of gender-based violence (see Figure 6.1).

Previous studies have shown the importance of psychological empowered females tends to influence the socio-culture domain in Pakistan. It has been reported that personality characteristics, development in gender, sociocultural support, and religious beliefs and values shape women's entrepreneur's psychological empowerment in Pakistan. Females possess competencies, knowledge, and skills to manage their businesses. Results of the study showed that females needed greater independence, support, and freedom to operate their businesses; the findings indicated that

FIGURE 6.1 Gender equality and women empowerment. (UN Report, 2023.)

women participate in entrepreneurial activities and feel psychological empowerment (Rehman & Basit, 2023). The utilization of social media and the increased number of female entrepreneurs can only be achieved through psychological empowerment for effective business outcomes. The involvement of social media has increased in recent years which influenced females to initiate their technological business ventures. It is demonstrated that female entrepreneurs' social media participation has a significant influence on digital entrepreneurship, which further influences psychological empowerment (Chakraborty & Biswal, 2023). Global sustainable development goals include gender equality. To achieve these goals, there is a need to focus on women's entrepreneurship, which further enhances women's empowerment. The women entrepreneurs' empowerment helps to achieve sustainable development goals like gender parity, decreasing unemployment, and social instability. It is reported that empowerment is defined by core dimensions, including clear goals and objectives, the implementation of effective control mechanisms, the development of essential competences, and the enhancement of self-esteem (Chakraborty & Biswal, 2022). The following hypothesized statement is formulated:

H3: Psychological empowerment behavior influences women's entrepreneurial success among Pakistani female entrepreneurs

6.6 FAMILY SUPPORT AND WES

Female entrepreneurs are progressing towards gender equality, which can be seen as a rise to take business initiatives among females to become entrepreneurs. Therefore, entrepreneurship in females plays a vital role in developing gender equality (Kyrgidou et al., 2021). The financial and economic strength of any country is increased by the contribution and participation of women's employment, formally or informally; growing contribution to the country will help to increase the economic development quickly rise in the household female's income and guarantee the equal distribution of income per capita. To accomplish aspiration number of programs have been introduced to increase women's economic status through creating awareness and encouraging and participating in entrepreneurship activities. Financial empowerment for women contributed overall to the country's development (Ismail et al., 2021).

The literature has reported emotional support as important factor in women's entrepreneurial success; family moral support can be psychological assistance to help female entrepreneurs deal with their business issues effectively (Agarwal et al., 2020). Family support helps to build confidence in women that they can easily manage their business responsibilities successfully. A positive association between families' support and business performance related to female entrepreneurial population was reported in China (Corrêa et al., 2022; Ojong et al., 2021). The significant challenges and problems women entrepreneurs currently face in small and medium-sized enterprises are family support, gender discrimination, difficulty getting loans and finance, and lack of motivation and encouragement, which have consistently decreased productivity and performance (Aljuwaiber, 2021).

Figure 6.3 presents the objective of the UN to reduce poverty through empowering and establishing businesses for females to fight against harsh economic conditions, and to meet the sustainable development goals. The objective is to reduce poverty in Pakistan by creating new jobs, and 88% of jobs for women are targeted. The UN supports the improvements in entrepreneurial services by enriching the knowledge, skills, and abilities of females in different districts of Pakistan to compete in the market. The UN contributes to capacity building, training, and education to enhance competencies and boost female employability by fostering entrepreneurship. In the agriculture sector, 55% of women out of 2,720 farmers are required to be trained for economic transformation in different provinces of Pakistan for poverty reduction. Entrepreneurial activities are necessary for the development of any country, it is a prime objective to increase the ratio of female entrepreneurs in Sindh by 55% of women possessing their businesses. Universities must provide entrepreneurial education and training to females for significant contributions to the market by owning and expanding home-based businesses (United Nations, 2023; see Figure 6.2).

FIGURE 6.2 Entrepreneurial objectives. (UN Report, 2023.)

Prior literature has discussed that financial support increases the performance of female entrepreneurs and helps to create job opportunities effectively. Management skills and family support also have a positive influence on performance, and these factors are essential for female entrepreneur development and improvement to enhance productivity and performance in the business (Lei et al., 2021).

The entrepreneurial performance was found to be low in the last two years due to the need for more management skills, limited access to loans and finance, and lack of family support, contributing to the low female entrepreneur performance. Reportedly, it is evident that female entrepreneur performance in the SME sector continuously decreased in terms of performance, which further reduced the development and growth of the Pakistani economy and SMEs. According to the report from 2015 to 2017, the performance of women entrepreneurs was high due to the support of different NGOs and financial support; after that, the progress and performance of the SMEs decreased due to the lack of family and financial support (Ariffin et al., 2020).

Researchers have focused to examine the impact of entrepreneurial intention on self-efficacy, further regret mediates the relationship. Family support is most crucial elements that tends to influence the relationship between exogenous constructs and female entrepreneurial success. It has been reported that family support is essential for females entrepreneurs for success as cultural support is necessary (Ahmed et al., 2021). Another research effort has been made to identify such factors that increase the performance female entrepreneurs. It was revealed that a higher level of emotional support is required for females to gain success. However, the lack of emotional support is a burning issue for women entrepreneurs, significantly hindering their productivity and performance (Neneh & Welsh, 2022). Other studies have considered the association among perceived family support and young individuals' perception related to the desirability and feasibility of starting a business; the study also explained the moderating role of cultural dimensions. Results of the study indicated that support of family members ensures the initiatives of businesses relying on national culture. It has been found that family support is essential in influencing entrepreneurial behavior in countries (Maleki et al., 2023).

The above literature helps in formulating the following hypothesis:

H4: Family support influences women's entrepreneurial success among Pakistani female entrepreneurs

6.7 MODERATING ROLE OF FAMILY SUPPORT

This section focuses on the moderation role of family support to explain the phenomenon of women's entrepreneurial success. The researcher argues that family support is an essential factor that predicts and plays a crucial role in business outcomes, including performance, success, and growth. This chapter entails the assessment of moderation effect of support of family to ensure innovative capabilities of females and women's entrepreneurial success, between risk-taking behavior and women's entrepreneurial success.

Family support is an essential factor that is referred to as the family dealing with good financial conditions and to determine profits. The unsupportive family members lower income and are considered as major challenge to deal with to achieve success. In Pakistan, most families need to be more supportive of women participating in business ventures, as the lack of support often forces women into harsh conditions and creates significant obstacles for them. Another issue is that the majority of female business owners in Pakistan are married; 51% of them have kids, 43% are found to be single, 8% are widowed, and 11% are separated from their spouses or divorced. The situation of Pakistani females is quite different from other women entrepreneurs in different countries (Qadri & Yan, 2023).

Previous research studies have highlighted various challenges and issues for females to conduct business activities, including limited resources, lack of family support, work-life balance, failure, social and cultural barriers, and societal issues. The pressure faced by family members and society prevents females from participating in business activities (Nawaz, 2018). Pakistan is a developing and masculine hegemonized country where females have no freedom in decisions related to everyday life sphere; while stressing that male members are responsible for financial support and women are required to take care of families, children, and other dependents (Rizvi et al., 2023). The research scholars have identified that family support is the most essential variable emotionally and financially linked to women's business success. Support from relatives and family members is vital in motivating female entrepreneurs. In Pakistan, it has been reported that females face strict restrictions from family members and from society that limit their freedom to participate in business activities, which is deeply rooted in Pakistani society. Female entrepreneurs face conflicts as compared to male members; due to societal approaches and cultural beliefs, female business owners receive a negative impact on investment decision-making. Females are also prevented from pursuing higher education and training for business initiatives (Meyer, 2019).

Family support is linked to the economic, instrumental, and emotional support that positively influences women's success in business ventures. Emotional support, encouragement, and guidance are required for females to achieve business success goals. Further, emotional stability and psychological capital are necessary for female entrepreneurs. Female entrepreneurs are restricted compared to male members and limit the opportunities in business management and technical skills hinder females. Further, male members are free to move in society, but females face diverse issues during mobility due to the cultural approach of the society. It is reported that in Pakistan, females are facing problems in commuting, meeting with counterparts, lack networking capabilities, and, most importantly, fulfilling the religious obligations that exert severe restrictions on the mobility of females and participation in business ventures (Yaqoob, 2020; Roomi & Parrott, 2008).

The above literature depicts that family support has a central role in female entrepreneurial success, the absence of family support leads to the negative impact and failure of business ventures; however, the support of family tends to influence the success. Thus, this research incorporates the moderation role of family support while arguing that family support increases the intensity of the relationship between

Effect of Family Support for Women's Entrepreneurial Success

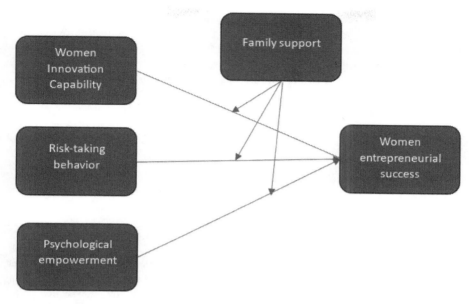

FIGURE 6.3 Research framework. (Developed by authors.)

exogenous and endogenous constructs. Therefore, the following moderating hypotheses are derived:

H5: Family support moderates the relationship between women's innovation capability and women's entrepreneurial success among Pakistani females
H6: Family support tends to moderate association between risk-taking behavior and WES among Pakistani females
H7: Family support tends to moderate association between psychological empowerment and WES among Pakistani females

6.8 RESEARCH FRAMEWORK

This section presents the research framework and hypothesized relationships between variables. There are four direct hypotheses and four moderating hypotheses. The direct hypotheses are between women's innovation capabilities and entrepreneurial success, risk-taking behavior and WES, and psychological empowerment and WES. Further, the moderation effect is also depicted in Figure 6.3.

6.9 RESEARCH METHODOLOGY

This section discusses the research approach adopted for data collection and analysis. This is a quantitative study, as data was collected cross-sectionally through a semi-structured questionnaire. The respondents for this study were females having businesses of any scale, and data were collected from 253 women entrepreneurs by employing purposive sampling technique from five major cities of Pakistan, including Karachi, Lahore, Rawalpindi, Faisalabad, and Sialkot. The reason for city selection is that these cities are major SME-based business hubs, and females are also engaged in businesses. The unit of analysis was an individual, as the sample was taken and supported by information from the Federation of Pakistan Chambers of Commerce and Industry, Export Promotion Bureau, SMEs Development Authority (SMEDA), First Women Bank, and from National Rural Support Program (NRSP) units that provide the micro-loans for small businesses to females specifically.

The measurement scales for each variable were adopted from prior literature having acceptability for data analysis. The dependent variable, women's entrepreneurial success, was borrowed from Basit et al. (2020) and had five items. The independent variable, women's innovation capability was from (Taleb et al., 2023), having three statements. The measurement scale of risk-taking behavior was from Khan et al. (2021) and has four items. The psychological empowerment consists of five items (Elshaer et al., 2021). Further, the moderating variable family support's measurement scale having four items is taken from the research of Goheer (2003). The five-point Likert scale was employed for assessment.

6.10 ANALYSIS AND DISCUSSION

This section entails the analysis of collected data: first a description respondents' is presented, and then the measurement model assessment and structural equation model are discussed.

6.11 DESCRIPTIVE ANALYSIS

Table 6.1 presents the descriptive statistics. As stated earlier, the data were collected from 253 individual female entrepreneurs from different cities. Table 6.1 shows the demographic description of the respondents.

6.12 MEASUREMENT MODEL ASSESSMENT

This measurement model assessment demonstrates the reliability and validity of the constructs based on the collected data. This section depicts the Cronbach alpha, composite reliability for determining the constructs' reliability that must remain higher than 0.70 minimum for acceptable reliability. The average variance extracted (AVE) demonstrates the convergent validity; the AVE must remain higher than 0.50 for achievement of convergent validity. The PLS algorithm method also assesses the discriminant validity by the Fornell and Larker method.

Table 6.2 satisfies reliability and validity, as values for Cronbach alpha, CR, and AVE are found to be in acceptable ranges.

TABLE 6.1
Demographic descriptive statistics for individual females

Age	
20–25	43
26–30	71
31–35	93
36–40	23
41–50	15
51–60	8
Education	
Under Matric	90
Intermediate	101
Bachelor	53
Masters	9
Above	0
Marital status	
Single	51
Married	142
Widowed	45
Divorced	15
Experience	
Less than 5 years	103
5–10 years	97
11–15 years	50
More than 20 years	03

TABLE 6.2
Reliability and validity analysis

S#	Constructs	α	CR	AVE
1	Women's entrepreneurial success	0.792	0.862	0.565
2	Women innovation capability	0.937	0.943	0.888
3	Risk-taking behavior	0.808	0.828	0.635
4	Psychological empowerment	0.895	0.906	0.706
5	Family support	0.945	0.946	0.859

Note: Women's Entrepreneurial Success (WES), Women's Innovation Capability (WIC), Risk-taking behavior (RTB), Psychological Empowerment (PE), Family Support (FS).

6.13 DISCRIMINANT VALIDITY

This section presents the discriminant validity. According to Fornell and Larcker's (1981) criteria, the square root of AVE must remain higher than correlational values with other latent variables.

TABLE 6.3
Discriminant validity

Constructs	FS	PE	RTB	WES	WIC
FS	**0.927**				
PE	0.453	**0.840**			
RTB	0.651	0.485	**0.797**		
WES	0.503	0.450	0.516	**0.752**	
WIC	0.693	0.467	0.611	0.461	**0.942**

Note: Women's Entrepreneurial Success (WES), Women's Innovation Capability (WIC), Risk-taking behavior (RTB), Psychological Empowerment (PE), Family Support (FS).

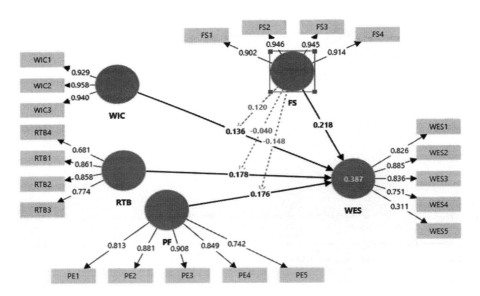

FIGURE 6.4 Measurement model assessment extracted from PLS.

In Table 6.3, discriminant validity is evident; the values show that the result achieved discriminant validity. The above statistics show that constructs are reliable and valid according to the PLS algorithm. Figure 6.4 shows the measurement model assessment extracted from PLS.

6.14 STRUCTURAL EQUATION MODEL (SEM)

This portion presents the SEM by investigating the relationship between variables, the four direct hypotheses, and three moderation effects. Table 6.4 presents the results.

TABLE 6.4
Direct effect

S#	Relationships	β	t-value	p-value
H1	WIC→WES	0.136	1.619	0.105
H2	RTB→WES	0.178	2.056	0.040
H3	PE→WES	0.176	2.612	0.009
H4	FS→WES	0.218	2.248	0.025

Note: Women's Entrepreneurial Success (WES), Women's Innovation Capability (WIC), Risk-taking behavior (RTB), Psychological Empowerment (PE), Family Support (FS).

6.15 DESCRIPTION

The results in Table 6.4 demonstrate that women's entrepreneurial success depends upon various factors; however, this study incorporated women's innovation capability, risk-taking behavior, psychological empowerment, and family support.

Hypothesis H1 investigates the relationship between women's innovation capabilities and women's entrepreneurial success. The relationship is insignificant due to the t-value and p-value. That means the females in the Pakistani business community are not innovative. The lack of education and training reduces the relationship and reduces the chances of becoming a successful businessperson. However, prior studies have reported that women's innovation significantly predicts WES (Riandika & Mulyani, 2020).

Hypothesis H2 investigated the relationship between risk-taking behavior and women's entrepreneurial success. The result depicted a significant relationship with WES. Risk-taking behavior is an essential trait of an entrepreneur's personality, and prior literature has identified that entrepreneurial success depends upon risk-taking initiatives (Mozumdar et al., 2022). However, it has also been reported that females are risk-averse compared to male businessmen (Amah & Okoisama, 2017).

Hypothesis H3 argues an association between psychological empowerment and WES. The results show that psychological empowerment is an essential element that plays a vital role in entrepreneurial success. According to resource-based theory (RBV), it is evident that resources such as managerial approach, personality, and psychological empowerment are the resources that contribute to organizational success. The results show that psychological empowerment significantly predicts women's entrepreneurial success.

Hypothesis H4 investigates the relationship between family support and women's entrepreneurial success. Family support is a prime concern for females when initiating business ventures. The Pakistani nation has an issue of family support, and females are restricted from participating in business activities. One of the major issues for Pakistani females is the lack of family support (Roomi & Parrott, 2008). Figure 6.5 shows the structural equation model extracted from PLS.

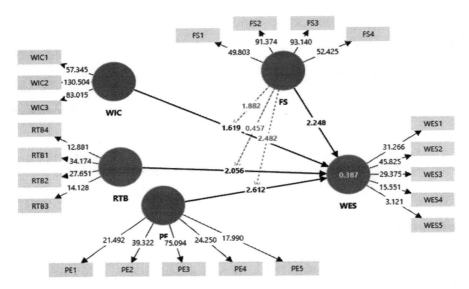

FIGURE 6.5 Structural equation model extracted from PLS.

6.16 MODERATING EFFECT

This research investigates the moderation effect of family support between exogenous and endogenous constructs. Table 6.5 presents the moderation analysis.

Hypothesis H5 addresses the moderation role of family support between women's innovation capabilities and WES. The result shows that family support has no moderation effect on the relationship between women's innovation capability and women's entrepreneurial success, which means the direct relationship between women's innovation capability and WES is insignificant. However, the moderation effect of family support does not support it as a moderating construct. This means family support is absent widely in Pakistani culture and needs to be addressed.

Hypothesis H6 also investigates the relationship between risk-taking behavior and WES and the moderating role of family support between these variables. The results show no moderation effect of family support between risk-taking behavior and WES. The direct relationship between risk-taking behavior and women's entrepreneurial success is reported, but the moderation role is insignificant. This shows that family support is absent, and a lack of family support prevents females from participating in business ventures.

Hypothesis H7 also investigated the moderation effect of family support on psychological empowerment and women's entrepreneurial success. The direct relationship is significant, and the moderation effect is also reported, which means that psychological empowerment in the presence of family support becomes significantly influential in predicting the WES.

TABLE 6.5
Moderation effect

S#	Relationships	β	t-value	p-value
H5	WIC*FS → WES	0.120	1.882	0.060
H6	RTB*FS → WES	0.040	0.457	0.648
H7	PE*FS → WES	0.148	2.482	0.013

Note: Women's Entrepreneurial Success (WES), Women's Innovation Capability (WIC), Risk-taking behavior (RTB), Psychological Empowerment (PE), Family Support (FS).

6.17 PRACTICAL IMPLICATIONS

This research effort provides guidelines for female entrepreneurs, women's development initiatives, and the government to make decisions and devise strategies for utilizing the female workforce.

1. There is a need to develop the curriculum and education for females to train them for effective business ventures.
2. There must be initiatives for female business centers to develop the capabilities, knowledge, skills, and abilities that ensure business success.
3. There is a dire need to create awareness about entrepreneurial education at every level. In Pakistan, it is another issue that females just take very basic education, and people culturally prevent females from getting enrolled in college or university. So, at the school level, there must be some training sessions for all females willing to start their businesses.
4. Females are generally risk-averse, so there is a need to create awareness about risk-taking initiatives and assess potential threats or benefits that can be harvested.
5. Empowerment is another issue in the Pakistani context for women, so governmental institutes must encourage and support female empowerment in making decisions during business activities.
6. The lack of family support is a prime concern and issue for females that drags them away from business activities. There is a need to conduct seminars, awareness sessions, and backup support for females.

6.18 LIMITATIONS AND FUTURE RESEARCH AVENUES

Every research has limitations; similarly, this research faced several limitations that must be addressed.

1. The study is limited in assessing the role of different exogenous constructs, and other variables that can be assessed for entrepreneurial success, such as personality traits, education level, awareness, and family business experience.

2. The study was conducted in five major cities of Pakistan, but industries are ignored in this study; future research may be conducted on diverse categories of small and medium-sized businesses.
3. The current study's sample was limited, and data were collected cross-sectionally; future research studies can consider the diverse sample size and may be compared to each other and male entrepreneurs. The longitudinal research approach can also be considered for explaining the phenomenon of women's entrepreneurial success.

6.19 CONCLUSION

Developing countries like Pakistan come across various complicated issues related to business initiatives taken into consideration by female entrepreneurs. The objective of the current chapter was to shed light on the entrepreneurial perspective of female business owners' success in the Pakistani market. Pakistan has various distinct issues associated with female business persons, such as lack of education, inappropriate behavior, knowledge, skills, abilities, capabilities, and empowerment. The most important factor is family support, which plays a vital role in business success. This research effort incorporated the emerging issues of women's innovative capabilities, risk-taking behavior, psychological empowerment, and family support to predict women's entrepreneurial success. The study's findings depicted that women's innovation capability needs to be improved among Pakistani female business owners; however, risk-taking behavior and psychological empowerment are significant to women's entrepreneurial success. Further, family support is required for business success for female entrepreneurs. Family support moderated the relationship between psychological empowerment and women's entrepreneurial success. However, no moderation role is evident in the relationship between women's innovation capability, risk-taking behavior, and women's entrepreneurial success. Governmental institutions must consider various guidelines for female business persons to take innovative initiatives that boost entrepreneurial activities.

REFERENCES

Agarwal, S., Lenka, U., Singh, K., Agrawal, V., & Agrawal, A. M. (2020). A qualitative approach towards crucial factors for sustainable development of women social entrepreneurship: Indian cases. *Journal of Cleaner Production*, 274, 123135.

Ahmed, I., Islam, T., & Usman, A. (2021). Predicting entrepreneurial intentions through self-efficacy, family support, and regret: A moderated mediation explanation. *Journal of Entrepreneurship in Emerging Economies*, 13(1), 26–38.

Aljuwaiber, A. (2021). Entrepreneurship research in the Middle East and North Africa: Trends, challenges, and sustainability issues. *Journal of Entrepreneurship in Emerging Economies*, 13(3), 380–426.

Alsaad, R. I., Hamdan, A., Binsaddig, R., & Kanan, M. A. (2023). Empowerment sustainability perspectives for Bahraini women as entrepreneurs. *International Journal of Innovation Studies*, 7(4), 245–262.

Amah, E., & Okoisama, T. (2017). Risk taking nd survival of small and medium scale enterprises in Nigeria. *Archives of Business Research*, 5(11), 35–42.

Ariffin, A. S., Baqutayan, S. M. S., & Mahdzir, A. M. (2020). Enhancing women entrepreneurship development framework: Policy & institution gap and challenges in the case of Malaysia. *Journal of Science, Technology and Innovation Policy*, 6(2), 22–33.

Basit, A., Hassan, Z., & Sethumadhavan, S. (2020). Entrepreneurial success: Key challenges faced by Malaysian women entrepreneurs in the 21st century. *International Journal of Business and Management*, 15(9), 122–138.

Boudreaux, C. J., Bennett, D. L., Lucas, D. S., & Nikolaev, B. N. (2023). Taking mental models seriously: Institutions, entrepreneurship, and the mediating role of socio-cognitive traits. *Small Business Economics*, 61(2), 465–493.

Brush, C. G., & Brush, C. G. (2006). *Growth-oriented women entrepreneurs and their businesses: A global research perspective*. Cheltenham: Edward Elgar Publishing.

Chakraborty, U., & Biswal, S. K. (2022). Psychological empowerment of women entrepreneurs: A netnographic study on Twitter. *Management Research Review*, 45(6), 717–734.

Chakraborty, U., & Biswal, S. K. (2023). Impact of social media participation on female entrepreneurs towards their digital entrepreneurship intention and psychological empowerment. *Journal of Research in Marketing and Entrepreneurship*, 25(3), 374–392.

Corrêa, V. S., Brito, F. R. D. S., Lima, R. M. De, & Queiroz, M. M. (2022). Female entrepreneurship in emerging and developing countries: A systematic literature review. *International Journal of Gender and Entrepreneurship*, 14(3), 300–322.

de Souza Barbosa, A., da Silva, M. C. B. C., da Silva, L. B., Morioka, S. N., & de Souza, V. F. (2023). Integration of Environmental, Social, and Governance (ESG) criteria: Their impacts on corporate sustainability performance. *Humanities and Social Sciences Communications*, 10(1), 1–18.

Diaz-Sarachaga, J. M., & Ariza-Montes, A. (2022). The role of social entrepreneurship in the attainment of the sustainable development goals. *Journal of Business Research*, 152, 242–250.

Dou, K., Wang, L.-X., Li, J.-B., Wang, G.-D., Li, Y.-Y., & Huang, Y.-T. (2020). Mobile phone addiction and risk-taking behavior among Chinese adolescents: A moderated mediation model. *International Journal of Environmental Research and Public Health*, 17(15), 5472.

Elshaer, I., Moustafa, M., Sobaih, A. E., Aliedan, M., & Azazz, A. M. S. (2021). The impact of women's empowerment on sustainable tourism development: Mediating role of tourism involvement. *Tourism Management Perspectives*, 38, 100815.

Fornell, C., & Larcker, D. F. (1981). Evaluating structural equation models with unobservable variables and measurement error. *Journal of Marketing Research*, 18(1), 39). https://doi.org/10.2307/3151312

Foss, L., Henry, C., Ahl, H., & Mikalsen, G. H. (2019). Women's entrepreneurship policy research: A 30-year review of the evidence. *Small Business Economics*, 53, 409–429.

Goheer, N. A. (2003). *Women entrepreneurs in Pakistan*. Geneva: International Labour Organization.

Hermayen, A., Tarique, A., Bhardwaj, A. B., Bishnoi, M. M., & Gupta, R. (2022). Family support and psychological empowerment: Women entrepreneurs in the UAE. *Journal of Positive School Psychology*, 6(9), 4244–4253.

Hibbs, L. (2022). "I could do that!"–The role of a women's non-governmental organization in increasing women's psychological empowerment and civic participation in Wales. *Women's Studies International Forum*, 90, 102557.

Ismail, N. N. H. M., Nasir, M. K. M., & Rahman, R. S. A. R. A. (2021). Factors that influence women to be involved in entrepreneurship: A case study in Malaysia. *Creative Education, 12*(4), 837.

Kadiyono, A. L., & Fathoni Cahyono, A. B. (2023). How does psychological empowerment affect entrepreneurial orientation at women-owned SMEs in Indonesia? *Global Journal of Business Social Sciences Review (GATR-GJBSSR), 11*(4), 93–106.

Khan, R. U., Salamzadeh, Y., Shah, S. Z. A., & Hussain, M. (2021). Factors affecting women entrepreneurs' success: A study of small and medium-sized enterprises in the emerging market of Pakistan. *Journal of Innovation and Entrepreneurship, 10*(1), 1–21.

Kongrode, J., Aujirapongpan, S., & Ru-Zhue, J. (2023). Exploring the impact of dynamic talent management capability on competitive performance: The mediating roles of dynamic marketing capability of startups. *Journal of Competitiveness, 1*.

Kyrgidou, L., Mylonas, N., Petridou, E., & Vacharoglou, E. (2021). Entrepreneurs' competencies and networking as determinants of women-owned ventures success in post-economic crisis era in Greece. *Journal of Research in Marketing and Entrepreneurship, 23*(2), 211–234.

Lei, H., Khamkhoutlavong, M., & Le, P. B. (2021). Fostering exploitative and exploratory innovation through HRM practices and knowledge management capability: The moderating effect of knowledge-centered culture. *Journal of Knowledge Management, 25*(8), 1926–1946.

Lim, S., & Envick, B. R. (2013). Gender and entrepreneurial orientation: A multi-country study. *International Entrepreneurship and Management Journal, 9*, 465–482.

Mahajan, R., & Bandyopadhyay, K. R. (2021). Women entrepreneurship and sustainable development: Select case studies from the sustainable energy sector. *Journal of Enterprising Communities: People and Places in the Global Economy, 15*(1), 42–75.

Maleki, A., Moghaddam, K., Cloninger, P., & Cullen, J. (2023). A cross-national study of youth entrepreneurship: The effect of family support. *The International Journal of Entrepreneurship and Innovation, 24*(1), 44–57.

Malik, M., Sarwar, S., & Orr, S. (2021). Agile practices and performance: Examining the role of psychological empowerment. *International Journal of Project Management, 39*(1), 10–20.

Man, S. S., Chan, A. H. S., Alabdulkarim, S., & Zhang, T. (2021). The effect of personal and organizational factors on the risk-taking behavior of Hong Kong construction workers. *Safety Science, 136*, 105155.

Mazhar, S., Sher, A., Abbas, A., Ghafoor, A., & Lin, G. (2022). Empowering shepreneurs to achieve the sustainable development goals: Exploring the impact of interest-free start-up credit, skill development and ICT use on entrepreneurial d. *Sustainable Development, 30*(5), 1235–1251.

Maziriri, E. T., Nyagadza, B., & Chuchu, T. (2024). Innovation conviction, innovation mindset and innovation creed as precursors for the need for achievement and women's entrepreneurial success in South Africa: Entrepreneurial education as a moderator. *European Journal of Innovation Management, 27*(4), 1225–1248.

Meyer, N. (2019). South African female entrepreneurs' business styles and their influence on various entrepreneurial factors. *Forum Scientiae Oeconomia, 7*(2), 25–35.

Mozumdar, L., Materia, V. C., Hagelaar, G., Islam, M. A., Velde, G. van der, & Omta, S. W. F. (2022). Contextuality of entrepreneurial orientation and business performance: The case of women entrepreneurs in Bangladesh. *Journal of Entrepreneurship and Innovation in Emerging Economies, 8*(1), 94–120.

Nair, S. R. (2020). The link between women entrepreneurship, innovation and stakeholder engagement: A review. *Journal of Business Research, 119*, 283–290.

Nawaz, A. (2018). Challenges faced by women entrepreneurs in Pakistan: A qualitative study. *Management and Organizational Studies, 5*(2), 13–26.

Neneh, B. N., & Welsh, D. H. B. (2022). Family support and business performance of South African female technology entrepreneurs. *International Journal of Entrepreneurial Behavior & Research, 28*(6), 1631–1652.

Neumeyer, X., Santos, S. C., Caetano, A., & Kalbfleisch, P. (2019). Entrepreneurship ecosystems and women entrepreneurs: A social capital and network approach. *Small Business Economics, 53*, 475–489.

Nguyen, H. T. N., Nguyen, H. T. T., Truong, A. T. L., Nguyen, T. T. P., & Nguyen, A. Van. (2023). Entrepreneurial culture and innovative work behaviour: The mediating effect of psychological empowerment. *Journal of Entrepreneurship in Emerging Economies, 15*(2), 254–277.

Ojong, N., Simba, A., & Dana, L.-P. (2021). Female entrepreneurship in Africa: A review, trends, and future research directions. *Journal of Business Research, 132*, 233–248.

Qadri, A., & Yan, H. (2023). To promote entrepreneurship: Factors that influence the success of women entrepreneurs in Pakistan. *Access Journal, 4*(2), 155–167.

Ratanavanich, M., & Charoensukmongkol, P. (2023). Effects of improvisational behavior on entrepreneurial activities and firm performance: The moderating roles of firm size and business experience. *Journal of Entrepreneurship in Emerging Economies. 16*(5), 1380–1408.

Rehman, T., & Basit, A. (2023). Sociocultural factors, religion and cognition development: An integrated view of psychological empowerment of female entrepreneurs. *Journal of Entrepreneurship and Innovation in Emerging Economies, 9*(2), 165–180.

Riandika, D., & Mulyani, E. (2020). The role of entrepreneurship development for women's welfare in rural areas. *Jurnal Ekonomi Pembangunan: Kajian Masalah Ekonomi Dan Pembangunan, 21*(1), 23–31.

Rizvi, S. A. A., Shah, S. J., Qureshi, M. A., Wasim, S., Aleemi, A. R., & Ali, M. (2023). Challenges and motivations for women entrepreneurs in the service sector of Pakistan. *Future Business Journal, 9*(1), 71.

Roomi, M. A., & Parrott, G. (2008). Barriers to development and progression of women entrepreneurs in Pakistan. *The Journal of Entrepreneurship, 17*(1), 59–72.

Sajjad, M., Kaleem, N., Chani, M. I., & Ahmed, M. (2020). Worldwide role of women entrepreneurs in economic development. *Asia Pacific Journal of Innovation and Entrepreneurship, 14*(2), 151–160.

Shaheen, N., Ahmad, N., & Hussain, S. (2022). Women entrepreneurship and empowerment in Pakistan: Gender, culture, education and policy in a broader perspective. *International Research Journal of Education & Social Sciences, 1*(1), 25–36.

Shahid, I., & Venturi, L. A. B. (2022). Women entrepreneurs problems in Pakistan (A study of District Mardan). *East Asian Journal of Multidisciplinary Research, 1*(7), 1405–1418.

Shahzad, M., Qu, Y., Zafar, A. U., Rehman, S. U., & Islam, T. (2020). Exploring the influence of knowledge management process on corporate sustainable performance through green innovation. *Journal of Knowledge Management, 24*(9), 2079–2106.

Shkabatur, J., Bar-El, R., & Schwartz, D. (2022). Innovation and entrepreneurship for sustainable development: Lessons from Ethiopia. *Progress in Planning, 160*, 100599.

Somwethee, P., Aujirapongpan, S., & Ru-Zhue, J. (2023). The influence of entrepreneurial capability and innovation capability on sustainable organization performance: Evidence of community enterprise in Thailand. *Journal of Open Innovation: Technology, Market, and Complexity, 9*(2), 100082.

Steenkamp, A., Meyer, N., & Bevan-Dye, A. L. (2024). Self-esteem, need for achievement, risk-taking propensity and consequent entrepreneurial intentions. *Southern African Journal of Entrepreneurship and Small Business Management, 16*(1), 1–11.

Strawser, J. A., Hechavarría, D. M., & Passerini, K. (2021). Gender and entrepreneurship: Research frameworks, barriers and opportunities for women entrepreneurship worldwide. *Journal of Small Business Management, 59*(1), S1–S15.

Taleb, T. S. T., Hashim, N., & Zakaria, N. (2023). Mediating effect of innovation capability between entrepreneurial resources and micro business performance. *The Bottom Line, 36*(1), 77–100.

Tanusia, A., Marthandan, G., & Subramaniam, I. D. (2016). Economic empowerment of Malaysian women through entrepreneurship: Barriers and enablers. *Asian Social Science, 12*(6), 81–94.

Thaddeus, K. J., Bih, D., Nebong, N. M., Ngong, C. A., Mongo, E. A., Akume, A. D., & Onwumere, J. U. J. (2022). Female labour force participation rate and economic growth in sub-Saharan Africa: "A liability or an asset." *Journal of Business and Socio-Economic Development, 2*(1), 34–48.

Trivedi, S. K., & Petkova, A. P. (2022). Women entrepreneurs journey from poverty to emancipation. *Journal of Management Inquiry, 31*(4), 358–385.

United Nations. (2023). *Key development trends in Pakistan 2023.* Retrieved from https://pakistan.un.org/en/270914-annual-report-2023

Yaqoob, S. (2020). The Emerging trend of women entrepreneurship in Pakistan. *Journal of Arts & Social Sciences, 7*(2), 217–230.

Zeidan, S., & Bahrami, S. (2011). Women entrepreneurship in GCC: A framework to address challenges and promote participation in a regional context. *International Journal of Business and Social Science, 2*(14), 100–107.

Zhang, Q., Wang, X., Miao, L., He, L., & Wang, H. (2022). The effect of Chronotype on risk-taking behavior: The chain mediation role of self-control and emotional stability. *International Journal of Environmental Research and Public Health, 19*(23), 16068.

7 Unveiling the Factors of Women Entrepreneurs on Social Media to Achieve Enterprise Sustainability

Anshu Rani, Vichitra Somshekar, Ramya U., and Mercy Toni

7.1 INTRODUCTION

The economic prosperity of a nation is significantly enhanced by the entrepreneurship paradigm (Saoula et al., 2023). Al-Mamary and Alshallaqi (2022) assert that entrepreneurship does not represent a quality or a particular way of thinking but rather the intentional pursuit of opportunities and the transformation of resources. Shane and Venkataraman (2000, p. 218) define it as "the means by which, by whom, and with what consequences opportunities to develop future goods and services are identified, assessed, and capitalized upon." It entails the identification of business potential by employing current, new, or a combined set of resources in an innovative and inventive manner (Ratten, 2023; Khoo et al., 2024).

The COVID-19 pandemic's disruption and displacement of businesses, persons, and markets have been associated with a higher rate of company failure (Rani & Shivaprasad, 2021). In 2021, women were 7.4% more likely than men to terminate their enterprises as a consequence of COVID-19 (Huang et al., 2022). Digital innovation has completely changed the world of entrepreneurship by providing previously unheard-of access to networks, markets, and materials (Khoo et al., 2024). Digital platforms have helped Indian women entrepreneurs overcome traditional barriers such limited mobility, poor market knowledge, and limited access to funding (Kreiterling, 2023). The democratization of entrepreneurship brought about by social media, mobile technology, and online marketplaces has allowed women to start and grow businesses at a much reduced cost and with a more flexible structure.

Social media is employed by numerous small business entrepreneurs in a variety of capacities (Ughetto et al., 2020). The marketing circumstances and conditions of small businesses are generally distinct from those of larger companies. Therefore, this phenomenon illustrates how small business owners have recognized the potential of social media as a powerful instrument, owing to its affordability, efficacy, and user-friendliness. It demonstrates the adaptability of organizations in the digital era, with respect to both technology and market conditions (Bican & Brem, 2020; Khoo et al., 2024).

The number of women-owned enterprises has been steadily increasing, and they have made significant contributions to economic development, innovation, and job creation (Suseno & Abbott 2021). The worldwide shift to sustainability has had an impact on entrepreneurial activities as well. It is becoming more well known that entrepreneurial women are capable of creating sustainable business models that give social and environmental goals equal weight with financial success (Bican & Brem, 2020; Saoula et al., 2023). Women have been heavily involved in campaigns in India that support social companies, organic products, and ecological technologies. Women-led businesses often give community development, moral behavior, and resource efficiency top priority, which aligns with the main goals of sustainable development (Kreiterling, 2023; Khoo et al., 2024). Still, challenges remain with regard to access to sustainable finance, capacity building, and joining international value chains. Nevertheless, they encounter distinctive obstacles in the effective promotion of their ventures, the development of networks, and the access to resources (Franzke et al., 2022; Ghouse et al., 2021; Kelly et al., 2020).

In recent years, social media platforms have emerged as a potent tool for business development and engagement, revolutionizing the manner in which companies engage with their target audiences (Kelly et al., 2020). This transformative impact is especially significant for women with entrepreneurial small enterprises, which are essential to the global economy. Although prior research has acknowledged the significance and value of social media adoption for small businesses (Chaker and Zouaoui, 2022; Mallios et al., 2023; Rehman et al., 2023; Alhajri & Aloud, 2024), there are numerous research gaps that require attention in order to develop a more comprehensive understanding of the ways in which women entrepreneurs specifically interact with and benefit from these platforms (Ughetto et al., 2020).

Additionally, even with the hopeful progress, women entrepreneurs in India still face a lot of recurring challenges. Gender prejudices and cultural conventions continue to hinder women's entrepreneurial dreams and progress. Finance access is still a major problem since women entrepreneurs sometimes have more trouble getting loans and investments than do their male colleagues (Ratten, 2023). Creating an atmosphere that supports women entrepreneurs requires focused legislative actions, encouraging ecosystems, and cultural shifts to overcome these obstacles (Alhajri & Aloud, 2024). The research studies on WE have drawn increasing attention throughout the last ten years because of their significant impact on the advancement of balanced development. This growing interest can be credited to a number of important elements that have profoundly changed the entrepreneurial environment in emerging nations, such India (Chaker & Zouaoui, 2022). On the one hand, the quick rise of digital innovation has completely changed the corporate landscape and given entrepreneurs new chances as well as difficulties. On the other hand, the increased support for corporate sustainability has highlighted the need of ethical and sustainable company practices even more (Kelly, et al., 2020). Notwithstanding these significant changes, earlier research has not been able to fully understand the role of women entrepreneurs in this changing environment.

One research gap concerns the researchers are identifying the factors related to women entrepreneurs in achieving enterprise sustainability. The studies that represent the distinctive obstacles and obstacles encountered by female entrepreneurs in the adoption and effective utilization of social media is limited. It is imperative to investigate factors which help women's entrepreneurship (WE) to flourish in emerging markets like India (Alhajri & Aloud, 2024). Another research gap is the examination of the specific benefits and outcomes that women-owned small businesses can achieve through social media, including business growth, network development, and empowerment. The precise mechanisms by which social media enables women entrepreneurs to access resources and establish networks are still unexplored. Therefore, this study will aim to identify and present the most relevant factors that influence the women entrepreneur's orientation on social media to develop sustainable enterprise.

The study compiles a summary of the research on factors influencing digital women's entrepreneurship to achieve sustainability. To accomplish this, the study organize a literature search and review to discover important research contributions. The goal in using search method was to find as many relevant studies as possible without missing any crucial article. After deciding on the Scopus and EBSCO database to use for review, the access to a huge library of academic journals was found on the topic.

To narrow the search pertaining to women's entrepreneurship, the topic filtering functions were applied.

The chapter narrowed the search to English-language publications covering the years 2012–2024, since this is a time of tremendous growth in the digitization and digital transformation of businesses. To ensure that my review included only the most pertinent papers, the study used strict inclusion and exclusion criteria. The following types of sources were not considered for this literature review: abstracts from conferences, papers without methodology, letters to editors, shorter pieces of writing, and grey literature. The study found 34 papers that were relevant to the topic after implementing this search method. Further, the reference of selected articles was also reviewed for better understanding of topic. This article describes the analysis of those papers and presents a conceptual model by using the relevant factors of WE which affects sustainability of business. Finally, this chapter will contribute on the factors that are linked with entrepreneurial orientation among women on social media and therefore will helps in gaining sustainability. This will comprehend our knowledge in the area of women entrepreneurial orientation (WEO). This study will further present implications, strategies and agenda for future research in the area of WE.

7.2 BACKGROUND OF STUDY

The nation's industry and economy have grown significantly as a result of the rising number of women who are entrepreneurs (Ughetto et al., 2020). Women-owned businesses are contributing significantly to society by creating jobs, changing the demographics of the nation and serving as an example for the next generation of female entrepreneurs (Alhajri & Aloud, 2024). Start-up India is dedicated to strengthening women's entrepreneurship in India through initiatives, schemes, the creation of enabling networks and communities and the activation of partnerships among diverse stakeholders in the start-up ecosystem. The organization's vision is to promote the sustainable development of women entrepreneurs for balanced growth in the nation. In every field, women are currently surpassing their male counterparts. Having advanced degrees is one of the most important traits that many successful female entrepreneurs share when it comes to schooling (Khoo et al., 2024). In addition, women business owners typically give greater health care benefit packages, on-the-job training and education, higher tuition reimbursement for employees pursuing postsecondary education, and more opportunities for paid time off and vacation, all of which contribute to the success of their companies.

7.2.1 Systems of Support and Interventions in Policy

One of the most important factors in achieving economic development that is sustainable, gender equality, and the reduction of poverty is the participation of women in the economy. According to the McKinsey Global Institute (MGI), if 68 million more women joined the workforce in India by 2025, the country's GDP might increase by USD 0.7 trillion (Sahu et al., 2024). If half of India's women were to enter the workforce, the country's GDP growth rate would rise by 1.5 percentage points, according to the World Bank (Franzke et al., 2022). Less than half of the world's GDP comes from women, and that figure is 17% in India. The Indian government has put many programs in place to support female entrepreneurs in association with a range of non-governmental organizations. Stand-Up India, the Mahila e-Haat platform, and the Pradhan Mantri Mudra Yojana (PMMY) are programs intended to provide female entrepreneurs with market access, skill development, and financial support (Khoo et al., 2024). Furthermore, women-led business-focused incubators and accelerators are starting to appear; they offer networking, resources, and mentoring. Closing the gap and creating a more inclusive entrepreneurship environment need these efforts.

7.2.2 Development of Women's Entrepreneurship in India

The number of women-owned enterprises has been rising consistently, and they significantly contribute to economic development, innovation, and the creation of jobs. For women-owned small enterprises, which are essential to the global economy, this transforming effect is especially noteworthy (Alhajri & Aloud, 2024). The adoption of social media by small businesses has been acknowledged as valuable and important by previous research. However, there are a number of research gaps that need to be filled in order to obtain a deeper understanding of how women entrepreneurs in particular interact with and benefit from these platforms (Franzke et al., 2022). It is imperative to investigate problems such as gender prejudices, resource scarcity, and industry-specific obstacles. Examining the particular advantages and results that female-owned small businesses can accomplish using social media (such as female empowerment, networking, and business expansion) is another research gap (Ughetto et al., 2020).

7.2.3 Digital Innovation and Development of Digital Entrepreneurship

Entrepreneurial success in today's competitive marketplaces is now mostly attributed to digital innovation. Adoption of digital technologies has shown to have enormous benefits for companies of all sizes and sectors, with important ramifications for society change and economic growth. One of the most effective forces promoting growth in GDP is digitalization. Streamlining operations using digital technologies helps firms save money and be more productive. Because of this, we can scale more rapidly and enjoy higher profit margins (Alhajri & Aloud, 2024). In addition, statistical evidence suggests that digitalization boosts average purchases, often tripling them. Several reasons have contributed to this, including the convenience and accessibility offered by digital platforms, which has led to an increase in consumers' intention to purchase. Digital technologies have not only increased consumers' desire to buy, but they have also greatly contributed to their overall happiness. Businesses may meet and surpass customer expectations by offering more personalized and seamless client experiences (Chaker & Zouaoui, 2022; Kelly, et al., 2020).

When customers are happy, they are more likely to return, tell their friends about their experience, and help spread the word about your brand. There are new obstacles and problems that entrepreneurs face as a result of the fast-paced digital innovation. The problems like outdated technology, socio-political conditions, funding issues etc. need special attention from policymakers (Kelly, et al., 2020). When companies take the time to learn about and resolve these issues, they will be able to fully utilize digital technology and digitalization to boost growth, purchase intent, and customer happiness.

7.2.4 Women's Entrepreneurship on Social Media

With limited resources and time, SM provide women with numerous and equal chances to work from home. Due to these reasons, it is noted that a large number of highly educated and competent women are currently enhancing the financial security of their families and countries (Sahu et al., 2024). Further empirical study on digital media, women entrepreneurs, and small enterprises is still needed. It is necessary to have a more thorough and comprehensive grasp of the roles and skills held by female entrepreneurs in performance (Parker and Smith, 2022; Mallios et al., 2023).

It is necessary to gain a more comprehensive and in-depth understanding of the role and skills of female entrepreneurs in SM-based enterprises. The body of knowledge on women entrepreneurs would be enhanced by researching the ways in which they employ social media to interact with clients and improve MSEs' financial success (Fatima and Ali, 2023). Technology development and the increase in social media users have spawned a number of new companies and female entrepreneurs. Social media is being employed by a lot of big businesses in addition to their current traditional marketing techniques. However, due to their limited financial and technical resources and capabilities, many MSEs use social media as their only marketing medium to carry out nearly all marketing activities (Fatima and Ali, 2023).

7.2.5 Green Product and Social Media Enterprise

People in traditional marketing often advertise to the public about their products and services. Product and service promotion takes place online in the cases of social and green marketing. It places a focus on developing environmentally friendly products, using environmentally friendly packaging, and putting marketing techniques into effect to promote sustainable business practices (Mallios et al., 2023).

In general, social media can be defined as online media, where users can share, engage in and produce content in the form of blogs, wikis, forums, social networks and technologically enabled virtual worlds through internet-based apps. The most popular and expanding social media platforms at the moment are blogs, wikis, and social networks.

In order to minimize their negative effects on the environment, green products must take environmental factors into account throughout the life cycle of the product. The goal of the minimization endeavor is to motivate all stakeholders to contribute to the advancement of technologies that will lead to environmentally friendly products. Utilizing a sustainable green product concept is one of the ways that the production sector can generate an environmentally friendly product.

7.2.6 Women's Entrepreneurship Orientation

In India, social considerations play a major role in the success of women entrepreneurs. A psychological component, including the capacity to accept setbacks, is common among Indian women business owners. Firm performance is a necessary condition for entrepreneurial satisfaction and the success of their businesses will be impacted by social, psychological, financial, and resource factors (Saoula et al., 2023; Khoo et al., 2024).

The government and other statutory entities might use some of the recommendations made by the writers to address problems that affect women entrepreneurs (Ratten, 2023; Khoo et al., 2024). Banks should support and encourage female businesses by providing affordable loans. In order for rural women entrepreneurs to use resources for constructive reasons, governments should direct their resource allocation towards them. It is imperative to construct rural areas in India with adequate infrastructure and market connectivity. It is necessary to enhance the understanding of female entrepreneurs in the field through a variety of programs, training, and financial assistance (Ingalagi, 2021).

7.2.7 Women's Entrepreneurship and Sustainability

Business sustainability has faced unprecedented challenges as a result of COVID-19. Previous research has identified a number of critical characteristics that women should cultivate in order to meet the challenges posed by COVID-19

(Ahmetaj et al., 2023; Huang et al., 2022). These included self-efficacy, a readiness to pick up new skills, share existing ones, and embrace newer technologies and procedures, professionalism in reporting and communication, and a sufficient level of competence in the legal and regulatory aspects of their line of work (Ingalagi, 2021). The GEM 2021 research revealed another important finding: women were much less skilled than males in the internet and communication technologies sector, which had the biggest gender disparity in early-stage entrepreneurial activity. These results point to crucial avenues for international development and policy study on strengthening women's emergency preparedness (Rehman et al., 2023).

In emerging markets, entrepreneurship is frequently characterized by a survivalist mindset. Women who were primarily from rural and already disadvantaged communities had fewer opportunities to start their own businesses as a result of COVID-19, which exposed them to even more difficult problems relating to unemployment and poverty. In entrepreneurial eco-systems, governments are expected to act as intermediaries and to foresee and prepare for emergencies and disruptions (Ahmetaj et al., 2023; Huang et al., 2022). The literature on the application of laws and regulations pertaining to female entrepreneurs was conspicuously lacking, particularly in regards to the degree of inspiration and effective support that these women offered during the COVID-19 pandemic in developing nations. It is clear from two successful models that organized and trained women entrepreneurs using a systems thinking approach that scalable solutions are needed to increase resilience in rural communities (Raman, 2022).

One of the most significant challenges faced by women entrepreneurs is a shortage of funding, which is made worse by COVID-19 (Huang et al., 2022). Access to financial resources is essential for innovation and growing business models. Crowdsourcing, microfinance, and risk investment supported by venture capitalists have all been mentioned in a small number of studies as possible survival strategies (Ratten, 2023). A key component of preventing disparities between women and business partners is financial literacy. Furthermore, creating long-term sustainable solutions for society and industry depends on investing in sustainable development (Raman, 2022).

7.3 MOST RELEVANT FACTORS AFFECTING WOMEN'S ENTREPRENEURSHIP ON SOCIAL MEDIA

An evolving and significant trend that has garnered considerable attention in recent years is women's entrepreneurship on social media. The digital landscape presents distinctive opportunities and challenges for female entrepreneurs, who utilize social media platforms to establish, expand, and maintain their enterprises (Emmanuel et al., 2022). A plethora of factors, such as accessibility, marketing tools, community development, educational resources, and socio-cultural dynamics, influence women's commerce on social media (Kreiterling, 2023). In the digital age, women can be even more empowered to succeed as entrepreneurs by addressing issues like funding, digital literacy, and cyberbullying.

7.3.1 Social Factors Affecting Women's Entrepreneurship on Social Media

Social factors are crucial in deciding how well women can use social media to launch and grow their companies.

Social acceptance: The mix of social and cultural elements that influence the success of female entrepreneurs is referred to as socio-cultural aspects. Women's entrepreneurial career decisions are influenced by and centered around a complex web of sociocultural variables (Dsouza & Panakaje, 2023). The way people view women in business is still influenced by traditional gender roles in many areas. Societies that are more accepting of gender equality typically provide a more encouraging atmosphere for female entrepreneurs.

Family support: Family members' emotional support can have a big influence on a woman's decision to become an entrepreneur and how persistent she is when faced with obstacles. Women's business journeys benefit from supportive and encouraging families that recognize their professional goals (Ughetto et al., 2021). Women frequently balance several responsibilities, such as being housewives and carers. Women who have family support in handling household chores have more time and energy to devote to their entrepreneurial endeavors. This kind of assistance can frequently mean the difference between a business's success and failure.

Motivation: One of the main drivers of women's social media entrepreneurship is intrinsic motivation, which is fueled by passion and personal fulfilment (Dsouza & Panakaje, 2023). Women who are driven to create, invent, and accomplish personal objectives are more likely to persevere in the face of difficulties. Social media gives them a forum to share their work, get feedback, and get noticed, which might increase their drive (Rao et al., 2023). Extrinsic elements are also important; these include things like social standing, financial security, and the desire to set an example for others. The need to ensure financial security for their children and themselves motivates a lot of women (Kreiterling, 2023). Social media's prominence can raise a person's social standing and present chances for women to take on leadership roles in their communities

7.3.2 Psychological Factors Affecting Women's Entrepreneurship on Social Media

Higher self-esteem, confidence, and a strong desire to learn are among the psychological characteristics that drive women's entrepreneurship on social media (Rao et al., 2023). These elements are vital in determining how women use social media platforms to take advantage of their benefits and overcome obstacles.

Self-worth and entrepreneurial identity: Increased self-worth has a big impact on how well female businesses do on social media. High self-esteem empowers women to see their value and potential, which is essential for success in the business world. According to Smith and Johnson (2023), women with higher self-esteem are more likely to engage in entrepreneurial activities because they feel more confident in their abilities to succeed. Social media platforms can enhance self-esteem by providing a space for women to receive positive feedback and validation (Rao et al., 2023). This feedback loop is essential as it boosts self-worth and reinforces entrepreneurial identity. For example, receiving likes, comments, and shares on social media can provide immediate validation, enhancing a woman's sense of achievement and encouraging further entrepreneurial efforts (Doe, 2023).

Continuous learning and adaptability: Continuous learning is crucial for staying relevant and competitive in the dynamic social media landscape. Women entrepreneurs who are committed to lifelong learning are better equipped to adapt to changes and innovations. Social media offers vast resources for learning, such as tutorials, webinars, and forums. Jones (2023) highlights that women who continuously seek new knowledge and skills are more likely to succeed in their entrepreneurial endeavors on social media. Women can join online communities, participate in discussions, and collaborate with other entrepreneurs to exchange insights and best practices. This collaborative learning environment significantly enhances their entrepreneurial skills and innovation (Brown & Davis, 2023).

7.3.3 Resource Factors Affecting Women's Entrepreneurship on Social Media

Women's entrepreneurship on social media is influenced significantly by resource factors such as product availability, demand for products, availability of digital facilities, and digital infrastructure. These factors shape the operational capabilities and market reach of women entrepreneurs, enabling or hindering their business success.

- *Product availability:* Product availability is critical for women entrepreneurs who rely on social media platforms to market and sell their goods. Efficient supply chain management ensures that products are available when needed, helping to meet customer demand promptly. A study by Chen and Li (2023) highlights that access to reliable suppliers and efficient inventory management systems are essential for maintaining product availability and customer satisfaction. Women entrepreneurs who manage their supply chains effectively can sustain their businesses by ensuring that their products are always in stock and ready for delivery.
- *Demand for product:* Understanding and predicting demand is crucial for the success of women entrepreneurs on social media. Accurate market research and consumer insights allow entrepreneurs to gauge the demand for their products and tailor their offerings accordingly. The study by Smith and Brown (2023), analyzed the data on consumer engagement and purchasing patterns, women entrepreneurs can better anticipate demand and adjust their inventory and marketing strategies. Lee and Kim (2022) found that real-time monitoring of social media trends helps entrepreneurs quickly adapt their product lines to meet changing consumer preferences, thereby maintaining a competitive edge in the market.
- *Availability of digital facilities:* The availability of digital facilities such as e-commerce platforms, payment gateways, and marketing tools is vital for women

entrepreneurs. These tools facilitate business operations, from setting up online stores to managing transactions and marketing campaigns. According to Garcia and Martinez (2023), women entrepreneurs who utilize comprehensive digital tools can streamline their operations, improve customer experiences, and enhance overall business efficiency.

Digital infrastructure: Reliable internet connectivity is a cornerstone of digital entrepreneurship. Women entrepreneurs need consistent and high-speed internet to manage their online businesses effectively. Poor digital infrastructure can hinder business operations, leading to delays and decreased customer satisfaction. According to Kaur and Singh (2023), regions with robust digital infrastructure see higher rates of entrepreneurial success among women due to better connectivity and access to online resources.

7.3.4 Financial Factor

Financial factors are critical to the success of women entrepreneurs on social media. Key financial factors include support from financial institutions, government support, and business income. These elements can significantly influence the capacity of women entrepreneurs to start, sustain, and grow their businesses.

Support from financial institutions: Access to credit and loans from financial institutions is a crucial factor for women entrepreneurs. Many women face challenges in securing traditional bank loans due to stringent requirements and gender biases. However, financial institutions that provide tailored loan products for women entrepreneurs can help bridge this gap. According to Williams and Thompson (2023), microfinance institutions and banks offering women-specific loan programs have significantly boosted female entrepreneurship by providing necessary capital.

Government support: Smith and Taylor (2023) emphasize that government grants specifically aimed at women entrepreneurs can provide a substantial boost, enabling them to invest in their businesses without the pressure of repayment that comes with loans.

Training and development programs: Governments also offer training and development programs that provide both financial support and skill development. These programs often include financial literacy training, which is crucial for effective business management. According to Lee and Kim (2022), such initiatives help women entrepreneurs better manage their finances, understand market dynamics, and develop sustainable business models.

Business income: The ability to generate consistent business income is fundamental for the sustainability of women-owned businesses. Higher business income not only ensures operational viability but also allows for reinvestment in growth initiatives. Johnson and Brown (2023) highlight that women entrepreneurs who effectively leverage social media platforms can achieve significant revenue growth through direct sales, marketing collaborations, and brand partnerships.

Diverse revenue streams: Diversifying revenue streams is a strategy that many successful women entrepreneurs use to stabilize their income. Social media platforms provide opportunities for multiple income sources, such as e-commerce sales, affiliate marketing, sponsored content, and online courses.

7.3.5 Firm-related Factor

Firm-related factors such as sales growth, profit growth, market share, and return on capital are crucial indicators of the success and sustainability of women entrepreneurs on social media. These factors reflect the financial health and competitive position of their businesses.

Sales growth: Sales growth is a primary indicator of a firm's market traction and customer acceptance. For women entrepreneurs on social media, leveraging digital marketing strategies and influencer collaborations can significantly enhance sales growth. According to Chen and Li (2023), businesses that actively engage in social media marketing experience higher sales growth due to increased visibility and customer engagement.

Customer acquisition and retention: Effective use of social media platforms for customer acquisition and retention strategies is essential for sales growth. Parker and Smith (2022) found that personalized marketing and customer relationship management on social media can lead to a substantial increase in repeat purchases and customer loyalty, driving continuous sales growth.

Profit growth: Profit growth reflects a firm's ability to manage costs and improve operational efficiency while increasing revenue (Suseno & Abbott 2021). Women entrepreneurs who utilize digital tools for inventory management, automation, and streamlined operations can significantly boost profit margins. According to Johnson and Brown (2023), efficient digital operations reduce overhead costs and improve profit growth for women-led businesses on social media.

Market share: Market share indicates a firm's competitive position in the industry. Women entrepreneurs who successfully differentiate their brands and build strong online communities can capture a larger market share. Nguyen et al. (2022) emphasize that unique branding and customer engagement on social media platforms are critical for gaining and maintaining market share.

Return on capital: Return on capital measures the efficiency of a firm in generating profits from its investments. Effective allocation of resources in high-impact areas such as digital marketing and product development can enhance return on capital.

Sustainable growth: Sustainable growth practices, including ethical sourcing and corporate social responsibility, can improve return on capital by attracting socially conscious consumers and investors. Williams and Thompson (2023) argue that sustainability initiatives not only enhance brand reputation but also contribute to long-term financial performance.

TABLE 7.1
Summary of factors affecting women's entrepreneurship on social media towards green product

Factors	Relevant study	Operational definition
Access to capital	Brush, C. G., Carter, N. M., Greene, P. G., et al. (2023)	The availability of financial resources, including loans and grants, to support business ventures on social media.
Networking and social capital	Burt, R. S. (2022)	The connections and relationships that provide support, information, and resources to women entrepreneurs.
Digital literacy	Martin, A., & Grudziecki, J. (2021)	The ability to effectively use digital tools and platforms to promote and manage entrepreneurial activities.
Marketing strategies	Gilmore, A., Carson, D., & Grant, K. (2022)	Techniques and approaches used to promote products or services on social media platforms.
Work–life balance	Jennings, J. E., & McDougald, M. S. (2022)	The equilibrium between personal life and business responsibilities, affecting women entrepreneurs' productivity.
Government and policy support	Minniti, M., & Naudé, W. (2023)	The role of governmental policies and initiatives in supporting or hindering women's entrepreneurship.
Education and training	Kelley, D. J., Singer, S., & Herrington, M. (2022)	Formal and informal educational programs that enhance entrepreneurial skills and knowledge.
Cultural and social norms	Brush, C. G., & Cooper, S. Y. (2022)	Societal attitudes and cultural beliefs that influence women's participation in entrepreneurship.
Technology access	Van Deursen, A. J. A. M., & Van Dijk, J. A. G. M. (2021)	The availability and affordability of technology tools and internet connectivity for entrepreneurial activities.
Role models and mentorship	Bosma, N., Hessels, J., Schutjens, V. (2022)	The influence of successful women entrepreneurs who can inspire and guide aspiring entrepreneurs.

Source: Authors.

7.3.6 SUSTAINABILITY FACTOR

Sustainability factors such as green products and addressing social and environmental concerns are increasingly important for women entrepreneurs, particularly those leveraging social media to promote and sell their products (Bican & Brem, 2020). These factors not only appeal to the growing segment of environmentally conscious consumers but also contribute to long-term business viability and positive social impact.

Green products and eco-friendly innovations: Green products, which are environmentally friendly and sustainably produced, have become a significant focus for many women entrepreneurs. These products appeal to consumers who prioritize sustainability in their purchasing decisions. Women entrepreneurs on social media can highlight the green aspects of their products to attract and retain customers who value sustainability.

Market demand for sustainable products: The demand for green products has been steadily rising, driven by increased consumer awareness and regulatory pressures. Women entrepreneurs who respond to this demand by offering sustainable products can tap into a lucrative market segment. A study by Nguyen et al. (2022) found that products marketed as eco-friendly saw a higher engagement rate on social media platforms, leading to increased sales and brand loyalty.

Corporate social responsibility (CSR): Addressing social and environmental concerns through CSR initiatives is another critical sustainability factor. Women entrepreneurs who engage in CSR activities, such as ethical sourcing, fair trade practices, and community support, can build a positive brand image and foster trust with their customers. Garcia and Martinez (2023) emphasize that CSR initiatives not only contribute to societal well-being but also enhance the reputation and success of businesses.

Environmental stewardship: Environmental stewardship involves taking proactive steps to minimize negative environmental impacts (Ghouse et al., 2021). This includes reducing carbon footprints, minimizing waste, and promoting recycling. Women entrepreneurs who prioritize environmental stewardship can differentiate their brands in a crowded market. According to Lee and Kim (2022), businesses that actively promote their environmental initiatives on social media tend to attract environmentally conscious consumers, thereby enhancing their market position and fostering long-term sustainability.

7.3.7 Moderating Role of Idealized Feminine Selves and Gender Role

The complex interactions between psychological, financial, firm-related, and sustainability aspects that influence the success of women entrepreneurs on social media are modified by the concepts of idealized feminine identities and gender roles (Amigot-Leache et al., 2023). Women are capable of effectively navigating the landscape of entrepreneurship due to psychological factors such as a commitment to lifelong learning, increased self-assurance, and a sense of self-worth. The resources necessary for expansion and sustainability are provided by institutional and governmental funding, as well as robust company income plans (Marlow & McAdam, 2012). The health and competitive position of a company are significantly influenced by market share, sales growth, profit growth, return on capital, and other firm-related metrics. Additionally, sustainability factors, such as the promotion of environmentally benign products and the integration of social and environmental concerns, promote consumer alignment and long-term viability.

Traditional gender roles and idealized feminine selves significantly moderate these characteristics. Female entrepreneurs frequently create their online personas to adhere to social norms, which can both facilitate and impede their business operations

(Amigot-Leache et al., 2023; Marlow & McAdam, 2012). By employing their idealized feminine selves, they can establish businesses that are both relatable and trustworthy to their target audiences. Nevertheless, the ability to fully leverage opportunities and resources may be diminished by the challenges that ingrained gender norms present, such as the need to balance business endeavors with family responsibilities.

The convergence of these variables, which are moderated by gender dynamics, underscores the intricate nature of women's social media entrepreneurship (Amigot-Leache et al., 2023). A nurturing environment that addresses the structural and psychological obstacles that women entrepreneurs face is essential for their empowerment, as is a comprehensive understanding of these factors. By challenging gender conventions that restrict women's business success and societal contributions in the digital era, we can establish conditions that promote financial independence, lifelong learning, and sustainable practices.

7.4 CONCEPTUAL MODEL DEVELOPMENT

The theory of planned behavior (TPB) can be utilized to investigate the characteristics that impact the adoption of social media by women business owners. Popular in the field of psychology, the TPB seeks to explain why people do things like adoption of new technologies. Perceived behavioral control, attitude, and subjective norms are the three components that make up the TPB (Rehman et al., 2023). When it comes to women-owned businesses, their perspective on digital platforms might be influenced by elements like their drive, ambition, and past experiences (Suseno & Abbott 2021). Furthermore, the impact of social factors, psychological factors, resource factors, financial factor, firm-performance related factor and technological factor is taken as determinant to investigate sustainability. Prior studies found that the social factors, psychological factors, resource factors, financial factor, firm-performance related factor and technological factor contribute to bring entrepreneurial orientation among women which in return bring sustainable practices in business. Figure 7.1 shows the conceptual model based on the current literature analysis in this study.

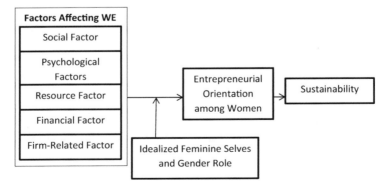

FIGURE 7.1 Factors affecting women's entrepreneurship on social media to achieve sustainability. (Developed by authors.)

It is implied that opportunities, resources, and knowledge can be accessed through social interactions and connections. Social media platforms allow for networking, collaboration, and the formation of social capital. This theory may be applied to the context of women entrepreneurs by exploring these ways in which these platforms work. It can also explain how social media can be useful for finding mentors, getting business advice, and connecting with support groups.

7.5 DISCUSSION

Over a period of last decade, it is observed that wide range of women's entrepreneurship is involved in different sectors. The factors that influence women's entrepreneurial orientation with their business and entrepreneurial performance are the focus of this research. The study examines the effect of these factors on enterprise sustainability. A driven entrepreneur puts in long hours, earns a good living wage, and provides for his or her family. Hence, society supports women business owners. The study's findings highlight the significance of financial variables on businesswomen's stainability practices (Ghouse et al., 2021). The success of enterprises is influenced by government financial help and family support. Having access to financial institutions or family members ensures that they can enjoy the monetary benefits of enterprises, which is a key indicator of their level of sustainability and, by extension, their continued relevance. Inadequate infrastructure, finding suitable storage facilities, obtaining information about the market and its worth, and sourcing raw materials are just a few of the many challenges faced by women entrepreneurs. When women business owners kept meticulous records of their company's progress—including sales growth, market share, profit, and capital return—they believed their ventures had succeeded.

7.5.1 Implications

This research adds to what is already known about women business owners, particularly in India. There are a lot of societal aspects that play a role in the success of female entrepreneurs in India. The capacity to deal with setbacks is a psychological component that is important for women business owners in India. The success of a business is crucial to bring out sustainability. In this conceptual study, the authors taken into account a number of factors that impact female entrepreneur's orientation. Women business owners' happiness and success are affected by monetary, social, psychological, and resource-related aspects, according to the research. In order to address the challenges that women business owners face, the authors have put up several recommendations that other statutory organizations and the government can implement (Bican & Brem, 2020). By providing affordable loans, banks can support female entrepreneurs and boost their businesses. Governments should direct their financial support towards rural women entrepreneurs so that they can put those funds to good use. Proper infrastructure facilities and market connectivity should be constructed in rural parts of India. It is necessary to equip female entrepreneurs with domain knowledge through training and other programs that provide financial support.

The uniqueness of this study lies in its evaluation of the impact of gender-role inclination and perceived cultural support on the progression of entrepreneurial ideal careers. The findings suggest that females who identify as having a masculine orientation are more inclined to pursue entrepreneurial endeavors, particularly in environments where they perceive a strong support for such endeavors. Thus, the study has proven that moderating role of gender orientations and idealized feminine orientation must be firmly included in the discussion of women's entrepreneurship if it is to be completely acknowledged. Researchers interested in women business owners will find this study useful.

7.5.2 Future Research Directions

The conceptual model presented in the chapter can be tested empirically by future researchers. Expanding the study to include additional regions of India by increasing the sample size will validate the model and we can then be surer about the relationship of WEO and Enterprise sustainability. Any other industry, like SMEs, might be used to expand the study. Since psychological traits are an expression of the desire for strength and belonging, they play an important role in satisfying businesswomen, which brings us to another aspect of women entrepreneurs: the relationship between psychological factors and their satisfaction can be studies at various cultural contexts. Because WE aren't afraid to try new things, entrepreneurs can put innovative ideas into action if they can just acquire new skills. An additional possibility is to conduct a comparative research by adding more variables that were not considered in the study might allow it to be more comprehensive. Psychologically, they are resilient enough to keep going even when things go wrong. However, the gender-role hinders their growth in emerging markets. Therefore future studies can take up the factors to study on various sectors and bring our useful results for policy makers.

7.6 CONCLUSION

The research study identifies social factors, psychological factors, resource factors, financial factor, firm-performance related factor and technological factor associated with WE to achieve sustainability. Realizing the full potential of women's entrepreneurship in India will need a multidimensional strategy. This means removing sociocultural obstacles, advancing access to markets and finance, and endorsing gender-sensitive laws. Promoting the adoption of technology and digital literacy of women entrepreneurs will speed up their business growth even more. Promotion of global cooperation and joint ventures between the public and commercial sectors can also provide women entrepreneurs with the necessary assistance to grow in a competitive and sustainable way. Finally, the development of women's entrepreneurship in India is being driven by digital innovation and a growing focus on sustainability. Though much has been achieved, understanding the complex roles and difficulties encountered by women entrepreneurs in an ever-changing environment is still essential. Future study should focus on analyzing these aspects and provide knowledge that may guide practice and policy to empower and assist women entrepreneurs. We shall then be able to support inclusive and sustainable economic growth.

REFERENCES

Ahmetaj, B., Kruja, A. D., & Hysa, E. (2023). Women entrepreneurship: Challenges and perspectives of an emerging economy. *Administrative Sciences*, 13(4), 111. https://doi.org/10.3390/admsci13040111

Alhajri, A., & Aloud, M. (2024). Female digital entrepreneurship: A structured literature review. *International Journal of Entrepreneurial Behavior & Research*, 30(2–3), 369–397. https://doi.org/10.1108/IJEBR-03-2024-062

Alhakimi, W., & Albashiri, S. (2023). Social media adoption by women entrepreneurial small businesses. *Asia Pacific Journal of Innovation and Entrepreneurship*, 17(3–4), 158–175. https://doi.org/10.1108/APJIE-03-2023-0060

Al-Mamary, Y. H., & Alshallaqi, M. (2022). Impact of autonomy, innovativeness, risk-taking, proactiveness, and competitive aggressiveness on students' intention to start a new venture. *Journal of Innovation and Knowledge*, 7(4), 100239. https://doi.org/10.1016/j.jik.2022.100239

Amigot-Leache, P., Carretero-García, C., & Serrano-Pascual, A. (2023). The limits of "no limits": Young women's entrepreneurial performance and the gendered conquest of the self. *Ethos*, 51(3), 285–304. https://doi.org/10.1111/etho.12398

Bican, P. M., & Brem, A. (2020). Digital business model, digital transformation, digital entrepreneurship: Is there a sustainable "digital"?. *Sustainability*, 12(13), 5239. https://doi.org/10.3390/su12135239

Brown, A., & Davis, L. (2023). Collaborative learning in digital entrepreneurship: The role of social media. *Journal of Business Research*, 136, 112–120. https://doi.org/10.51505/IJEBMR.2022.6608

Chaker, H., & Zouaoui, S. (2022). Meeting the challenge of entrepreneurship with social media: The case of Tunisian women entrepreneurs. *Journal of Entrepreneurship and Innovation in Emerging Economies*, 9(1), 33–61. https://doi.org/10.1177/23939575221138439

Chen, X., & Li, Y. (2023). Supply chain management for small businesses: Ensuring product availability. *Journal of Business Logistics*, 44(1), 67–85. https://doi.org/10.3390/logistics7040070

Doe, J. (2023). Empowerment through social media: Enhancing self-esteem among women entrepreneurs. *Journal of Digital Marketing*, 17(3), 56–72. https://doi.org/10.3390/su12135239

Dsouza, A., & Panakaje, N. (2023). Factors affecting women entrepreneurs' success: A study of small and medium-sized enterprises–A review. *International Journal of Case Studies in Business, IT and Education (IJCSBE)*, 7(2), 51–89. https://doi.org/10.47992/IJCSBE.2581.6942.0260

Emmanuel, C. P., Qin, S., Hossain, S. F. A., & Hussain, K. (2022). Factors influencing social-media-based entrepreneurship prospect among female students in China. *Heliyon*, 8(12). https://doi.org/10.1016/j.heliyon.2022.e12041

Fatima, N., & Ali, R. (2023). How businesswomen engage customers on social media?. *Spanish Journal of Marketing-ESIC*, 27(2), 221–240. https://doi.org/10.1108/SJME-09-2021-0172

Franzke, S., Wu, J., Froese, F. J., & Chan, Z. X. (2022). Female entrepreneurship in Asia: A critical review and future directions. *Asian Business and Management*, 21(3), 343–372. https://doi.org/10.1057/s41291-022-00186-2

Garcia, R., & Martinez, S. (2023). The impact of corporate social responsibility on brand loyalty among women entrepreneurs. *Journal of Business Ethics*, 175(1), 98–115. https://doi.org/10.3390/su15031781

Ghouse, S. M., Durrah, O., & McElwee, G. (2021). Rural women entrepreneurs in Oman: Problems and opportunities. *International Journal of Entrepreneurial Behavior and Research*, 27(7), 1674–1695. https://doi.org/10.1108/ijebr-03-2021-0209

Huang, Y., Li, P., Wang, J., & Li, K. (2022). Innovativeness and entrepreneurial performance of female entrepreneurs. *Journal of Innovation and Knowledge*, 7(4), 100257. https://doi.org/10.1016/j.jik.2022.100257

Ingalagi, S. S., Nawaz, N., Rahiman, H. U., Hariharasudan, A., & Hundekar, V. (2021). Unveiling the crucial factors of women entrepreneurship in the 21st century. *Social Sciences*, 10(5), 153. https://doi.org/10.3390/socsci10050153

Johnson, E., & Brown, P. (2023). Profit growth strategies for women entrepreneurs on social media. *Journal of Digital Entrepreneurship*, 11(1), 67–83. https://doi.org/10.1145/3579619

Jones, R. (2023). Lifelong learning and entrepreneurial success: Insights from women on social media. *Entrepreneurship & Innovation*, 24(2), 98–115. https://doi.org/10.47992/IJCSBE.2581.6942.0260

Kaur, P., & Singh, R. (2023). The impact of digital infrastructure on entrepreneurial success. *Digital Business Review*, 10(1), 45–62. https://doi.org/10.1016/j.jretconser.2023.103395

Kelly, L., Kimakwa, S., & Brecht, S. (2020). Building entrepreneurial community: A collaborative benefit corporation for women empowering women. In Poonamallee, L., Scillitoe, J., & Joy, S. (eds) *Socio-Tech Innovation*. Cham: Palgrave Macmillan, pp. 239–260. https://doi.org/10.1007/978-3-030-39554-4_13

Khoo, C., Yang, E. C. L., Tan, R. Y. Y., Alonso-Vazquez, M., Ricaurte-Quijano, C., Pécot, M., & Barahona-Canales, D. (2024). Opportunities and challenges of digital competencies for women tourism entrepreneurs in Latin America: A gendered perspective. *Journal of Sustainable Tourism*, 32(3), 519–539.

Kreiterling, C. (2023). Digital innovation and entrepreneurship: A review of challenges in competitive markets. *Journal of Innovation and Entrepreneurship*, 12(1), 49. https://doi.org/10.1186/s13731-023-00320-0

Lee, H., & Kim, S. (2022). Environmental stewardship and market positioning for women-led businesses. *Journal of Environmental Management*, 304, 113–129. https://doi.org/10.1177/19389655231164064

Lee, H., & Kim, S. (2023). Product diversification and profitability in women-owned businesses. *Journal of Business Strategy*, 34(2), 76–92. https://doi.org/10.3390/su11113076

Mallios, P., Zampetakis, L., & Moustakis, V. (2023). Social media impact on entrepreneurship intention: Lessons learned from business startuppers. *Journal of Business and Entrepreneurship*, 32(2), Article 2. https://repository.ulm.edu/jbe/vol32/iss2/2

Marlow, S., & McAdam, M. (2012). Analyzing the influence of gender upon high–technology venturing within the context of business incubation. *Entrepreneurship Theory and Practice*, 36(4), 655–676.

Nguyen, T., & Thompson, L. (2022). Consumer engagement with sustainable products on social media. *Journal of Marketing Science*, 40(3), 225–243.

Olsson, A. K., & Bernhard, I. (2021). Keeping up the pace of digitalization in small businesses–Women entrepreneurs' knowledge and use of social media. *International Journal of Entrepreneurial Behavior & Research*, 27(2), 378–396. https://doi.org/10.1108/IJEBR-10-2019-0615

Okadiani, N. L. B., Mitariani, N. W. E., & Imbayani, I. G. A. (2019). Green product, social media marketing and its influence on purchasing decisions. *International Journal of Applied Business and International Management (IJABIM)*, 4(3), 69–74. https://doi.org/10.32535/ijabim.v4i3.684

Parker, J., & Smith, K. (2022). Customer retention strategies for sales growth on social media. *Journal of Business Management*, 15(4), 202–218.

Raman, R., Subramaniam, N., Nair, V. K., Shivdas, A., Achuthan, K., & Nedungadi, P. (2022). Women entrepreneurship and sustainable development: Bibliometric analysis and emerging research trends. *Sustainability*, 14(15), 9160. https://doi.org/10.3390/su14159160

Rani, A., & Shivaprasad, H. N. (2021). Revisiting the antecedent of electronic word-of-mouth (eWOM) during COVID-19 Pandemic. *Decision*, 48(4), 419–432. https://doi.org/10.1007/s40622-021-00298-2

Rao, S. A., Abdul, W. K., Kadam, R., & Singh, A. (2023). Factors affecting the performance of micro-level women entrepreneurs: A comparative study between UAE and India. *Measuring Business Excellence*, 27(3), 460–482. https://doi.org/10.1108/MBE-02-2022-0034

Ratten, V. (2023). Entrepreneurship: Definitions, opportunities, challenges, and future directions. *Global Business and Organizational Excellence*, 42(5), 79–90. https://doi.org/10.1002/joe.22217

Rehman, W., Yosra, A., Khattak, M. S., & Fatima, G. (2023). Antecedents and boundary conditions of entrepreneurial intentions: Perspective of theory of planned behavior. *Asia Pacific Journal of Innovation and Entrepreneurship*, 17(1), 46–63. https://doi.org/10.1108/apjie-05-2022-0047

Sahu, T. N., Agarwala, V., & Maity, S. (2024). Effectiveness of microcredit in employment generation and livelihood transformation of tribal women entrepreneurs: Evidence from PMMY. *Journal of Small Business & Entrepreneurship*, 36(1), 53–74. https://doi.org/10.1080/08276331.2021.1928847

Saoula, O., Shamim, A., Ahmad, M. J., & Abid, M. F. (2023). Do entrepreneurial self-efficacy, entrepreneurial motivation, and family support enhance entrepreneurial intention? The mediating role of entrepreneurial education. *Asia Pacific Journal of Innovation and Entrepreneurship*, 17(1), 20–45. https://doi.org/10.1108/apjie-06-2022-0055

Shane, S., & Venkataraman, S. (2000). The promise of entrepreneurship as a field of research. *Academy of Management Review*, 25(1), 217–226. https://doi.org/10.5465/amr.2000.2791611

Smith, A., & Brown, K. (2023). Leveraging social media analytics for market research. *Journal of Marketing Analytics*, 14(1), 32–50. https://doi.org/10.1111/isj.12327

Suseno, Y., & Abbott, L. (2021). Women entrepreneurs' digital social innovation: Linking gender, entrepreneurship, social innovation and information systems. *Information Systems Journal*, 31(5), 717–744. https://doi.org/10.1111/isj.12327

Thompson, M., Parker, R., & Richards, S. (2023). Psychological traits and entrepreneurial performance: The moderating role of social media. *Journal of Business Psychology*, 38(1), 75–93. https://doi.org/10.1002/job.1794

Ughetto, E., Rossi, M., Audretsch, D., & Lehmann, E. E. (2020). Female entrepreneurship in the digital era. *Small Business Economics*, 55, 305–312. https://doi.org/10.1007/s11187-019-00298-8

Williams, D., & Thompson, B. (2023). Financial institution support for female entrepreneurs: Challenges and opportunities. *Journal of Banking & Finance*, 140, 105–120. https://doi.org/10.3390/su15031783

8 Women's Entrepreneurship and Changing Landscape in Africa
Strategies and Solutions

Benneth Uchenna Eze, Festus Ekechi, Oluyemisi Agboola, Iyabode Abisola Adelugba, and Adeola Abosede Oyewole

8.1 INTRODUCTION

Emerging businesses, a burgeoning spirit of entrepreneurship, and a rise in innovation are all signs of a fundamental transition in Africa's women entrepreneurial scene. People all around the continent are grabbing the chance to start companies that not only solve regional problems but also advance the continent's economic growth (Odeyemi et al., 2024; Mlambo et al., 2022). Africa has a thriving and diversified entrepreneurial landscape, encompassing both tech-driven startups and businesses with roots in conventional sectors.

This increase is especially noticeable in light of demographic changes, as a growing number of young women and active people see entrepreneurship as a means of achieving economic independence (Odeyemi et al., 2024). Gobena and Kant (2022) posit that a massive wave of innovative thinking and creativity is transforming the face of business throughout the continent as urbanization picks up speed and technological advances becomes more widely available. In light of the current state of the global economy, studying entrepreneurial activities of women in Africa is crucial.

The continent offers special possibilities and problems that demand careful analysis because of its various markets, unrealized potential, and resilient women entrepreneurs. Beyond its effects on the economy, African entrepreneurship is essential for tackling social problems, promoting inclusive growth, and realizing the potential of regional communities. Furthermore, the story of African women's entrepreneurship dispels myths and emphasizes how crucial it is to comprehend business dynamics in the context of various cultural, economic, and infrastructure contexts. Acknowledging and understanding the complexities of African women's entrepreneurship is crucial for a comprehensive grasp of the global economic tapestry at a period where economies around the globe are becoming more intertwined (Odeyemi et al., 2024).

The story of African women entrepreneurs demonstrates how this demographic is propelling the nano, micro, and small enterprises sector of the informal economy. In Africa, women are perceived as having roles in both reproduction and productivity, which support the well-being of the family and the country's economy, respectively. They continue to experience discrimination in the job and stark gender disparities, despite the recognition of their positive contributions to society. However, as more women take on bigger positions in big businesses and even global corporations, the scene in Africa is shifting.

This research endeavors to elucidate the reasons behind the evolving African entrepreneurship scene while also putting forward tactics and fixes to deal with the shifting African business environment. The research aims to address the following queries: What factors are causing the landscape of African women entrepreneurs to change? What challenges do Africa women entrepreneurs face? Which strategies and solutions are available for African women entrepreneurs to use?

8.2 LITERATURE REVIEW

8.2.1 African Entrepreneurship Landscape

Africa's entrepreneurship has undergone a paradigm change, driven by a number of variables that come together to support its rapid expansion. The dynamics of population and urbanization, government programs, foreign alliances, and the prevailing innovation culture all contribute to this rise. The young demography and fast urbanization of Africa's people play a major role in propelling the rise of entrepreneurship throughout the continent. The growing population of the continent results in a broad market with a varied customer base. More people are moving into cities as urbanization picks up speed, creating congested markets with rising demand for products and services. According to Odeyemi et al. (2024), business owners are taking advantage of this demographic benefit by figuring out what the demands of this growing urban population are.

Policies and programs from the government are essential for encouraging the expansion of women's entrepreneurship. Numerous African governments actively assist new and small women enterprises by offering incentives and specialized programs. Financial assistance, programs for mentoring, and links to incubation facilities are common forms of this help. Governments encourage innovation and risk-taking by empowering entrepreneurs to overcome early hurdles via the creation of enabling environments. A robust women entrepreneurial ecosystem is fostered by policies that facilitate firm registration, give grants, and offer tax advantages (Odeyemi et al., 2024; Wei, 2022). The government's attempts to draw in investments serve as further fuel for the expansion of women's entrepreneurship. Tax rebates, less regulatory barriers, and investment-friendly policies are some of the incentives that draw in both local and overseas investors.

In addition, governments are essential in building infrastructure, improving connectivity, and fostering business-friendly environments. These initiatives help to create an environment that supports entrepreneurship and lowers operating expenses. The rise of entrepreneurship in Africa is mostly driven by international relationships. The region's entrepreneurial landscape benefits from the many viewpoints, resources, and skills that partnerships with foreign firms and organizations bring to the table (Odeyemi et al., 2024). Forming strategic partnerships with foreign companies opens up new markets, gives access to international networks, and makes it easier to share best practices and information.

The absorption of Africa's expanding labor force is largely dependent on the growth of entrepreneurship-driven nano, micro, small, and medium-sized companies (NMSMEs). Policymakers and other stakeholders may further amplify the good effects of entrepreneurship on employment creation, especially for the continent's rapidly growing youth population, by creating a climate that is favorable to company development (Oriji et al., 2023). Growth in women's entrepreneurship in Africa pushes economies away from an over-reliance on conventional industries and promotes economic diversification.

Although traditionally agriculture and the extractive industries have dominated most economies in the continent, diversification into new sectors is currently being led by entrepreneurial activity. The creative and technological industry is expanding rapidly as a result of startups and tech-driven corporations upending established sectors and establishing new ones. Notable enterprises in financial technology, electronic commerce, and green energy contribute to a more diverse economic environment (Odeyemi et al., 2024). In addition to increasing economic resilience, this diversification puts African nations in a position to profit from new international trends.

The creative industries, which include the arts, entertainment, and fashion are seeing an increase in the number of entrepreneurs. These endeavors act as cultural ambassadors, showcasing Africa's rich legacy and inventiveness on a worldwide scale in addition to aiding in economic diversification. It is also important to state that there is high presence of women entrepreneurs in the creative industry (Odeyemi et al., 2024; Adaga et al., 2024). African entrepreneurs may help create a more sustainable and well-balanced economic portfolio by fostering these businesses. Beyond its effects on the economy, entrepreneurship is a potent force for community development and social empowerment in Africa. When business people are successful, they not only better themselves but also have a good knock-on impact on the communities in which they operate (Rayburn & Ochieng, 2022).

By enabling people to take charge of their financial futures, entrepreneurship promotes a culture of independence and resiliency. In underprivileged groups, where entrepreneurship may be a means of reducing poverty and promoting social mobility, this empowerment is especially important for women. Through offering guidance, coaching, and resource access, entrepreneurs help create a strong ecosystem that backs community-based projects.

Narrowing gender inequalities and promoting inclusion are two things that entrepreneurship helps with. As role models, successful female entrepreneurs push social conventions and encourage other women to follow their dreams of becoming business owners. Through proactive promotion of diversity and inclusiveness, business transforms into a catalyst for social transformation, dismantling obstacles and cultivating a fairer community. The rise of entrepreneurship in Africa has a revolutionary effect on the continent's economy and society. Building robust and dynamic societies is facilitated by the expansion of job possibilities, economic diversity, and social empowerment taken together. The crucial role that women's entrepreneurship plays in promoting sustainable development throughout the continent must be acknowledged and supported by policymakers, stakeholders, and the international community (Odeyemi et al., 2024; Danladi et al., 2023).

8.3 CHALLENGES FACING AFRICAN WOMEN ENTREPRENEURS

African women entrepreneurs work in a dynamic environment with enormous promise and difficult obstacles. This assessment examines the various obstacles that African women entrepreneurs must overcome, including deficiencies in infrastructure, restricted access to capital, regulatory impediments, and the effects of political instability. For African business owners, inadequate infrastructure, especially in the areas of transportation and electricity supply, presents serious difficulties. The entrepreneurial activities of women entrepreneurs are hampered by erratic electricity supplies, poor transportation networks, a lack of capital, educational inequality, gender bias, and the underrepresentation of women in leadership roles across a range of industries and governments in Africa (Odeyemi et al., 2024; Ndubuisi-Okolo et al., 2023; Eze, 2022).

One of the main issues that women in business confront is gender bias and stereotyping, which can take many various forms, including discrimination, prejudice, and unconscious prejudices. These prejudices frequently support gender-based stereotypes, which hinders women's access to resources, job development prospects, and professional growth. The underrepresentation of females in top roles throughout Africa's governments and businesses presents another difficulty. The phenomenon known as the "glass ceiling" describes the imperceptible obstacles that keep women from rising to positions of high leadership. The perception that women are less competent in leadership posts is reinforced by this lack of representation, which also restricts the recognition of women role models (Ndubuisi-Okolo et al., 2023).

Ndubuisi-Okolo et al. (2023) opine that one of the biggest obstacles facing women in business today is finding a work-life balance. Women are frequently put in a tough situation when juggling their desire for a job with their obligations to their families and society expectations. This can lead to them having to make decisions that could limit their ability to advance in their careers (Glass & Cook, 2016). This problem is made worse by the absence of supporting policies as well as flexible work schedules, which commonly compel women to choose between advancing their careers and finding personal contentment. For women in business, juggling caregiving and family duties with career obligations presents serious obstacles. Work–life conflicts are a common result of the traditional assumption that women should take main responsibility for childcare, which makes it challenging for women to devote themselves entirely to their job.

Unreliable power supplies cause delays and higher expenses by upsetting manufacturing schedules. Inadequate transportation infrastructure complicates logistics and impedes the flow of commodities, which impacts the supply chain as a whole. The efficiency of operation of enterprises is a direct consequence of infrastructural deficiencies. Investments in alternate energy sources are required due to frequent power outages, which puts a burden on limited financial resources. Increased wait times resulting from transportation constraints have an impact on inventory management and production schedules (Eze, 2018).

All of these difficulties add up to increased manufacturing costs, diminished competitiveness, and inefficient operations. A prevalent obstacle for African businesses is the restricted availability of funding. Enterprising individuals, especially those from underprivileged backgrounds, have obstacles while trying to obtain funding for their fledgling businesses. Many prospective entrepreneurs that lack these requirements are turned away by financial institutions because they frequently demand collateral and evidence of trustworthiness. Building business ownership in Africa requires financial inclusion. The gap may be closed by putting policies that support inclusive funding into practice, such as government-backed guarantee programs and microfinance efforts (Eze et al., 2019a; Eze et al., 2019b).

Giving women business owners access to finance and capital promotes not just the expansion of their enterprises but also the creation of jobs and economic growth (Adenutsi, 2023; Eze et al., 2019b). For entrepreneurs in Africa, navigating complicated regulatory regimes is a major problem. Uncertainty is increased and company development is impeded by onerous bureaucratic procedures, unclear laws, and uneven enforcement (Oladimeji et al., 2018; Oladimeji & Eze, 2017). It may be difficult for women entrepreneurs to understand and follow regulations, which can cause legal issues and perhaps impede corporate operations. It takes coordinated efforts to simplify procedures for ease of doing business in order to remove regulatory impediments.

8.4 STRATEGIES AND SOLUTIONS TOWARDS ADDRESSING AFRICAN ENTREPRENEURSHIP CHALLENGE

Odeyemi et al., 2024 argue that in order to cut down on bureaucratic red tape, governments throughout Africa should conduct thorough examinations of their current rules and simplify them. Entrepreneurship will be promoted by streamlining the licensing, compliance, and business registration procedures. Enacting laws that entice both international and domestic investors will boost the economy. Tax discounts, subsidies, and less regulatory barriers are examples of incentives that can create an atmosphere that is more favorable to company development.

The growth of microfinance organizations should be encouraged to give modest loans to female business owners who may be without access to conventional banking systems. To reach a wider range of entrepreneurs, governments along with lenders may work with fintech businesses to build creative solutions like digital lending platforms and mobile banking. By reducing the risks involved, government-backed guarantee schemes can be established to incentivize financial institutions to provide financing to nano, micro, small, and medium-sized enterprises (NMSMEs).

Career training programs should be improved to emphasize women's entrepreneurial abilities, encouraging practical learning and problem-solving in the actual world. To guarantee that women's entrepreneurship education stays pertinent to the changing business environment, cooperation should be promoted between academic institutions and industry participants. Laws should be adopted that explicitly address gender inequality to guarantee women business owners equitable access to capital, resources, and opportunities.

Mentoring programs should be created to pair up prospective female business owners from underrepresented groups with knowledgeable mentors who can offer advice and assistance. Diversity should be promoted by organizing networking events as well as platforms that provide entrepreneurs from all backgrounds the chance to collaborate and develop partnerships. African countries may foster an atmosphere that is more conducive to the success of women entrepreneurs by putting these principles into practice. In addition to addressing current issues, these actions create the groundwork for an inclusive and sustainable business environment that will promote social and economic advancement throughout Africa (Oladimeji et al., 2018).

Governments can save administrative costs by implementing clear and uncomplicated regulatory frameworks. According to Safitra et al. (2023), periodic evaluations and modifications of current legislation guarantee their pertinence and conformity with changing commercial situations. Investing in online tools for regulatory compliance may speed up procedures even more, creating an atmosphere that is more conducive to business. In Africa, political instability is a constant danger to investor trust and business continuity. An unpredictable business climate is brought about by frequent changes in administration, civil instability, and geopolitical risks.

Long-term planning presents difficulties for entrepreneurs, and unstable political environments may make investors reluctant to make financial commitments, which might impede overall economic growth. Adaptive tactics are necessary for enterprises to handle political risks. According to Tohİnean et al. (2020), it is imperative to diversify activities across geographies, maintain adaptable business models, and conduct comprehensive risk assessments. Resilience can be improved by creating backup plans that take political unrest-related interruptions into consideration. Initiatives in the areas of infrastructure development, equitable financing, fewer restrictions, and adaptable strategies in the face of political instability are crucial if the continent is to fully exploit its vast entrepreneurial potential.

Furthermore, proactive interaction with stakeholders and local communities promotes resilience against the effects of political unpredictability by creating a network of support. African entrepreneurs face a wide range of complex issues. Deficits in infrastructure, restricted financial resources, regulatory obstacles, and unstable political environments all provide challenges requiring calculated actions. To overcome these issues and create an atmosphere where entrepreneurs may flourish, innovate, and make a substantial contribution development of Africa's economic growth, governments, the private sector, and international players must work together (Opute et al., 2021).

8.5 CONCLUSION

A number of variables have combined to create a remarkable increase in Africa's entrepreneurial landscape. Significant business potential has been created by the continent's growing population and rising urbanization. The rise of entrepreneurship has also been greatly aided by international cooperation, government programs, and a developing innovation culture, notably in the technology industry. But this expansion is accompanied by a number of difficulties that African women entrepreneurs are confronted with.

Strong barriers are created by poor infrastructure, restricted financial options, complicated regulations, gender discrimination, and unstable political environments. Aspiring women entrepreneurs confront additional difficulties due to socio-cultural variables including gender and educational inequality. Africa may fully realize the endless possibilities of its entrepreneurial landscape by putting into practice customized strategies that take into account the various socio-economic and cultural factors.

By implementing these strategies, which include diversity initiatives, family-friendly policies, and focused support programs, we can create an inclusive workplace that advances gender equality, encourages innovation, spurs economic growth, and gives women the tools they need to succeed in their entrepreneurial activities and contribute to a more inclusive future. Furthermore, to fully realize the continent's enormous entrepreneurial potential, initiatives in the areas of infrastructure development, equitable finance, reduced regulations, and flexible tactics in the face of political unpredictability are essential.

REFERENCES

Adaga, E.M., Egieya, Z.E., Ewuga, S.K., Abdul, A.A., & Abrahams, T.O. (2024). Philosophy in business analytics: A review of sustainable and ethical approaches. *International Journal of Management & Entrepreneurship Research*, 6(1), 69–86.

Adenutsi, D.E. (2023). Entrepreneurship, job creation, income empowerment and poverty reduction in low-income economies. *Theoretical Economics Letters*, 13(6), 1579–1598.

Danladi, S., Prasad, M.S.V., Modibbo, U.M., Ahmadi, S.A., & Ghasemi, P. (2023). Attaining sustainable development goals through financial inclusion: Exploring collaborative approaches to fintech adoption in developing economies. *Sustainability*, 15(17), 1–14. https://doi.org/10.3390/su151713039

Eze, B.U. (2018). Corporate entrepreneurship and manufacturing firms' performance. *Emerging Market Journal*, 8(1), 12–17.

Eze, B.U. (2022). Effect of innovation on agricultural-based micro, small and medium enterprises survival in Nigeria. *World Review of Entrepreneurship, Management and Sustainable Development*, 18(5–6), 500–513.

Eze, B.U., Oladimeji, M.S., & Fayose, J. (2019a). Entrepreneurial orientation and micro, small and medium enterprises (MSMEs) performance in Abia State, Nigeria. *Covenant Journal of Entrepreneurship*, 3(1), 19–35.

Eze, B.U., Adelekan, A.S., Udoh, I.P., & Sunday, D.K. (2019b). Determinants of women entrepreneurship in South-West, Nigeria. *NOUN Journal of Management and International Development*, 5(1).

Glass, J., & Cook, A. (2016). Leading with their hearts? How gender stereotypes of emotion lead to biased evaluations of female leaders. *The Leadership Quarterly, 27*(3), 415–428.

Gobena, A.E., & Kant, S. (2022). Assessing the effect of endogenous culture, local resources, eco-friendly environment and modern strategy development on entrepreneurial development. *Journal of Entrepreneurship, Management, and Innovation, 4*(1), 118–135.

Mlambo, D.N., Manganyi, T.P., & Mphurpi, J.H. (2022). Re-examining the notion of local economic development (LED) post democratization: Anticipated outputs, impediments and future expectation (s). *EUREKA: Social and Humanities, 1*, 31–42.

Ndubuisi-Okolo, P.U., Dibua, E.C., & Akaegbobi, G.N. (2023). Challenges faced by women in business and possible solutions. *African Banking and Finance Review Journal, 5*(5), 211–216.

Odeyemi, O., Oyewole, A.T., Adeoye, O.B., Ofodile, O.C., Addy, W.A., Okoye, C.C., & Ololade, Y.J. (2024). Entrepreneurship in Africa: A review of growth and challenges. *International Journal of Management & Entrepreneurship Research, 6*(3), 608–622.

Oladimeji, M.S., & Eze, B.U. (2017). Determinants of born-global firms: Evidence from Nigeria. *Izvestiya Journal of Varna University of Economics, 61*(4), 377–392.

Oladimeji, M.S., Eze, B.U., & Adebayo, A.A. (2018). Strategic renewal, corporate venturing and internationalization of banks in Nigeria. *Lapai Journal of Management Science, 8*(1), 63–72.

Opute, A.P., Kalu, K.I., Adeola, O., & Iwu, C.G. (2021). Steering sustainable economic growth: Entrepreneurial ecosystem approach. *Journal of Entrepreneurship and Innovation in Emerging Economies, 7*(2), 216–245.

Oriji, O., Shonibare, M.A., Daraojimba, R.E., Abitoye, O., Daraojimba, C. (2023). Financial technology evolution in Africa: A comprehensive review of legal frameworks and implications for AI-driven financial services. International Journal of Management & Entrepreneurship Research, 5(12), 929–951. https://doi.org/10.51594/ijmer.v5i12.627

Rayburn, S.W., & Ochieng, G. (2022). Instigating transformative entrepreneurship in subsistence communities: Supporting leaders' transcendence and self-determination. *Africa Journal of Management, 8*(3), 271–297.

Safitra, M.F., Lubis, M., & Fakhrurroja, H. (2023). Counterattacking cyber threats: A framework for the future of cybersecurity. *Sustainability, 15*(18), 13369.

Tohănean, D., Buzatu, A.I., Baba, C.A., & Georgescu, B. (2020). Business model innovation through the use of digital technologies: Managing risks and creating sustainability. *Amfiteatru Economic, 22*(55), 758–774.

Wei, Y. (2022). Regional governments and opportunity entrepreneurship in underdeveloped institutional environments: An entrepreneurial ecosystem perspective. *Research Policy, 51*(1), 104380.

9 An Overall Evaluation of Women's Entrepreneurship

Mustafa Altintaş

9.1 INTRODUCTION

Entrepreneurship emerges as a concept that progresses on the foundations of innovation, creativity, and risk-taking, aiming to create new opportunities in the business world throughout the historical process. Entrepreneurship, which promotes economic growth and development, is also a crucial element supporting social transformation. Offering various benefits for individuals and societies, entrepreneurship is a structure that can be established not only by organizations but also by individuals with their own ideas. Individuals with an entrepreneurial spirit can become entrepreneurs to demonstrate their abilities and tackle challenges by leveraging opportunities arising from specific experiences, educational levels, and financial resources.

Individual entrepreneurship refers to the process where individuals develop, establish, and manage their own business ideas. The entrepreneurial individual initiates the process with an idea and then seeks the necessary resources. Regardless of the sector, risk is the most crucial factor in starting a business, and the entrepreneur undertakes these risks by presenting and developing their idea. Entrepreneurship facilitates the realization of innovative ideas and personal desires in the business world. On an individual level, entrepreneurship offers the opportunity to demonstrate one's potential, become one's own boss, and manage one's future. In doing so, the entrepreneurial individual develops business ideas according to personal preferences, showcasing creativity and leadership skills during the implementation of these ideas.

When examining the historical content of the concept of entrepreneurship, one important issue that emerges is women's entrepreneurship, as it is well known that women constitute a significant portion of the population both in Turkey and around the world. Despite this, from an entrepreneurial perspective, women are not at the desired levels either as entrepreneurs or in economic life. There are many reasons for this, and the reason for addressing 'women's entrepreneurship' as a separate category is due to the specific barriers that exist in this field worldwide. Women's entrepreneurship aims to mobilize dormant potential, thereby contributing more significantly to economic development and growth objectives (Sırkıntıoğlu-Yıldırım and Şencan-Kukuş, 2022).

An Overall Evaluation of Women's Entrepreneurship

Although women's entrepreneurship faces numerous obstacles, there is a noticeable increase in entrepreneurial tendencies among women in today's world. This increase is driven by factors such as the desire to be one's own boss, willingness to take risks, aspiration for independent work, and the urge to innovate and satisfy the need for change. Broadly speaking, the rise in the number of female entrepreneurs can be attributed to socio-cultural transformations and concerns within the working environment. Additionally, the concept frequently discussed in the literature known as the "glass ceiling"—the perception that women cannot advance beyond a certain point in organizations—plays a significant role in the initiation of entrepreneurial activities (Gürol, 2000). Therefore, the form and structure of entrepreneurial activities that women may be inclined to engage in are shaped both officially through laws, regulations, or policies, and informally by elements such as the cultural values of society and the family structure. In this determination, economic conditions and societal expectations are prominent (Al-Dajani et al., 2015; Welter, 2020).

This study focuses on women's entrepreneurship and aims to determine whether it arises as a choice or due to dilemmas, utilizing insights from existing literature. We draw upon current literature and research findings to achieve this goal. In particular, we attempt to define the perspective on women's entrepreneurship in light of societal changes and developments from the past to the present and into the future. We believe that this study will contribute to the body of knowledge in the literature.

9.2 WOMEN'S ENTREPRENEURSHIP

Entrepreneurship, when viewed broadly, emerges as a concept that has developed both locally and internationally with the transition from an industrial society to an information society and has evolved alongside technology. Currently, entrepreneurship continues to evolve and has rapidly become significant for many sectors in both social and economic dimensions (Börü, 2006). The concept of entrepreneurship is known to be closely related to notions such as change, creativity, and innovation, requiring a departure from existing conditions. Entrepreneurship aims to take advantage of emerging opportunities or to create new ones. It can be said that entrepreneurship progresses independently of the control of resources. Additionally, it involves a process aimed at generating and following up on opportunities (Başar, 2005).

While entrepreneurship is a highly valued and rapidly changing phenomenon, women's entrepreneurship is equally important. One of the most significant factors affecting entrepreneurship is suggested to be the concept of gender. According to research conducted by the Global Entrepreneurship Monitor in 2001, which included 29 countries, women engaged in entrepreneurial activities more frequently than men (Çetinkaya et al., 2012). Studies investigating entrepreneurship from the perspective of gender roles indicate that, within the framework of technological advancements in today's world, women are taking on more roles in the business world (Thompson et al., 2010). Changing and developing living conditions provide more space for the involvement of women. The evolving socio-demographic structures of societies have clarified the position of women entrepreneurs in the business world (Karabat and

Sönmez, 2012). Thus, this suggests that in the future, the term "women's entrepreneurship" might be used without any gender connotation.

From a gender perspective, it is known that both female and male entrepreneurs may have similarities as well as differences. The literature indicates that factors such as the desire to utilize their abilities, the idea of establishing an independent business, the aspiration to be their own boss/manager, and economic challenges are influential for both genders (Çelebi, 1997). On the other hand, while entrepreneurs share common tendencies in realizing their entrepreneurial pursuits, significant differences exist between men and women in the initial stages of concretizing these ideas. Men's entry into entrepreneurship is often driven by concerns about the future and a desire for control, whereas women are motivated by the desire to achieve something and the pursuit of freedom. Additionally, in terms of professional life, factors such as fear of career stagnation, lack of promotion opportunities, and professional problems are primary drivers pushing women towards entrepreneurship. Men typically utilize savings and bank loans to start their ventures, whereas women tend to rely more on individual resources. Furthermore, sectoral differences reveal that men usually have experience in finance, manufacturing, and technical fields, while women are more specialized in education and retail sectors (Arıkan, 2004).

Although women's entrepreneurship is examined from a gender perspective, the societal context is highly significant. The social position of women and the entrepreneurial profile within the social structure are also important. The very presence of the term "women's entrepreneurship" in the literature indicates that women do not have equal opportunities with men. In underdeveloped societies, women still face obstacles regarding inheritance rights, contract-making, and property ownership (OECD, 2004). Entrepreneurship is crucial for the development and economic advancement of a society, especially on a national level, and it is suggested that sufficient entrepreneurial initiatives are lacking in underdeveloped and developing countries. Limited educational opportunities, cultural prejudices, and gender roles confining women to the home have hindered their contribution to economic development. It is asserted that overcoming these fundamental issues will be achieved through economic and entrepreneurial activities (Lashgarara et al., 2011).

In entrepreneurship, where the significance of gender differences remains, Hisrich and Peters (2002) stated that there are distinct differences between male and female entrepreneurs. The views of researchers who make a distinction in this regard are presented below:

Motivational characteristics (male entrepreneurs):

- Achievement resulting from task completion
- Personal independence
- Job satisfaction due to having control

Motivational characteristics (female entrepreneurs):

- Achievement resulting from fulfilling a purpose
- Independence derived from performing the work

Starting point characteristics (male entrepreneurs):

- Dissatisfaction with current job
- Having been involved in a new job elsewhere
- Being laid off or leaving a job
- Desire to achieve something

Starting point characteristics (female entrepreneurs):

- Job-related frustration
- Ability to recognize opportunities in the field

Funding sources characteristics (male entrepreneurs):

- Personal assets and savings
- Banks
- Investors
- Loans from acquaintances (friends/family)

Funding sources characteristics (female entrepreneurs):

- Personal assets and savings
- Personal loans

Personal characteristics (male entrepreneurs):

- Starting a business between the ages of 25–35
- Having a parent who owns a business
- University graduate (business/engineering)
- Being the first child in the family

Personal characteristics (female entrepreneurs):

- Starting a business between the ages of 25 and 35
- Having a parent who owns a business
- University graduate
- Being the first child in the family

Support groups (male entrepreneurs):

- Friends, accountants, and lawyers
- Business acquaintances
- Spouses

Support groups (female entrepreneurs):

- Close friends
- Spouses
- Family
- Professional women's groups
- Trade associations

Type of first business founded (male entrepreneurs):

- Construction or manufacturing

Type of first business founded (female entrepreneurs):

- Service sector (education, consulting, public relations)

Hisrich and Peters (2002) provided detailed information on the distinguishing characteristics of male and female entrepreneurs. Although the researchers outlined a general framework, it can be said that the same factors may not be experienced in every society. This is because each society's economic conditions, cultural lifestyles, and perspectives on life differ. For example, when examining the National Action Plan for Gender Equality published by the Ministry of Development in Turkey in 2009, measures such as increasing childcare and eldercare services, reducing the prevalence of traditional norms, and targeting women in rural areas were emphasized. This demonstrates that women's entrepreneurship in Turkey is influenced by various socio-demographic factors.

9.2.1 Historical Background of Women's Entrepreneurship

Examining the historical background of women's entrepreneurship reveals that it has always been a concept throughout human history. The literature indicates that it is positioned on two main bases: before and after the Industrial Revolution (Özkaya, 2020). Essentially, entrepreneurship is a part of the management phenomenon, and it can be said that where there are humans, these concepts are very old and never outdated. Özyılmaz (2016) states that entrepreneurship was not fully used in its scientific sense before the Industrial Revolution. The reason for this is that people in earlier periods did not feel the need for risk-taking and creativity.

Although referred to as the periods before and after the Industrial Revolution, both entrepreneurship and women's participation in the workforce can be said to date back to more recent times. In the more recent past, it is observed that women's integration into the workforce in the 1980s significantly supported small-scale businesses. The 1980s marked an extraordinary transformation worldwide, and during this period, small business entrepreneurship emerged as a solution to poverty in developing countries (Yetim, 2002). Moreover, the Industrial Revolution is seen as a crucial focal point for women's entrepreneurship and their participation in the workforce. Women's participation in the workforce in the 20th century represented social and economic development, and rapid demographic changes worldwide made women's involvement in the workforce inevitable. Until the Industrial Revolution, women were primarily confined to the home, but afterward, they began to be present outside the home (Arslan and Toksoy, 2017).

Examining the literature, both globally and in Turkey, it is observed that during the periods of the Industrial Revolution before the establishment of the republic, there were empires and states ruled by sultans. These circumstances tend to perpetuate a traditional approach from social, political, and cultural perspectives. During these years, characterized by a patriarchal structure, women's exclusion from the workforce and limitations on entrepreneurship led to a concentration of such occurrences in the recent past (Ersoy, 2018).

Another important piece of information, especially concerning Turkey, is the existence of significant developments related to women during the Second Constitutional Period. Despite the prevailing patriarchal societal structure, it can be said that opportunities for education for women and young girls improved during this period. Furthermore, vocational training similar to trade schools was provided for women's employment, and efforts were initiated to make them active in the workforce. This ongoing process contributed to raising awareness about women's rights and paved the way for the emergence of women's movements (Kaymaz Mert and Kaymaz Gençal, 2023). The foundation of women's progress in entrepreneurship began with their integration into daily life. Initially, issues such as suffrage and employability in businesses were significant steps toward achieving gender equality in social life. A fundamental factor contributing to the distinction of women in the entrepreneurial scene is their delayed integration into societal life.

Both the democratic state understanding and the progression of women's employment and entrepreneurship, which had undergone certain stages previously, continued to develop and increase by the 1990s. Examining Turkey specifically, the policies put forth by political authorities have been oriented towards women's employment and entrepreneurship. Economic crises in the 1970s and subsequent ones underscored the necessity of women's participation in the workforce. Women's entrepreneurship has been supported by every political authority, and the necessity of establishing a balanced structure for development has been realized (Sallan Gül and Altındal, 2016).

Subsequently, as the process advanced, women's employment in jobs traditionally dominated by men, their engagement in entrepreneurial activities in sectors such as public relations, consultancy, and sales, signify an important dimension of women's entrepreneurship. Assessing the 21st century and the present, Gürol (2000) points to reasons for the increase in women's entrepreneurship such as social and cultural changes, the rise in women's education levels, the continued increase in the number of women rising in professional management ranks, and the growing number of women entrepreneurs serving as role models, along with the prevalence of the glass ceiling syndrome. In this context, considering that everyone is somehow part of a consumer society in the economy, it can be said that both today and in the future, it is imperative for both men and women to work. While traditional views still persist in many countries, it is known that women are somehow involved in entrepreneurial activities and are part of the economy.

9.2.2 Barriers Faced by Women Entrepreneurs

Women entrepreneurs generally lag behind their male counterparts in the business world, often encountering significant obstacles on the path to success. These obstacles are usually attributed to economic, social, and cultural factors. Prejudices identified as societal gender roles can be considered the most significant challenge women face on the path to entrepreneurship. Such biases, occurring in paid employment roles held by women, lead to questioning their ability to establish and manage independent businesses, thus perceptually constraining entrepreneurial activities (Brush et al., 2009).

Issues surrounding women's past entrepreneurial activities, such as leadership skills, their impact on business performance, and differences in organizational culture, have also been the subject of various studies (Çelik and Özdevecioğlu, 2001). Research suggests that men have more experience and infrastructure related to entrepreneurship than women, and female entrepreneurs often have to take greater risks due to their lack of readiness for business establishment (Kutanis, 2003). Marlow and Patton (2005) have indicated the economic challenges women entrepreneurs face in accessing capital, highlighting that this difficulty, particularly at the inception of a business, poses a significant disadvantage in expanding and ensuring the sustainability of enterprises. Additionally, Carter et al. (2003) have expressed the lack of adequate consultancy and collaboration for women in financial matters, asserting that this deficiency hampers women entrepreneurs in identifying and evaluating opportunities and obtaining necessary support.

On the other hand, women entrepreneurs encounter cultural barriers. Cultural barriers, which play a significant role, are used to denote situations in societal structures that cannot be overcome. The family and household responsibilities imposed on women, along with family expectations, hinder their ability to be active and entrepreneurial in the workplace. These circumstances, which increase the likelihood of work–life imbalance, make it challenging for women to devote their time fully to their work, even in paid employment, let alone engaging in entrepreneurial activities (Jennings and Mcdougald, 2007). Eddleston and Powell (2012) suggest that women's inability to balance work and life can lead to stress and burnout. Naguib and Jamali (2015) state that women face criticism from their cultural environment for starting and managing businesses, feeling underestimated, which creates bias against women's entrepreneurial activities. Furthermore, they highlight that even women occupying leadership positions in the workplace is considered unusual, and sexist attitudes restrict women's advancement.

If various examples of women's entrepreneurship from a cultural perspective are to be given, the presence of an informal culture is cited as a deficiency of women's entrepreneurship in Ghana. Certain societal pressures contribute to this. It is stated that the deterrent factors in social life directly affect women (Adom and Anambane, 2019). On the other hand, the relationships that female entrepreneurs have with institutions are considered important, and it is noted that change is closely linked to entrepreneurship. In fact, female entrepreneurs are even given the nickname 'agents of change.' It is reported that this could create a negative environment for institutions (Ojediran and Anderson, 2020).

The study by Bullough et al. (2022) presents a general framework that encompasses both women's entrepreneurship and culture. This framework includes socio-cultural dimensions, gender role expectations and identities, and the entrepreneurial environment. Within the socio-cultural dimensions, aspects such as uncertainty avoidance, gender egalitarianism, power distance, humane orientation, assertiveness, future orientation, and in-group collectivism are included. The dimension of gender role expectations and identities covers elements such as stereotypes, team processes, risk-taking behaviors, perceptions of women entrepreneurs, family commitment, gender roles, social role theory, family firm, gender role theory, gender role orientation, traditional gender norms, and collaborative and supportive elements. Lastly, the entrepreneurial environment dimension includes social entrepreneurship, socially supportive cultures, regulations, entrepreneurship-supportive factors, the entrepreneurial process, and implicit theories of entrepreneurship. Overall, environmental pressures are frequently discussed and debated as the most significant factor globally.

In Turkey, it can be said that the problems faced by women entrepreneurs parallel the literature. Issues such as women not receiving adequate and quality education, their societal roles not being considered suitable for the business environment, the lack of connections to sectors due to the roles imposed on them, family power dynamics and reactions, and work–life balance are all highlighted (Özkaya Onay, 2009).

9.2.3 General Status of Women's Entrepreneurship in Turkey

Studies on women's entrepreneurship in Turkey gained significant importance, especially after the 1990s. Women's entrepreneurship, with its strong historical background, has risen to higher levels thanks to government policies. Particularly in the 1990s, the establishment of the Directorate General on the Status of Women, affiliated with the Ministry of Labor and Social Security at the time, provided an official identity for women's entrepreneurship and women's studies (Çakıcı, 2004). One of the most significant obstacles of gender discrimination, the requirement for spousal permission for married women to engage in business and establish enterprises, was abolished in the Turkish Civil Code. Since this requirement contradicted the notion of the "free individual" in the law, its removal was a commendable step. Chronologically, studies conducted on women and women's entrepreneurship in Turkey are listed below (Çakıcı, 2004; Suğur and Demiray, 2009):

The first seminar on "Women's entrepreneurship in Turkey" and the organization of a panel on "Encouragement and Support for Women's entrepreneurship" discussed:

- Conducting the first research on "Family and Business Relationships of Independent Female Business Owners;"
- Initiation of the "Women Entrepreneur" credit program by Halk Bankası (People's Bank);
- Comprehensive research titled "Small Entrepreneurship Project" aimed at examining micro and small businesses owned by women, conducted by the Directorate General on the Status of Women and Women's Issues in 1995;

- Establishment of the Women Entrepreneurs Association (KAGİDER) in 2002 with the aim of promoting women's entrepreneurship and strengthening the position of women in economic and social life;
- Launch of the "Support for Women Entrepreneurs" project by the European Union and the Confederation of Turkish Tradesmen and Craftsmen (TESK);
- Implementation of the "Women's entrepreneurship" project through the partnership of the Southeastern Anatolia Project, Entrepreneur Support and Guidance Center, and Women's Center (GAP, GİDEM, KAMER).

The advancement of women's entrepreneurship through the initiatives mentioned above is considered significant for women's presence in the business world. Support organizations for women and women's entrepreneurship in Turkey are listed below (Erdemir, 2018):

- Small and Medium Enterprises Development and Support Administration;
- Women's Labor Foundation;
- Women Entrepreneurs Association;
- Women Entrepreneurs Board of Turkey Chambers and Commodity Exchanges Union;
- Directorate General on the Status of Women;
- Turkish Employment Agency;
- Turkish Young Businessmen's Association;
- Turkey Development Foundation.

Since the 1990s, various organizations in Turkey have shown increasing interest in women's entrepreneurship, as evidenced by numerous studies and supporting institutions. In addition, the adequacy of these organizations and studies regarding women's employment and labor force participation in Turkey has been examined through the latest data provided by the TurkStat (TUIK, 2023). According to TUIK's 2023 data, the number of employed women aged 15 and over was 24,851 in 2005, while by March 2024, this figure had risen to 33,235. In 2005, the labor force participation rate for women was 20.8%, whereas by March 2024, this rate increased to 36.4%. These indicators demonstrate that, along with population growth, the employment rate of women has increased from 2005 to March 2024. It can be said that the role of women in the workforce has grown over the years. Therefore, it can be concluded that global developments and public policies implemented by the authorities in the 1990s have produced favorable outcomes over the past 30 years.

9.3 CONCLUSION

This study was developed with a focus on women's entrepreneurship, specifically aiming to address the general situation of women's entrepreneurship in Turkey. The aim of the study is to summarize the general state of women's entrepreneurship and provide actionable recommendations. Particularly in this section, approaches explaining women's entrepreneurship will be elucidated. Women's entrepreneurship is an important topic in today's business world and is especially problematic. This is because, from a gender perspective, women's participation in the workforce, their employment, and entrepreneurship are subject to unofficial restrictions economically, socially, and culturally. Despite women's participation in the workforce, advancement within organizations is often hindered, and the presence of women in various sectors of society is sometimes perceived as unsettling. The roles attributed to women tend to deter them from the business world. For instance, although there is suffrage, opportunities for women to be elected as members of parliament or mayors are limited. Furthermore, the appointments to significant positions within public authority administrations often feature a scarcity of women.

In the literature, there are various approaches explaining women's entrepreneurship. However, within the scope of this study, certain theories related to women employment and entrepreneurship have been considered. Particularly, approaches emphasizing women's economic and sociological participation, along with feminist theories, are considered significant in explaining women employment and entrepreneurship. Examination of research on women's entrepreneurship in the literature generally reveals analyses of the general situation and explanations within frameworks of various approaches. For instance, De Vita et al. (2014), in their study examining research on women's entrepreneurship in developing countries, reviewed 70 articles and concluded that theoretical sociological frameworks were not addressed and feminist theories were not considered in the research. Additionally, the articles reviewed mostly involved survey applications, with very few articles based on international examples. In addition, Prasastyoga et al. (2021) state that entrepreneurial activities have a complex structure that cannot be handled alone and are affected by cognitive conditions. Stating that very few studies in the literature address cognitive issues, researchers refer to theories.

One of the economic theories explaining women employment and entrepreneurship is development theory. Seers (1969) asserts that development should revolve around the individual and define it as the provision of conditions necessary for the realization of an individual's personality. Flammang (1979) suggests that development encompasses more output and encompasses technical and theoretical structural changes. Sen (2004) argues that development should focus on the central role of human life. Development theory, not being a standalone theory but rather a framework covering economic, social, technical, and many other aspects, suggests that women's participation in development activities should be inclusive in all respects. The approach advocating that development will occur through women's labor force participation advocates for women's participation in the economy, enabling them to share in growth and development, leading to an improvement in their position within the family. Women's participation in employment is necessary for development activities to occur at the local, regional, national, and international levels in a balanced manner (Toksöz, 2011).

Another approach referring to women's entrepreneurship is the social capital approach. Social capital approach defines potential benefits provided by relationships and communication based on long-term and purposeful social networks (Bourdieu, 1986). In another definition, the social capital approach is defined as a concept that provides resources for collaborative behavior among groups (Coleman, 1988). Social capital is essential in interpersonal relationships, integration, and collaboration. The social capital approach advocates that individuals should be independent of class and gender, suggesting that women may have less social capital than men due to the characteristics they possess or lack. For instance, a woman may reject the business world due to marriage, children, or family reasons (Kaynar, 2022). Paxton (1999) argues that social capital has a significant impact on necessary social processes and is essential for open-system businesses. Therefore, women entrepreneurs can turn this situation in their favor by expanding their social networks, seeking counseling, and establishing business connections.

Finally, perhaps the most important approach in explaining women's entrepreneurship is feminist approaches. Feminist approaches examine how gender inequalities arise and are created to explain women's experiences in various social, cultural, economic, and similar fields. Like development theories, feminist theories, which emerge in different ways, offer solutions to the inequalities and discriminations applied to women. Feminist approaches, put forth in various forms such as liberal, radical, socialist, postmodern, seek solutions to the inequalities applied to women from different perspectives. For example, Tong (2009) states that liberal feminism advocates for the implementation of political and legal reforms to achieve gender equality. Butler (1990) suggests that the postmodern feminist approach has a historical and cultural basis and argues that women should be advocated for individually rather than with universal discourses. From the perspective of feminist approaches to women's entrepreneurship, the necessity for women to be present in all segments of society emerges. Both in employment and entrepreneurship, feminist approaches strive for equality.

This study is planned as a review study and, based on the topics discussed, evaluations made, and considering the literature, it provides the following recommendations, especially from the perspective of public authorities:

- Special entrepreneurship training and consultancy programs should be organized for women entrepreneurs to not only increase their educational level but also enhance their knowledge and skills.
- Incentives should be provided by organizing tax deductions and exemption programs specifically for women entrepreneurs.
- Returns should be increased by providing low-interest loans, grant programs, and even non-repayable loans specifically for women entrepreneurs.
- Legal regulations should be made, especially in favor of women entrepreneurs.
- In addition to all these, events should be organized where women entrepreneurs and potential women entrepreneurs can meet, create collaboration networks, and programs should be arranged to obtain tangible results.
- Campaigns promoting women's entrepreneurship and reaching the entire country should be organized to create awareness.
- Considering the boundaries disappearing and the widespread digitization in the technology world, integration of technology with women's entrepreneurship is

crucial, as it presents a significant opportunity. Therefore, integration of technology with women's entrepreneurship is necessary.

Considering the recommendations mentioned above, our aim is to support women's entrepreneurship and to help women become stronger and more competitive in the business world through comprehensive measures. The primary purpose of presenting these recommendations to public authorities is to ensure the feasibility and implementation of various state-led policies. With the enhancement of education, practical skills, mentorship programs, credit support and incentives, technological advancements, and numerous other opportunities, it is anticipated that women's entrepreneurship will be in a better position in the future. Moreover, it can be said that this study, supported by various theories, aims to maximize the social and economic contributions of women entrepreneurs.

REFERENCES

Adom, K. and Anambane, G. (2019). Understanding the role of culture and gender stereotypes in women's entrepreneurship through the lens of the stereotype threat theory. *Journal of Entrepreneurship in Emerging Economies*, 12(1), 100–124.

Al-Dajani, H., Carter, S., Shaw, E. and Marlow, S. (2015). Entrepreneurship among the displaced and dispossessed: Exploring the limits of emancipatory entrepreneuring. *British Journal of Management*, 26(4), 713–730. https://doi.org/10.1111/1467-8551.12119

Arıkan, S. (2004). *Girişimcilik Temel Kavramlar ve Bazı Güncel Konular*. Siyasal Kitabevi, Ankara.

Arslan, İ. K. and Toksoy, M. D. (2017). Türkiye'de Kadınları Girişimciliğe Yönelten Faktörler Karşılaştıkları Sorunlar ve Çözüm Önerileri. *İstanbul Ticaret Üniversitesi Girişimcilik Dergisi*, 1(1), 123–148.

Başar, M. (2005). Girişimcilik ve Girişimcinin Özellikleri. In Yavuz Odabaşı (Ed.), *Girişimcilik içinde* (pp. 4–10). Eskişehir: T.C. Anadolu Üniversitesi Yayını No:1567 Açıköğretim Fakültesi Yayını.

Bourdieu, P. (1986). The forms of capital on CALL. In J. G. Richardson (Eds.), *CALL Handbook of theory and research for the sociology of education* (pp. 241–258). New York: Greenwood.

Börü, D. (2006). *Girişimcilik Eğilimi Marmara Üniversitesi İşletme Bölümü Öğrencileri Üzerine Bir Araştırma*. İstanbul: Marmara Üniversitesi Yayını.

Brush, C. G., de Bruin, A. and Welter, F. (2009). A gender-aware framework for women's entrepreneurship. *International Journal of Gender and Entrepreneurship*, 1(1), 8–24. https://doi.org/10.1108/17566260910942318

Bullough, A., Guelich, U., Manolova, T. S. and Schjoedt, L. (2022). Women's entrepreneurship and culture: Gender role expectations and identities, societal culture, and the entrepreneurial environment. *Small Business Economics*, 58(2), 985–996. https://doi.org/10.1007/s11187-020-00429-6.

Butler, J. (1990). *Gender trouble: Feminism and the subversion of identity*. New York: Routledge.

Carter, N. M., Brush, C. G., Greene, P. G., Gatewood, E. J. and Hart, M. M. (2003). Women entrepreneurs who break through to equity financing: The influence of human, social and financial capital. *Venture Capital: An International Journal of Entrepreneurial Finance*, 5(1), 1–28. https://doi.org/10.1080/1369106032000082586

Coleman, J. (1988). Social capital in the creation of human capital. *American Journal of Sociology*, 94(1), 95–120.

Çakıcı, A. (2004). Kadın Girişimcilerin İşletme Fonksiyonlarındaki Etkisinin Belirlenmesine Yönelik Bir Araştırma. *Yönetim Bilimleri Dergisi*, 2(1), 129–142.

Çelebi, N. (1997). *Turizm Sektöründeki Küçük İşyeri Örgütlerinde Kadın Girişimciler, TC.* Ankara: Başbakanlık Kadın Statüsü ve Sorunları Genel Müdürlüğü Projesi.

Çelik, C. and Özdevecioğlu, M. (2001). "Kadın Girişimcilerin Demografik Özellikleri ve Karşılaştıkları sorunlara İlişkin Nevşehir İlinde Bir Araştırma," 1. *Orta Anadolu Kongresi (Nevşehir)*, 487–498.

Çetinkaya Bozkurt, Ö., Kalkan, A., Koyuncu, O., Alparslan, A. M. (2012). Türkiye'de Girişimciliğin Gelişimi: Girişimciler Üzerinde Nitel Bir Araştırma. *Süleyman Demirel Üniversitesi Sosyal Bilimler Enstitüsü Dergisi*, (15), 229–247.

De Vita, L., Mari, M. and Poggesi, S. (2014). Women entrepreneurs in and from developing countries: Evidences from the literatüre. *European Management Journal*, 32, 451–460. https://doi.org/10.1016/j.emj.2013.07.009

Eddleston, K. A. and Powell, G. N. (2012). Nurturing entrepreneurs' work–family balance: A gendered perspective. *Entrepreneurship Theory and Practice*, 36(3), 513–541. https://doi.org/10.1111/j.1540-6520.2012.00506.x

Erdemir, T. (2018). *Türkiye'de Kadın Girişimciliği ve Kadın Girişimciliğini Etkileyen Engeller ve Fırsatlar Üzerine Bir Araştırma.* İstanbul: Yayımlanmamış Yüksek Lisans Tezi, İstanbul Gelişim Üniversitesi Sosyal Bilimler Enstitüsü.

Ersoy, A. İ. (2018). *Kadın Girişimci Kobilerde Eğitim İhtiyaç Algısının Finans ve Diğer Fonksiyonlar Açısından Karşılaştırmalı Analizi.* İstanbul: Yayımlanmamış Yüksek Lisans Tezi, İstanbul Medipol Üniversitesi Sosyal Bilimler Enstitüsü.

Flammang, R. A. (1979). Economic growth and economic development; counterparts or competitors?. *Economic Development and Cultural Change*, 28(1), 47–61.

Gürol, A. (2000). *Türkiye'de Kadın Girişimci ve Küçük işletmesi: Fırsatlar, Sorunlar, Beklentiler ve Öneriler.* Ankara: Atılım Üniversitesi Yayını.

Hisrich, R. D. and Peters, M. P. (2002). *Entrepreneurship* (5th Edition). USA: McGraw-Hill International.

Jennings, J. E. and McDougald, M. S. (2007). Work-family interface experiences and coping strategies: Implications for entrepreneurship research and practice. *The Academy of Management Review*, 32(3), 747–760. https://doi.org/10.2307/20159332

Karabat, B. Ç. and Sönmez, Ö. A. (2012). Bölgesel Kalkınmada Kadın Girişimciliğinin Rolü. *Balkan Journal of Social Sciences Balkan Sosyal Bilimler Dergisi*, 1(2), 1–12.

Kaymaz Mert, M., and Kaymaz Gençal, E. (2023). II. Meşrutiyet'ten Cumhuriyet'in İlanına Kadar İstanbul'da Kamusal Alanda Kadınlar. *Adam Academy Journal of Social Sciences*, 13(1), 25–50. https://doi.org/10.31679/adamakademi.1147676

Kaynar, G. (2022). *Kadın Girişimciliğinin Başarısını Etkileyen Faktörler; Mersin İlinde Bir Araştırma.* Mersin: Yayımlanmamış Yüksek Lisans Tezi, Toros Üniversitesi Lisansüstü Eğitim Enstitüsü.

Kutanis, R. Ö. (2003). Girişimcilikte Cinsiyet Faktörü: Kadın Girişimciler. 11. Yönetim ve Organizasyon Kongresi Bildiriler Kitabı, *Afyon Kocatepe Üniversitesi İ.İ.B.F. Dergisi*, 8(2), 139–153.

Lashgarara, F., Roshani, N. and Najafabadi, M. O. (2011). Influencing factors on entrepreneurial skills of rural women in Ilam City, Iran. *African Journal of Business Management*, 5(14), 5536–5540.

Marlow, S., and Patton, D. (2005). All credit to men? Entrepreneurship, finance, and gender. *Entrepreneurship Theory and Practice*, 29(6), 717–735. https://doi.org/10.1111/j.1540-6520.2005.00105.x

Naguib, R. and Jamali, D. (2015). Female entrepreneurship in the UAE: A multi-level integrative lens. *Gender in Management: An International Journal*, 30(2), 135–161. https://doi.org/10.1108/GM-12-2013-0142

OECD. (2004). Kadın Girişimciliği: Sorunlar ve Politikalar. 2. *OECD Küçük ve Orta Ölçekli İşletmelerden (KOBİ'ler) Sorumlu Bakanlar Konferansı (İstanbul, Türkiye).*

Ojediran, F. O. and Anderson, A. (2020). Women's entrepreneurship in the global South: Empowering and emancipating? *Administrative Sciences*, 10(4), 87–109

Özkaya Onay, M. (2009). Kadın Girişimcilere Yönelik "Strateji Geliştirmede" Yerel Yönetimlerle İşbirliği İçinde Olmak, Mümkün Mü?. *Journal of Management and Economics Research*, 7(11), 56–72.

Özkaya, G. (2020). *Kadın Girişimcilerin Girişimci Olma Nedenleri ve Girişimcilik Algılarının Sosyo-Demografik Değişkenler Açısından Değerlendirilmesi: İzmir İlinde Bir Araştırma.* İzmir: Yayımlanmamış Yüksek Lisans Tezi, İzmir Kâtip Çelebi Üniversitesi Sosyal Bilimler Enstitüsü.

Özyılmaz, A. M. (2016). *Türkiye'de Kadın Girişimciliği ve Girişimci Kadınların Karşılaştıkları Sorunlar Üzerine Bir Araştırma.* Nevşehir: Yayımlanmamış Yüksek Lisans Tezi, Nevşehir Hacı Bektaş Veli Üniversitesi Sosyal Bilimler Enstitüsü.

Paxton, P. (1999). Is social capital declining in the United States? A multiple indicator assessment. *The American Journal of Sociology*, 105(1), 88–127.

Prasastyoga, B., Harinck, F. and van Leeuwen, E. (2021). The role of perceived value of entrepreneurial identity in growth motivation. *International Journal of Entrepreneurial Behavior & Research*, 27(4), 989–1010. https://doi.org/10.1108/IJEBR-03-2020-0170

Sallan Gül, S. and Altındal, Y. (2016). Türkiye'de Kadın Girişimciliğinin Serüveni: Başarı Mümkün Mü?. *Süleyman Demirel Üniversitesi İktisadi ve İdari Bilimler Fakültesi Dergisi*, 21(4), 1361–1377.

Seers, D. (1969). *The meaning of development.* Institute of Development Studies.

Sen, A. K. (2004). *Özgürlükle Kalkınma, (Çeviren: Y. Alogan).* İstanbul: Ayrıntı Yayınları.

Sırkıntıoğlu Yıldırım, Ş. and Şencan Kukuş, T. (2022). Kadın Girişimci Profili Üzerine Bir Analiz: Amasya İli Örneği". *Nişantaşı Üniversitesi Sosyal Bilimler Dergisi*, 2(10) 286–300.

Suğur, S., Demiray, E., Eşkinat, R. and Ağaoğlu, E. (2009). *Esmahan, Toplumsal Yaşamda Kadın.* Eskişehir: Anadolu Üniversitesi Yayınları.

Thompson, P., Evans, D. J. and Kwong, C. C. Y. (2010). Education and entrepreneurial activity: A comparison of White and South Asian Men. *International Small Business Journal*, 28(2), 147–162.

Toksöz, G. (2011). *Kalkınma da Kadın Emeği.* Varlık Yayınları İstanbul.

Tong, R. (2009). *Feminist thought: A more comprehensive introduction.* Boulder, CO: Westview Press.

TurkStat. (2023). Temel işgücü göstergeleri. https://data.tuik.gov.tr/Bulten/Index?p=Isgucu-Istatistikleri-Aralik-2023-49379. (Accessed 01.06.2024).

Welter, F. (2020). Contexts and gender – looking back and thinking forward. *International Journal of Gender and Entrepreneurship*, 12(1), 27–38. https://doi.org/10.1108/IJGE-04-2019-0082

Yetim, N. (2002). Sosyal Sermaye Olarak Kadın Girişimciler: Mersin Örneği. *Ege Akademik Bakış*, 2(2), 79–92.

10 An Analysis of Women's Entrepreneurship in Turkey
Bibliometric Methods

Meltem Ince Yenilmez

10.1 INTRODUCTION

The entrepreneurial endeavours of women are vital to the social and economic development of societies. In addition to having a significant positive impact on economic growth and job creation, it also empowers women and improves gender equality, leading to more inclusive and sustainable development. Societies gain from a variety of viewpoints and creative ideas, which are crucial for tackling difficult problems in the modern global economy when women are encouraged to succeed as entrepreneurs.

Women's entrepreneurship has a huge economic impact. Women-owned companies contribute significantly to employment and income generation, which lowers poverty and boosts the economy. In many nations, female entrepreneurs have played a major role in the growth of small and medium-sized businesses (SMEs), which are the foundation of many economies. Governments can promote women's entrepreneurship to increase economic resilience and diversification in underdeveloped countries where women are often essential to informal economies.

Beyond its economic benefits, women's entrepreneurship has societal effects. Strong economic standing encourages women to invest in their families, communities, and overall well-being, which improves their chances for education and general well-being. This ripple effect can help break the cycle of poverty and promote wealth across generations. Furthermore, successful female entrepreneurs inspire the next generation of girls and women to follow their passions and challenge societal norms that restrict their potential by serving as role models.

Women's entrepreneurship, which defies traditional gender conventions and reduces inequities in the labour market, further advances gender equality. The relationships in families and communities can shift when women own and operate businesses because they can become financially independent and have more decision-making power. This empowerment helps to undermine patriarchal norms and promotes a more equitable distribution of opportunities and resources. Furthermore, diverse leadership in organisations can lead to inclusive policies and practices that benefit stakeholders and employees equally.

Competitiveness and innovation need to support women who want to launch their enterprises. Women entrepreneurs often identify gaps in the market and develop products and services that appeal to a range of consumers. They frequently offer unique perspectives and approaches to the business community. This diversity of perspectives encourages innovation and creativity, two traits necessary to be competitive in a global economy that is changing swiftly. Moreover, inclusive business ecosystems that support women can attract talent and investment, enhancing a country's economic potential.

Women's entrepreneurship is essential for achieving gender equality, social progress, and economic success. Societies that encourage women entrepreneurs can unlock their potential to promote innovation, create jobs, and achieve inclusive development. Policymakers, business executives, and communities must work together to remove barriers and support women in their entrepreneurial aspirations to guarantee a more prosperous and equitable future for everybody.

10.2 INTERNATIONAL WOMEN'S ENTREPRENEURSHIP

The term "female entrepreneurship" refers to the expanding global phenomenon of businesswomen and their enterprises (Hechavarria et al., 2019). According to Elam et al. (2019), the Global Entrepreneurship Monitor 2018/2019 women's entrepreneurship report, around 231 million women were starting or managing new businesses worldwide in 2019. Studies on female entrepreneurship have been conducted more frequently in tandem with this development. These studies have been summed up in review papers. For instance, Brush (1992) examined empirical research studies on female entrepreneurship and recommended a new "integrated perspective" for future female entrepreneurship research based on 57 studies published between 1975 and 1991. Ahl (2006) conducted an assessment of 81 research publications that were published in prestigious magazines on management and entrepreneurship between 1982 and 2000. He criticised various shortcomings in the literature on female entrepreneurship, pointed out a tendency to perpetuate the idea that women are less valuable than males, and suggested new research directions based on discourse analysis.

Hughes et al. (2012) demonstrated how subsequent research on female entrepreneurship has taken Ahl's (2006) critiques into account. Jennings and Brush (2013) evaluated how, during the preceding 30 years, research on female entrepreneurship has enhanced broader notions about entrepreneurship. While the previous reviews provided readers with a basic overview of the subject of female entrepreneurship and were insightful, a more in-depth analysis of this academic field is required in light of the current substantial growth in female entrepreneurship studies.

Furthermore, there is a growing global acknowledgement that women's entrepreneurship plays a significant role in driving social and economic progress. Around the world, more women are launching and operating businesses, which promotes creativity, creates jobs, and lessens poverty. In order to achieve the goals of sustainable development, female entrepreneurs are essential, as recognised by the World Bank and the International Monetary Fund, among other international organisations. In spite of these developments, female entrepreneurs often face unique challenges that could impede the growth of their companies, like limited access to resources, finance, and education.

The difference in access to financial resources based on gender is one of the major issues affecting female entrepreneurs worldwide. According to Singh and Dash (2021), women are less likely than men to be able to secure loans or draw investment for their businesses due to a variety of variables, including discriminatory lending policies and lower levels of collateral. In addition, female entrepreneurs may lack the financial knowledge necessary to manage complex financial systems. A few of the targeted interventions required to close these inequalities are microfinance programmes, financial literacy training, and legislation that promotes gender equity in lending practices (Leitch et al., 2018).

Education and training are also critical for empowering women entrepreneurs. Globally, women often have less access to business education and vocational training than men do. This educational gap may make it more difficult for them to learn the skills necessary for their company to succeed, like marketing, financial management, and strategic planning (Meyer and Hamilton, 2020). The ability to start and build successful enterprises can be achieved by implementing measures that eliminate this gap and give women access to high-quality education and training. Women entrepreneurs benefit immensely from networks and mentorship. The most common grievances voiced by female entrepreneurs are loneliness and a lack of chances, guidance, and support from professional networks. Creating networks, mentorship programmes, and women-only business associations can help foster a cooperative, information-sharing, and mentoring-focused environment (Theaker, 2023). These networks can also help women overcome gender bias and cultural conventions that could be impeding their ability to follow their entrepreneurial dreams.

Global cultural and societal norms have a significant impact on women's business. Traditional gender conventions and expectations in many nations may make it more difficult for women to pursue entrepreneurship. The way that society views women in the workplace can have an impact on their confidence and propensity for taking risks (Ghimire, 2024). We must push for laws that encourage work–life balance, dispel stereotypes, and advance gender equality to foster an environment where women may thrive as entrepreneurs. It will take much work to change cultural perceptions and emphasise the value women bring to the economy to achieve gender parity in entrepreneurship.

10.3 TURKISH WOMEN ENTREPRENEURS

Turkey has seen a rise in the number of successful female entrepreneurs in recent years, which is a sign of more significant social and economic shifts. The goal of the policies and initiatives put in place by the government and numerous organisations is to encourage female entrepreneurs. Through initiatives like the Small and Medium Enterprises Development Organisation (KOSGEB), female entrepreneurs can obtain financing, support, and training. Turkish women still face several challenges in spite of their efforts, including limited access to funding, cultural barriers, and a lack of business networks (Tunç and Alkan, 2019).

In Turkey, traditional gender roles and cultural norms often dictate the kinds of businesses that women start and how much growth they can achieve with them. Many female entrepreneurs engage in retail, textiles, and personal services industries since these are perceived as extensions of their domestic responsibilities. This sectoral specialisation may limit their companies' ability to expand and compete in other lucrative industries (Ince Yenilmez, 2021). For female entrepreneurs to achieve economic emancipation, their industries must be diversified.

Obtaining funding is one of the major obstacles Turkish women entrepreneurs face. Women usually have trouble getting loans and investing because of discriminatory rules and a lack of collateral. Although their efficacy and scope may be limited, government grants and microfinance initiatives have been put in place to assist in addressing these problems. Enhancing women's financial literacy and providing customised financial products that satisfy their requirements are two ways to overcome these barriers (Ince Yenilmez, 2022). Furthermore, it is imperative to have a more varied financial environment in order to provide female entrepreneurs with the financing they require in order to grow their enterprises.

There are still gaps in the education and training alternatives available to women entrepreneurs in Turkey, despite increased options. Programmes that advance business acumen and entrepreneurial education are essential if women are to be given the resources they need to succeed. Academic institutions, industries, and governmental entities can work together to increase the availability and quality of these programmes (Mert, 2021). Additionally, by fostering a culture of continuous learning and professional development, women entrepreneurs can preserve their competitiveness in a market that is always changing.

The significance of networks and mentorship for female entrepreneurs in Turkey cannot be overstated. Women can interact, share experiences, and gain support through networks, mentorship programmes, and women's business associations. According to Raza et al. (2024), these networks can help women overcome the challenges of entrepreneurship, improve their self-esteem, and increase their commercial prospects. By strengthening these support networks and increasing the visibility of successful female entrepreneurs, more women can be inspired to undertake entrepreneurial endeavours and contribute to Turkey's economic development (Kalemci Tüzün and Araz, 2017).

10.4 FACTORS INFLUENCING WOMEN'S ENTREPRENEURSHIP

The entrepreneurship of women is influenced by various factors, including personal incentives and systemic constraints. Personal characteristics including financial independence, flexibility, and the desire to manage work and family obligations are common drivers of female entrepreneurship. According to Turan Torun (2022), these incentives have the potential to be powerful motivators for launching a business, but they must be supported by an environment that facilitates access to resources, knowledge, and support networks. It is critical to understand these traits to develop initiatives and policies that effectively promote women entrepreneurs.

The availability of capital is one of the key factors influencing women's entrepreneurship. Women business owners usually have trouble getting funding because of discriminatory lending practices, a lack of collateral, and a lack of financial literacy. Funding services and products tailored to women, such as microloans, venture capital, and grants, are required to get beyond these barriers (Ghosh and Ghosh, 2017). Women can obtain capital for business starts and expansion, as well as more effective money management, by improving their financial literacy through courses and training programmes.

The form of women's entrepreneurship is largely controlled by their access to education and training opportunities. Women must have access to top-notch education and training programmes that provide them with the necessary information and skills to excel in the business world. A variety of themes related to entrepreneurship should be covered in these courses, including financial management, marketing, leadership, and business planning (Meyer and Hamilton, 2020; Lee et al., 2024). Partnerships amongst educational institutions, governmental agencies, and the private sector could increase the number and quality of these programmes and ensure that women are ready to take on the challenges of startup.

Social and cultural norms have a big influence on women's entrepreneurship since they limit their access to opportunities and resources. In many countries, women's entrepreneurial potential can be restricted by traditional gender conventions and expectations. It is essential to dispel these myths and advance gender equality in order to foster an atmosphere that is encouraging to female entrepreneurs. By dismantling obstacles and supporting laws that promote gender equality and work–life balance, women can fully participate in the entrepreneurial environment (Sarı and Karabulut, 2023). This can be accomplished by advocating for legislation that modifies public perceptions.

Systems of support and mentoring are essential for female entrepreneurs. By having access to business groups, professional networks, and mentorship programmes, women can receive the guidance, support, and opportunities required for success. According to Bianco et al. (2017), women can gain valuable knowledge, boost their confidence, and overcome hurdles in the corporate world with the help of these networks. Rebuilding these support systems and increasing the visibility of successful female entrepreneurs are critical to inspiring more women to choose entrepreneurship and contributing to economic growth and development.

10.5 BIBLIOMETRIC ANALYSIS

For some reason, this study (Block and Fisch 2020; Donthu et al. 2021; Zupic and Čater, 2015) employed bibliometric analysis to determine the framework of prior research on entrepreneurial communication. The 383 articles that make up the current study attest to the benefits of bibliometrics for a wide range of research areas, as suggested by Block and Fisch (2020) and Donthu et al. (2021). Additionally, this field of study includes subcategories of communication such as investor relations, public relations, and employee relations. Second, although earlier research was restricted to working with print journals, there are now a great deal more digital publications, which makes it more challenging to handle sizable publishing volumes (Kraus et al.,

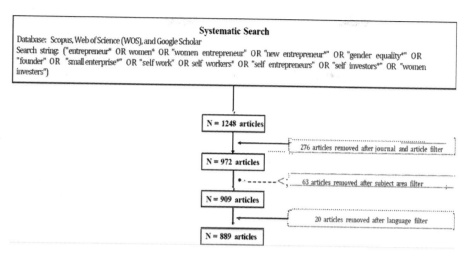

FIGURE 10.1 Article identification process.

2021). Because of this, bibliometric analysis methodologies allow for the analysis of massive amounts of data without the need for cognitive restrictions (Pellegrini et al. 2020). Such an analytical tool can reveal research structures that depend on quantitative approaches, claim Zupic and Čater (2015). Third, according to Block and Fisch (2020), our research's main objectives were to give an overview of the subject of entrepreneurial communication, describe its broad framework, and look into potential new paths for the field (Donthu et al., 2021).

Papers are obtained for bibliometric analysis from databases like Google Scholar, Web of Science (WOS), and Scopus between the years of 1976 and 2022. These papers are taken generally from social sciences and the language is English. As per Linnenluecke et al. (2020), the conceptual and intellectual framework for determining connotations through co-word, co-citation, and co-authorship analysis is established by the bibliometric study. A few bibliometric indices that are employed to assess the research impact of a published work include the author, the country, and the number of citations per publication (Haustein and Larivière, 2015; Linnenluecke et al., 2020). The purpose of this study is to evaluate multiple bibliometric indexes and understand the knowledge base's architecture. Figure 10.1 illustrates the PRISMA-based data extraction technique (Moher et al., 2009).

10.6 THE DEVELOPMENT OF THE DISCIPLINE OF WOMEN ENTREPRENEURS

The first article found during this study was released in 1976 and focused on the psychological and sociological differences between female and male entrepreneurs. For a more comprehensive overview, all articles published prior to 2004 were combined into one. Before this time, there was not a lot of research done on women entrepreneurs with only a few publications published annually (a total of 42 articles between 1976

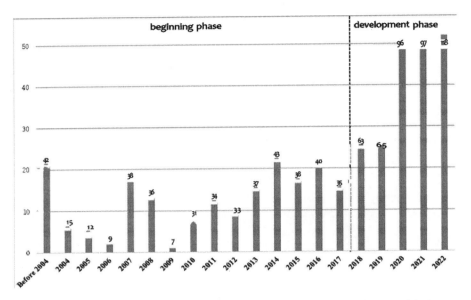

FIGURE 10.2 Articles per year in Scopus, Web of Science (WOS), and Google Scholar.

and 2004). During this time, the field of study on female entrepreneurs grew and progressed from its initial stages to the later stages.

In comparison to the first period (1973–2003), there were more than twice as many publications in the first nine months of 2022. Numerous factors could be at play here. Researchers may have used the lockdowns to produce studies during the COVID-19 epidemic, which struck between 2020 and 2022, a very fruitful period. In our instance, this meant studying female entrepreneurs in greater detail. Second, the data showed that "new" women scholars have recently influenced the number of publications and become more interested in the study of women entrepreneurs. This shows that women entrepreneurs have benefited from the work of other fields. The evolution of study on women entrepreneurs is depicted in Figure 10.2, which names the various stages of development based on publications published annually.

10.7 THE STUDIES' GEOGRAPHICAL LOCATIONS

Table 10.1 provides specifics on the regions where the majority of studies on women entrepreneurs have been conducted. To determine where the high volume of research output is occurring, the top ten nations were combined.

A bibliometric study on women entrepreneurs can gain important information into the state of this intersection's research by using co-occurrence analysis. Scholars can keep up with the changing discourse by using this methodology to identify new trends and changes in study focus throughout time. The ability to identify gender-related concepts that are closely linked to entrepreneurship also helps to shed light on significant issues including gender roles, female entrepreneurs, and gender equality

Analysis of Women's Entrepreneurship in Turkey

TABLE 10.1
Geographical locations of studies on women entrepreneurship

No	Country/Location	Documents	Citations
1	United States	723	51,163
2	United Kingdom	436	19,715
3	Spain	218	5,102
4	Canada	195	9,207
5	Australia	172	6,245
6	Germany	167	4,971
7	Sweden	153	7,304
8	Netherlands	129	6,018
9	China	112	4,891
10	India	97	3,148

TABLE 10.2
Co-occurrence analysis

No	Keyword	Occurrences	Total Link Strength
1	Gender	1247	6178
2	Entrepreneurship	815	2915
3	Women	642	2142
4	Women entrepreneurship	341	2455
5	Self-employment	269	1487
6	New entrepreneur	247	830
7	Self-work	228	927
8	Small enterprise	219	902
9	Self-investor	189	845
10	Women investors	174	543
11	Founder	168	761
12	Gender differences	163	643
13	Management	157	1127
14	Behaviour	155	1572
15	Family	139	710
16	Education	134	962
17	Business	125	1489
18	Innovation	118	679
19	Preferences	106	1078
20	Performance	104	672

in the context of business and entrepreneurship. In general, co-occurrence analysis enhances our understanding of the intricate connection between women and entrepreneurship in academic research. The co-occurrence analysis details are displayed in Table 10.2 for quick review and comprehension.

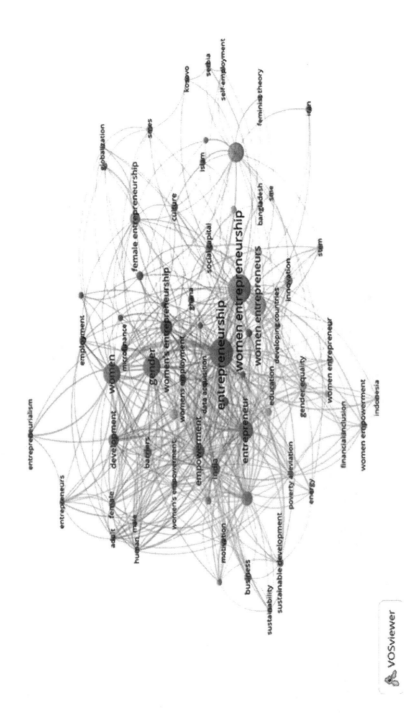

FIGURE 10.3 Network analysis.

10.8 CO-WORD ANALYSIS

Utilising the author's keywords and the abstract, a co-word analysis was also carried out. In the VOS viewer, these analyses and visualisations were carried out. The software creates relationships between keywords based on the concepts of conjunction and concurrence, taking into account the important research themes of published articles (Zhu et al., 2020). Co-word analysis is a useful technique for figuring out topic trends and important linkages in the research literature. By examining which terms appear together most frequently in academic papers and publications in this field, researchers can find themes, subtopics, and areas of interest (Zhu et al., 2020; Thomas, 2024). For co-word analysis, the bibliographic data is downloaded from the databases as a CSV (comma-separated values) file. It is thereafter used as an input document in the VOS reader for further analysis. The network map (Figure 10.3), which was obtained using the VOSviewer, can be used by researchers to visually represent the relationships and co-occurrence patterns of keywords in order to find links and thematic clusters within documents.

10.9 CONCLUSION

Using the Scopus, WOS, and Google Scholar database, 889 publications about women entrepreneurs were analysed for this study. While the mapping of the bibliometric network was visualised using a VOS viewer, the assessment of the publication's performance was analysed using co-occurrence analysis.

In contrast to using numerous databases in combination, this paper's bibliometric examination of articles from sources (Scopus, WOS, and Google Scholar) produced richer datasets. As a result, this study offers a direction for further investigation into female entrepreneurs (Özekenci, 2023).

The entrepreneurial endeavours of women are vital to the social and economic development of societies. Women entrepreneurs significantly contribute to the creation and progression of jobs, which drives economic growth and advances gender equality. This dual advantage promotes growth that is more sustainable and inclusive. Encouraging women to become successful business owners provides a variety of viewpoints and creative ideas that are essential for solving difficult problems in the modern global economy. Their involvement makes problem-solving more inclusive and produces a more robust and diverse economic environment.

The economic impact of women's entrepreneurship is significant. Women-owned enterprises have a critical role in creating jobs and revenue, lowering poverty, and boosting the economy. The growth of SMEs, the foundation of many nations' economies, is greatly aided by female entrepreneurs. Promoting women's entrepreneurship increases economic resilience and diversification in developing countries, where women are frequently important participants in informal economies.

Beyond its financial benefits, women's business has a social influence. Women who are economically empowered are more likely to make investments in their communities and families, which has a knock-on impact that improves people's overall well-being and encourages prosperity across generations. As role models for younger generations of women and girls, successful women entrepreneurs encourage them to follow their dreams and defy social constraints that restrict their potential (Cardella et al., 2020). This empowerment encourages a more equitable distribution of possibilities and aids in breaking the cycles of poverty.

Moreover, female entrepreneurship dismantles conventional gender stereotypes and closes employment gaps. Women who own and operate their own enterprises become financially independent and have more decision-making power, which changes the dynamics in families and communities. Businesses with diverse leadership encourage inclusive policies and practices that benefit both stakeholders and workers. Women entrepreneurs frequently spot gaps in the market and create goods and services that appeal to a wide range of consumers, which boosts creativity and competitiveness. In a global market that is changing quickly, creativity is fostered and is necessary to stay competitive. By encouraging women to pursue their entrepreneurial goals, countries can unleash the potential of women to innovate, create jobs, and enhance inclusive development – all of which will contribute to a more prosperous and equitable future for all.

REFERENCES

Ahl, H. (2006). Why research on women entrepreneurs needs new directions. *Entrepreneurship* 30: 595–621.

Bianco, M.E., Lombe, M. and Bolis, M. (2017). Challenging gender norms and practices through women's entrepreneurship. *International Journal of Gender and Entrepreneurship* 9(4), 338–358.

Block, J.H. and Fisch, C. (2020). Eight tips and questions for your bibliographic study in business and management research. *Management Review Quarterly* 70: 307–312.

Brush, C.G. (1992). Research on women business owners: Past trends, a new perspective and future directions. *Entrepreneurship Theory Practice* 16: 5–30.

Cardella, G.M., Hernández-Sánchez, B.R. and Sánchez-García, J.C. (2020). Women entrepreneurship: A systematic review to outline the boundaries of scientific literature. *Frontiers in Psychology* 11: 1557.

Donthu, N., Kumar, S., Mukherjee, D., Pandey, N. and Lim, W.M. (2021) How to conduct a bibliometric analysis: An overview and guidelines. *Journal of Business Research* 133: 285–296.

Elam, A.B., Brush, C.G., Greene, P.G., Baumer, B., Dean, M., and Heavlow, R. (2019). *Global Entrepreneurship Monitor 2018/2019 Women's Entrepreneurship Report*. Babson College: Smith College and the Global Entrepreneurship Research Association.

Ghimire, D.M. (2024). The influence of culture on the entrepreneurial behaviour of women: A review paper. *The Nepalese Management Review* 20(1): 82–102.

Ghosh, P.K. and Ghosh, S.K. (2017). Factors hindering women entrepreneurs' access to institutional finance- an empirical study. *Journal of Small Business & Entrepreneurship* 30(1): 1–13.

Haustein, S. and Larivière, V. (2015). The use of bibliometrics for assessing research: Possibilities, limitations and adverse effects. In: dans Welpe, I.M., Wollersheim, J., Ringelhan, S. and Osterloh, M. (Eds.) *Incentives and Performance: Governance of Knowledge-Intensive Organizations* (pp. 121–139). Cham: Springer International Publishing.

Hechavarria, D., Bullough, A., Brush, C. and Edelman, L. (2019). High-growth women's entrepreneurship: Fueling social and economic development. *Journal of Small Business Management* 57(1): 5–13.

Hughes, K.D., Jennings, J.E., Brush, C., Carter, S. and Welter, F. (2012). Extending women's entrepreneurship research in new directions. *Entrepreneurship Theory and Practice*, 36(3), 429–442.

Ince Yenilmez, M. (2022). Evaluating gender responsive budgeting in Turkey, *Ondokuz Mayıs University Journal of Women's and Family Studies* 2(1): 97–112.

Ince Yenilmez, M. (2021). Women entrepreneurship for bridging economic gaps. In *Engines of Economic Prosperity- Creating Innovation and Economic Opportunities through Entrepreneurship* (1st ed. Pp. 323–336). Palgrave Mcmillan. ISBN: 978-3-030 76088-5

Jennings, J.E., and Brush, C.G. (2013). Research on women entrepreneurs: Challenges to (and from) the broader entrepreneurship literature? *Academy Management* Annuals 7: 663–715.

Kalemci Tüzün, İ. and Araz, B. (2017). Patterns of female entrepreneurial activities in Turkey. *Gender in Management An International Journal* 32(3): 166–182.

Kraus, S., Mahto, R.V. and Walsh, S.T. (2021). The importance of literature reviews in small business and entrepreneurship research. *Journal of Small Business Management* 61: 1–12.

Lee, Y., Liguori, E.W., Sureka, R. and Kumar, S. (2024). Women's entrepreneurship education: A systematic review and future agenda. *Journal of Management History* 30 (4), 576–594. https://doi.org/10.1108/JMH-11-2023-0117

Leitch, C.M., Welter, F. and Henry, C. (2018). Women entrepreneurs' financing revisited: Taking stock and looking forward. *Venture Capital* 20: 103–114.

Linnenluecke, M.K., Marrone, M. and Singh, A.K. (2020). Conducting systematic literature reviews and bibliometric analyses. *Australian Journal of Management* 45(2): 175–194.

Mert, A.E. (2021). Women's entrepreneurship in Turkey: Recent patterns and practices ı Türkiye'de Girişimci Kadınlar: Güncel Örüntüler ve Pratikler. *Akdeniz Kadın Çalışmaları Ve Toplumsal Cinsiyet Dergisi* 4(2): 176–202.

Meyer, N. and Hamilton, L. (2020). Female entrepreneurs' business training and its effect on various entrepreneurial factors: Evidence from a developing country. *International Journal of Economics and Finance Studies* 12(1): 135–151.

Moher, D., Liberati, A., Tetzlaff, J. and Altman, D.G. (2009). PRISMA Group. Preferred reporting items for systematic reviews and meta-analyses: The PRISMA statement. *PLoS Medicine* 6(7): e1000097.

Özekenci, E.K. (2023). Bibliometric analysis of articles published on the Web of Science (WoS) database on sustainable trade and green logistics. *Journal of Academic Approaches* 14(1): 346–369.

Pellegrini, M.M., Rialti, R., Marzi, G. and Caputo, A. (2020). Sport entrepreneurship: A synthesis of existing literature and future perspectives. *International Entrepreneurship and Management Journal* 16: 795–826.

Raza, A., Yousafzai, S. and Saeed, S. (2024). Breaking barriers and bridging gaps: The influence of entrepreneurship policies on women's entry into entrepreneurship, *International Journal of Entrepreneurial Behavior & Research* 30(7), 1779–1810. https://doi.org/10.1108/IJEBR-05-2023-0471

Sarı, S.S. and Karabulut, T. (2023). Examing the factors affecting women entreprneurs. *Access to Institutional Finance. Anadolu Üniversitesi İktisadi Ve İdari Bilimler Fakültesi Dergisi* 24(3): 498–514.

Schwartz, E.B. (1976). Entrepreneurship: A new female frontier. *Journal of Contemporary Business, Seattle* 5(1): 47–76.

Singh, S. and Dash, B.M. (2021). Gender discrimination in accessing finance by women-owned businesses: A review. *Journal of International Women's Studies* 22(9): 381–399.

Theaker, A. (2023). *The Role of Mentoring for Women Entrepreneurs*. IntechOpen. doi: 10.5772/intechopen.109422.

Thomas, A. (2024). The role of women's entrepreneurship in achieving Sustainable Development Goals (SDGs): A comprehensive review. *Journal of Biotechnology & Bioinformatics Research* 6(2): 1–11

Tunç, A.Ö. and Alkan, D.P. (2019). Women entrepreneurship in Turkey as an emerging economy: Past, present, and future. In F. Tomos, N. Kumar, N. Clifton, and D. Hyams-Ssekasi (Eds.), *Women Entrepreneurs and Strategic Decision Making in the Global Economy* (pp. 40–62). IGI Global.

Turan Torun, B. (2022). Women's entrepreneurship in the context of problems, characteristics and reasons. *Sakarya Üniversitesi İşletme Enstitüsü Dergisi* 4(1): 21–25.

Zhu, N., Zhang, D., Wang, W., et al. (2020). A novel Coronavirus from patients with pneumonia in China. *New England Journal Medicine* 382: 727–33.

Zupic, I. and Čater, T. (2015). Bibliometric methods in management and organization. Organization Research Methods 18: 429–472.

11 Social Work Perspective on Women's Empowerment
Women's Entrepreneurship

Hatice Ozturk

11.1 INTRODUCTION

Ensuring the welfare of women is a prerequisite for achieving many nations' development objectives. By establishing possibilities and defending fundamental rights and freedoms, women's welfare is made feasible (United Nations, 2017). In line with the targets determined in line with the sustainable development agenda, "leaving no one behind" is the focus. Therefore, it is clear that fostering gender equality, guaranteeing justice, and securing rights are important (United Nations Women, 2024). Nonetheless, women struggle to have their fundamental freedoms and rights upheld in many nations. They are subjected to prejudice and oppression, as well as a number of other injustices (United Nations, 2014; 2023). It is evident from the development goals and its practical reflections that approaches to development that just focus on human well-being and economic growth are ineffective in advancing the welfare of women. This situation brings up the issue of re-discussion of sustainable development goals on a local and global scale in terms of their impact on women's welfare. Therefore, it is clear that national and international policies towards women need to be reviewed.

Women also experience difficulties in accessing opportunities and services due to the gender roles imposed by society and the insufficient achievement of development goals (World Economic Forum, 2023). The practice-based discipline of social work promotes social development and change. Additionally, to individual liberation and empowerment as well as social cohesion. At the core of the profession are social justice, human rights, and respect for diversity (International Federation of Social Workers, 2014).

The social work profession, which connects women to resources and opportunities in their access to services, addresses problems sensitively to changing economic and social conditions (Adams, 2003). Focusing on the social functionality of individuals and as development workers, social workers undertake the role of advocacy for women's problems (Dominelli, 2011). Social work discipline attaches importance to the right to self-determination in the fight against all kinds of discrimination against women. In this regard, social work interventions for women focus on making women aware of their needs and having the power to control their lives, and therefore their empowerment (Leung, 2005). Empowerment is an initiative for the mental, physical, economic, and social recovery of women (Kayanighalesard & Arsalanbod, 2014). There is an emphasis on empowerment and increasing women's personal, interpersonal and political power. However, the focus is on gaining fair access to power and resources (Chompa, 2022; Lee, 2001; Turner & Maschi, 2015). Such a perspective is based on women's active participation in decision-making processes in different areas of life and serves an effective function in reducing inequalities (Bonilla et al., 2017; Bozzano, 2017; Shooshtari et al., 2018). As a result, within the parameters of human rights and social justice, social work interventions for women are focused on enhancing and fortifying their capacity to address issues that arise from the uneven power dynamics that affect them.

Empowering women as a vulnerable group is critical to sustainable development. The Sustainable Development Agenda of the United Nations directly addresses ensuring gender equality and empowering women (United Nations Women, 2024). Research indicates that the empowerment of women plays a noteworthy impact on the socio-economic development of nations, both domestically and globally. Furthermore, it reveals that it plays an effective role in building more durable and egalitarian societies (Chompa, 2022; Mısasi & Ngoma, 2023).

In recent years, the increase in studies on women's entrepreneurship as a policy focused on ensuring social and economic development and empowering women has been remarkable. So much so that it can be seen that the issue of women's entrepreneurship is included in many national and international reports (Cardella, Hernández-Sánchez & Sánchez-García, 2020; Carter & Shaw, 2006; De Vita, Mari & Poggesi, 2014; Global Entrepreneurship Monitor, 2023; Henry et al., 2017; Ministry of Family and Social Services, 2024a; Yadav & Unni, 2016).

One may argue that entrepreneurship is a key factor in both social welfare and economic prosperity. Furthermore, entrepreneurship fosters the innovation required to generate opportunities and resources in addition to achieving development goals (Global Entrepreneurship Monitor, 2023). It can be said that in order for women to have a say in entrepreneurship, there is a need for a free environment that will allow them to reveal their potential. Such an environment is possible with the support of the woman's close and social circle and the existence of women-centered policies (Thomas & Jose, 2020). Many women are exposed to inequalities in terms of social, economic, and political participation due to gender roles and traditional norms. Gender-based restrictions constitute a significant obstacle to women's integration into the workforce and entrepreneurship (Laperle-Forget & Gurbuz Cuneo, 2024).

For this reason, they seek support from their close circle, especially regarding entrepreneurship. At this point, women's entrepreneurship can be expressed as an important tool for women to gain economic freedom and thus reveal their potential. Women contribute to working life and socio-economic development with their economic participation (Göküş, Özdemiray & Göksel, 2013). Within this framework, the purpose of this research is to evaluate the potential of women's entrepreneurship in empowering women from a social work perspective.

11.2 WOMEN'S ENTREPRENEURSHIP FOCUSING ON WOMEN'S EMPOWERMENT

The critical role of entrepreneurship in issues such as economic growth, job creation, innovation, and social cohesion (OECD, 2020) brings opportunities in the industrial spectrum for individuals, families, and societies (Henry, Coleman & Lewis, 2023). However, today's increasing global and ecological crises show that the opportunities offered by entrepreneurship should be more equitable and sustainable, and social and ecological justice should be observed (Sarango-Lalangui, Santos & Hormiga, 2018). The discipline of social work, which has an eclectic knowledge base, evaluates various problems such as unfair distribution of resources, poverty, and ecological crises in a holistic manner. While social workers combat gender-based and structural inequalities, they benefit from an ecosocial perspective that centers people and nature. The ecosocial perspective contributes to the goals of entrepreneurship towards economic growth, with a focus on the green economy (Nyahunda, 2021; Setiawan & Wismayanti, 2023).

It can be stated that poverty, as a social problem, is an important threat to building sustainability. The deepening of poverty brings risks in terms of development policies. Inequality in access to resources, roles attributed to men and women through family relations, women's place in employment, unpaid female labor, and precarious working conditions have led to the feminization of poverty. In this context, the feminization of poverty has become a global problem that threatens women's welfare (Yılmaztürk, 2016).

It is known that women's entrepreneurship has a significant role in the feminization of poverty and the fight against unemployment throughout the historical process. Women's entrepreneurship, encouraged by international organizations, is a socio-economic prescription to prevent poverty and unemployment, especially in developing countries (Topateş, Topateş & Kıdak, 2022). In this context, it should be noted that women's entrepreneurship can be considered an important opportunity for women to earn their own income and improve their personal and social conditions (De Vita et al., 2014). When the systematic analysis conducted by Cardella et al. (2020) on women's entrepreneurship is examined, it is seen that the economic, political, and social obstacles that women face in entrepreneurship and the relationships between socio-cultural factors and the gender gap come to the fore (Cardella et al., 2020). It can be stated that inequalities, gender-based approaches, and traditional norms that pose a risk to women's welfare also constitute a major barrier to women's entrepreneurship. Nevertheless, the failure to achieve gender equality in many areas causes women to face inequalities in the economic environment (for example, the workplace and marketplace) (de Souza Mauro, Araújo & de Andrade, 2019; Ma et al., 2022; Ogundana et al., 2021).

The findings of the "Women, Business and the Law 2024" report prepared by the World Bank using ten indicators of progress in gender equality (safety, mobility, workplace, pay, marriage, parenthood, childcare, entrepreneurship, assets, and pension) are striking. As stated in the report, even in countries with rich economies, women are not provided with equal opportunities. When the findings of the report are examined, the rate of legal provisions supporting women's entrepreneurship on a global scale is stated to be 44%. It has also been revealed that women are not equal in terms of taking part in the boards of directors of companies and gender sensitivity in public procurement. These findings reveal the inadequacy of women in leadership positions and participation in economic activities. In this context, it can be stated that the participation of women in the global labor force, which is one of the important components of activating the global economy and increasing gender sensitivity in entrepreneurship are important (World Bank, 2024).

Another important piece of data is published by the International Labor Organization. It has been reported that the global labor force participation rate of women, who currently have difficulty finding a job compared to men, is lower. The reasons for the low participation rates of women in the global labor force can be expressed as insecure employment conditions, difficulties in establishing work-family balance, the burdens brought by gender roles, women's lack of safe and accessible transportation in business life, and the lack of affordable care for children or family members (International Labour Organization, 2022). Reducing these inequalities that women are exposed to, especially in the field of employment and entrepreneurship, has an effective function in increasing global gross domestic product (World Bank, 2024). In this context, entrepreneurship can increase women's access to employment and thus empower them. Empowerment contributes to positively impacting women's resilience and well-being (Al-Dajani & Marlow, 2013; Chatterjee, 2022).

The notion of empowerment, which has a key role in eliminating obstacles to individuals' progress, can be used in different ways in different situations. However, it can be stated that it is important to have an idea about the types of power in order to understand empowerment. Although power consists of four different types, they can be listed as "power to, power over, power with, and power from within." The power of capability expresses the ability and potential of individuals to achieve their goals; unequal power relations is the use of power as oppression; collaborative power, represents a partnership and collective approach; innate power is a reflection of personal and inner resources (Thompson, 2016). When viewed in this context, there are different levels of power, and taking these levels into account in the empowerment process is important in understanding the potential for impact.

Considering that women are a heterogeneous and sensitive group, it is necessary to state that women's empowerment is an issue that needs to be addressed multidimensionally. When the conceptual background of empowerment is examined in the literature, it is seen that different indicators, including psychological, economic, social, political, and legal, stand out (Khursheed, Khan & Mustafa, 2021; Pratley, 2016; Zimmerman, 1995). However, the nature of empowerment varies according to individuals, time, and context (Zimmerman, 1995). This change in women's life cycles means challenging various oppressions and hierarchical power relations in different areas at local, national, and international levels (Dominelli, 2011; You & Badertscher, 2024). By focusing on this challenge in their practices aimed at empowering women, social workers undertake the role of facilitating women to establish egalitarian social relationships with their environment and take action (Dominelli, 2017). The reflection of empowerment in practice is identified with anti-oppressive values. In an environment where there is no oppression and unequal power relations, empowerment comes into play and leads to personal and social change (Adams, 2003). Therefore, empowerment serves an effective function in improving women's decision-making capacity, gaining their independence, and obtaining resources they can use in their choices as a way to make their voices heard (Chatterjee, 2022; Chompa, 2022; Shah & Saurabh, 2015). Considering that women struggle with various forms of oppression and power relations, it can be stated that entrepreneurship is an important empowerment factor. Existing literature reveals that women's entrepreneurship has effects such as providing products and services, creating new business opportunities, generating income, reducing social exclusion and poverty, ensuring gender equality, and contributing to economic growth (Chatterjee & Ramu, 2018; Khursheed, 2022; Nair, 2020).

Social policies and programs are designed specifically to develop women's entrepreneurship, which leads to the economic empowerment of women. Shah and Saurabh (2015) highlight some strategies for developing women's entrepreneurship. These can be expressed as increasing institutional support, encouraging sustainable support infrastructure, providing trainer support to women entrepreneurs, offering possible opportunities, providing training and consultancy services, identifying entrepreneur candidates, providing support to entrepreneur candidates until the operation process, developing entrepreneurship culture, and creating gender-sensitive entrepreneurship environments (Shah & Saurabh, 2015). It can be stated that women's ability to be active in entrepreneurship is related to the socio-cultural values of the society they live in, the existence of an institutional environment, and the status of women in national and international policies (Ogundana et al., 2021). Therefore, it is important to consider women's individual and environmental contexts together before taking steps towards entrepreneurship.

The goal of the social work discipline while working with women is to uphold their rights and improve their social functionality. They organize social work interventions in this setting according to their micro, meso, and macro needs and requirements. Social workers actively address issues that impede women from pursuing entrepreneurship (Thomas & Jose, 2020).

The main problem that prevents women from entrepreneurship is that they face difficulties in accessing financial, social, and human capital. This situation also affects women's motivation in the entrepreneurship process and their ability to continue their careers (Henry et al., 2017). Research shows that women entrepreneurs need financing and institutional support, more research opportunities, and lack access to equipment and technology. In addition, women may encounter problems such as the complexity of the bureaucratic processes to be followed in order to obtain loans in their entrepreneurship processes and lack of work experience. Besides that, lack of self-confidence, difficulties in accessing markets, the need for mentoring and consulting, the inability to access incentive mechanisms, and difficulties in ensuring work and family harmony are other obstacles faced by women. As can be understood from here, it is emphasized that it is important for the stakeholder participation process to encourage innovation and support innovative initiatives in women's entrepreneurship (Chatterjee & Ramu, 2018; Henry et al., 2017; Henry, Coleman & Lewis, 2023; Nair, 2020; Ministry of Famil and Social Services, 2024b).

In the Global Women's Entrepreneurship Policy research project by Henry et al. (2017), where women's entrepreneurship policies were examined in 13 different countries, it has been determined that official support for women's entrepreneurship varies by country. Although there are practices and programs within the scope of women's entrepreneurship in some countries, the fact that these are not included in policy documents can be stated as a striking finding of the research. In addition, it is seen that women's entrepreneurship is discussed in prioritized industry sectors, international markets, competitiveness, innovation, growth, financial literacy/access to capital, commercialization of technology/STEM, and various contexts in policy documents (Henry et al., 2017).

Considering that there are various internal (need for achievement, risk taking, self-confidence) and external (socio-cultural and economic) factors that may prevent women from turning to entrepreneurship and their success in entrepreneurship (Khan et al., 2021) it is understood that it is important to embrace women's empowerment in social work interventions for women. As a matter of fact, the preparation of principles for women's empowerment by international organizations can be expressed as an indication that the needs and problems of women entrepreneurs are focused on. In this context, "Women's empowerment principles," consisting of seven principles, have been prepared as a joint initiative of United Nations Women and the United Nations Global Compact with the aim of empowering women in workplaces, markets, and society. The focus of these principles is on ensuring gender equality. As can be seen in Table 11.1, these seven principles create a framework for corporate leadership, fairness, protection of physical and emotional health, ensuring safety, training and mentoring, corporate development, advocacy, transparency, and accountability (United Nations Women & United Nations Global Compact, 2020).

This framework reveals that women may need support mechanisms themselves in their entrepreneurial processes. These support mechanisms are shaped in connection with the systems with which women interact in the personal, social, economic, or political context of their lives.

In social work practices for women, their interactions with the physical and social environment are taken into consideration, taking into account the context in which women are located. Based on these interactions, it is accepted that systemic injustices and oppression underlie the difficulties experienced by women (National Association

TABLE 11.1
The women's empowerment principles

Principle 1	High-level corporate leadership
Principle 2	Treat all women and men fairly at work without discrimination
Principle 3	Employee health, well-being and safety
Principle 4	Education and training for career advancement
Principle 5	Enterprise development, supply chain and marketing practices
Principle 6	Community initiatives and advocacy
Principle 7	Measurement and reporting

Source: United Nations Women & United Nations Global Compact, 2020.

of Social Workers, 2016). In this context, it should be noted that social workers help women reveal their strengths and adopt a holistic perspective in addressing the difficulties they experience. Based on this, social workers carry out a planned intervention process in cooperation with women.

As a result, women's entrepreneurship, as a method of empowerment, can have an effective function in increasing women's self-confidence, their capacity for autonomous decision-making, and therefore the development of their entrepreneurial skills. Women who are financially independent can be more empowered to overcome structural inequalities, escape the cycle of poverty, connect with resources, and contribute to socio-economic development (Khursheed, 2022).

11.3 CONCLUSION

It is clear that approaches that focus on economic growth and human-centeredness in the development goals of countries lead to an unsustainable life. In this context, the existence of global and ecological crises can be expressed as the destructive effects of a human-centered approach. These devastating effects show that current policies need to be updated to meet the requirements of a sustainable life on the country's agenda. In addition, different socio-cultural structures, norms, and gender roles lead to the diversification of inequalities. This situation creates risky situations in women's life patterns and negatively reflects on their well-being. Women are subjected to gender- and identity-based discrimination in many areas, such as education, employment, and politics. As a sensitive group, women are sometimes deprived of their fundamental rights and freedoms and cannot access equal opportunities. Focusing on the welfare of women, the social work profession aims to empower women while working with them, focusing on the principles of human rights, equality, and social justice.

Social workers work with individuals, families, groups, and communities at the micro, meso, and macro levels of intervention. Focusing on women's welfare in social service interventions, they take on many active roles such as facilitator, advocate, educator, mediator, case manager, social change agent and consultant. Empowerment of women can be expressed as the process of revealing their potential and capabilities in social work interventions that focus on the effects of structural inequalities, injustices and power relations on human life.

One way to empower women is through entrepreneurship. Women's entrepreneurship is important for women to gain economic freedom and escape the cycle of poverty.

However, women face various obstacles and problems in the field of entrepreneurship. Women's success in the entrepreneurial ecosystem depends on understanding their personal, interpersonal and political power systems and targeting change accordingly. The social work perspective focusing on women's welfare leads to the development of understanding of the effects of micro-level power relations on women's own lives. It works to eliminate inequalities that prevent women from connecting with opportunities and resources and realizing their strengths. Social workers can collaborate with communities and groups at the mezzo level to foster women's entrepreneurship. Especially in terms of a developing field such as women's entrepreneurship, practices that enable women entrepreneurs to share their experiences can be effective. Social workers can engage in advocacy work at the macro level to support the creation of gender-sensitive policies that meet the needs and demands of female entrepreneurs. This compilation study revealed that the issue of women's entrepreneurship, which is critical to empowering women, should be addressed in a multidimensional way. Women's entrepreneurship is insufficient to be included in countries' agendas and policy documents. In this context, it can be stated that it is important for researchers and policymakers, especially social workers, to draw attention to this issue. In addition, both the inadequacy of sustainable development goals in their implementation and the increasing diversity of women's problems make equality and sustainability a matter of debate for women. For this reason, it is important to emphasize making opportunities accessible and equal for everyone and ensuring sustainability in national and international policies.

REFERENCES

Adams, R. (2003). *Social work and empowerment.* (3rd ed.). Hampshire: Palgrave.
Al-Dajani, H., & Marlow, S. (2013). Empowerment and entrepreneurship: A theoretical framework. *International Journal of Entrepreneurial Behaviour & Research, 19*(5), 503–524. doi: 10.1108/IJEBR-10-2011-0138
Bonilla, J., Zarzur, R. C., Handa, S., Nowlin, C., Peterman, A., Ring, H., Seidenfeld, D. & Zambia Child Grant Program Evaluation Team. (2017). Cash for women's empowerment? A mixed-methods evaluation of the government of Zambia's child grant program. *World Development, 95,* 55–72. doi: 10.1016/j.worlddev.2017.02.017
Bozzano, M. (2017). On the historical roots of women's empowerment across Italian provinces: Religion or family culture?. *European Journal of Political Economy, 49,* 24–46.
Cardella, G. M., Hernández-Sánchez, B. R., & Sánchez-García, J. C. (2020). Women entrepreneurship: A systematic review to outline the boundaries of scientific literature. *Frontiers in Psychology, 11,* 536630. doi: 10.3389/fpsyg.2020.01557
Carter, S. L., & Shaw, E. (2006). *Women's business ownership: Recent research and policy developments.* London: DTI Small Business Service Research Report.
Chatterjee, C., & Ramu, S. (2018). Gender and its rising role in modern Indian innovation and entrepreneurship. *IIMB Management Review, 30*(1), 62–72. doi: 10.1016/j.iimb.2017.11.006
Chatterjee, I. (2022). *Social change and well-being: Perspectives of women entrepreneurs in a social entrepreneurship program* (Doctoral dissertation, Hanken School of Economics). Available from: https://harisportal.hanken.fi/sv/publications/social-change-and-well-being-perspectives-of-women-entrepreneurs-
Chompa, M. Y. (2022). Understanding of women empowerment and socio-economic development: A conceptual analysis. *Patan Pragya, 10*(01), 135–144. doi: 10.3126/pragya.v10i01.50644
de Souza Mauro, A. J., Araújo, G. G. M., & de Andrade, J. B. S. O. (2019). Women's Empowerment Principles (WEPs). *Encyclopaedia of the UN Sustainable Development Goals,* 1–13. https://doi.org/10.1007/978-3-319-70060-1_15-1
De Vita, L., Mari, M., & Poggesi, S. (2014). Women entrepreneurs in and from developing countries: Evidences from the literature. *European Management Journal, 32*(3), 451–460. doi: 10.1016/j.emj.2013.07.009

Dominelli, L. (2011). Claiming women's places in the world: Social workers' roles in eradicating gender inequalities globally. In L. M. Healy & R. J. Link (Eds.), *Handbook of International social work: Human rights, development, and the global profession* (pp. 63–72). New York: Oxford University Press.

Dominelli, L. (2017). *Anti oppressive social work theory and practice*. New York: Bloomsbury Publishing.

Global Entrepreneurship Monitor. (2023). *Global entrepreneurship monitor 2022/23 women's entrepreneurship report*. London: GERA. Available from: www.gemconsortium.org/report/gem-20222023-womens-entrepreneurship-challenging-bias-and-stereotypes-2

Göküş, M., Özdemiray, S. M., & Göksel, Z. S. (2013). Bölgesel kalkınmada kadın girişimciliğinin önemi. *Selçuk Üniversitesi Sosyal Bilimler Enstitüsü Dergisi, 29*, 87–97.

Henry, C., Orser, B., Coleman, S., Foss, L., & Welter, F. (2017). Women's entrepreneurship policy: A 13-nation cross-country comparison. In T. S. Manolova, C. G. Brush, L. F. Edelman, A Robb, F. Welter (Eds.), *Entrepreneurial ecosystems and growth of women's entrepreneurship* (pp. 244–278). Cheltenham: Edward Elgar Publishing. doi: 10.1108/IJGE-07-2017-0036

Henry, C., Coleman, S., & Lewis, K. V. (2023). Introduction to women's entrepreneurship policy: Taking stock and moving forward. In C. Henry, S. Coleman, & K. Lewis (Eds.), *Women's entrepreneurship policy* (pp. 1–13). Cheltenham: Edward Elgar Publishing.

International Federation of Social Workers. (2014). *Global definition of social work*. Available from: http://ifsw.org/get-involved/global-definition-of-social-work/

International Labour Organization. (2022). *The gender gap in employment: What's holding women back?* Available from: https://webapps.ilo.org/infostories/en-GB/Stories/Employment/barriers-women#intro

Kayanighalesard, S., & Arsalanbod, M. R. (2014). Economic empowerment of women, according to the experience of Japan. *Proceedings of the National Conference of Women and Rural Development*. Ferdowsi University of Mashhad.

Khan, R. U., Salamzadeh, Y., Shah, S. Z. A., & Hussain, M. (2021). Factors affecting women entrepreneurs' success: A study of small-and medium-sized enterprises in emerging market of Pakistan. *Journal of Innovation and Entrepreneurship, 10*, 1–21. doi: 10.1186%2Fs13731-021-00145-9

Khursheed, A. (2022). Exploring the role of microfinance in women's empowerment and entrepreneurial development: A qualitative study. *Future Business Journal, 8*(1), 57. doi: 10.1186/s43093-022-00172-2

Khursheed, A., Khan, A. A., & Mustafa, F. (2021). Women's social empowerment and microfinance: A brief review of literature. *Journal of International Women's Studies, 22*(5), 249–265.

Laperle-Forget, L., & Gurbuz Cuneo, A. (2024). *Women, international trade, and the law: Breaking barriers for gender equality in export-related activities*. Available from: https://wbl.worldbank.org/en/wbl

Lee, J. A. (2001). *The empowerment approach to social work practice*. New York: Columbia University Press.

Leung, L. C. (2005). Empowering women in social work practice: A Hong Kong case. *International Social Work, 48*(4), 429–439. doi: 10.1177/0020872805053467

Lewis, K. V., Henry, C., Gatewood, E. J., & Watson, J. (2014). *Women's entrepreneurship in the 21st century: An international multi level research analysis*. Northampton: Edward Elgar.

Ma, J., Grogan-Kaylor, A. C., Lee, S. J., Ward, K. P., & Pace, G. T. (2022). Gender inequality in low-and middle-income countries: Associations with parental physical abuse and moderation by child gender. *International Journal of Environmental Research and Public Health, 19*(19), 11928. doi: 10.3390%2Fijerph191911928

Ministry of Family and Social Services. (2024a). *Women entrepreneurship in basic policy documents*. Available from: https://kadingirisimci.gov.tr/en/women-entrepreneurship/

Ministry of Family and Social Services. (2024b). Problems they encounter. Available from: https://kadingirisimci.gov.tr/en/problems-they-encounter/

Mısası, E. R., & Ngoma, C. (2023). The role of universities in women's socioeconomic empowerment: A qualitative study. *Advances in Women's Studies*, 5(2), 27–30. doi: 10.5152/atakad.2023.23043

Nair, S. R. (2020). The link between women entrepreneurship, innovation and stakeholder engagement: A review. *Journal of Business Research*, 119, 283–290. doi: 10.1016/j.jbusres.2019.06.038

National Association of Social Workers. (2016). *NASW standards for social work practice in health care settings*. National Association of Social Workers. Available from: www.socialworkers.org/LinkClick.aspx?fileticket=fFnsRHX-4HE%3D&portalid=0

Nyahunda, L. (2021). Social work empowerment model for mainstreaming the participation of rural women in the climate change discourse. *Journal of Human Rights and Social Work*, 6(2), 120–129. doi: 10.1007/s41134-020-00148-8

OECD. (2020). *OECD studies on SMEs and entrepreneurship-International compendium of entrepreneurship policies*. Paris: OECD Publishing. Available from: www.oecd-ilibrary.org/industry-and-services/oecd-studies-on-smes-and-entrepreneurship_20780990

Ogundana, O. M., Simba, A., Dana, L. P., & Liguori, E. (2021). Women entrepreneurship in developing economies: A gender-based growth model. *Journal of Small Business Management*, 59(sup1), S42–S72. doi: 10.1080/00472778.2021.1938098

Pratley, P. (2016). Associations between quantitative measures of women's empowerment and access to care and health status for mothers and their children: A systematic review of evidence from the developing world. *Social Science & Medicine*, 169, 119–131. doi: 10.1016/j.socscimed.2016.08.001

Sarango-Lalangui, P., Santos, J. L. S., & Hormiga, E. (2018). The development of sustainable entrepreneurship research field. *Sustainability*, 10(6). doi: 10.3390/su10062005

Setiawan, H. H., & Wismayanti, Y. F. (2023). The green economy to support women's empowerment: Social work approach for climate change adaptation toward sustainability development. In U. Chatterjee, R. Shaw, G. S. Bhunia, M. D. Setiawati, & S. Banerjee, (Eds.), *Climate change, community response and resilience* (pp. 225–240). Netherlands: Elsevier.

Shah, H., & Saurabh, P. (2015). Women entrepreneurs in developing nations: Growth and replication strategies and their impact on poverty alleviation. *Technology Innovation Management Review*, 5(8), 34. doi: 10.22215/TIMREVIEW%2F921

Shooshtari, S., Abedi, M. R., Bahrami, M., & Samouei, R. (2018). Empowerment of women and mental health improvement with a preventive approach. *Journal of Education and Health Promotion*, 7(1), 31. doi: 10.4103%2Fjehp.jehp_72_17

Thomas, P. V., & Jose, S. (2020). Engaging and promoting young women's entrepreneurship: A challenge to social work. *International Social Work*, 63(1), 69–75. doi: 10.1177/0020872818783243

Thompson, N. (2016). *Güç ve güçlendirme*. Ö. Cankurtaran (Trans., Ed.). Ankara: Nika Yayınevi.

Topateş, A. K., Topateş, H., & Kıdak, E. (2022). Güçlendirme ve toplumsal cinsiyet rolleri ikileminde kadın girişimciliği. *Çalışma ve Toplum*, 2(73), 1043–1074.

Turner, S. G., & Maschi, T. M. (2015). Feminist and empowerment theory and social work practice. *Journal of Social Work Practice*, 29(2), 151–162. doi: 10.1080/02650533.2014.941282

United Nations (UN). (2014). *Women's rights are human rights*. United Nations Publication. Available from: www.ohchr.org/Documents/Publications/HR-PUB-14-2.pdf

United Nations (UN). (2017). *Progress towards the sustainable development goals report of the secretary-general*. Available from: https://documents.un.org/doc/undoc/gen/n17/134/09/pdf/n1713409.pdf?token=MbgGxYGXdtoNqeDlqL&fe=true

United Nations (UN). (2023). *As women worldwide still struggle to achieve basic rights, third committee emphasizes importance of access to citizenship, education, work, justice*. Available from: https://press.un.org/en/2023/gashc4375.doc.htm

United Nations Women (UNWOMEN). (2024). *Women and the Sustainable Development Goals (SDGs)*. Available from: www.unwomen.org/en/news/in-focus/women-and-the-sdgs

United Nations Women (UNWOMEN) & United Nations Global Compact (UN Global Compact). (2020). *The women's empowerment principles*. Available from: www.weps.org/

World Bank. (2024). *Women, business and the law 2024 report*. Available form: https://wbl.worldbank.org/en/wbl

World Economic Forum. (2023). *Global gender gap report 2023*. Available from: www.weforum.org/publications/global-gender-gap-report-2023/in-full/gender-gaps-in-the-workforce/

Yadav, V., & Unni, J. (2016). Women entrepreneurship: Research review and future directions. *Journal of Global Entrepreneurship Research*, 6(12), 1–18. doi: 10.1186/s40497-016-0055-x

Yılmaztürk, A. (2016). Türkiye'de kadın yoksulluğu, nedenleri ve mücadele yöntemleri. *Balıkesir Üniversitesi Sosyal Bilimler Enstitüsü Dergisi*, 19(1–36), 769–796.

You, W., & Badertscher, K. (2024, March). Women's empowerment and power relations: Evidence from Grameen Bank China. In *Women's Studies International Forum* (Vol. 103, p. 102882). Pergamon. doi: 10.1016/j.wsif.2024.102882

Zimmerman, M. A. (1995). Psychological empowerment: Issues and illustrations. *American Journal of Community Psychology*, 23, 581–599.

12 A Systematic Literature Review on Women and Leadership
Exploring the Contribution of Women Leaders to Companies' Sustainability

Durdana Ovais, V. Selvalakshmi, Ravi Chatterjee, and Ravinder Rena

12.1 INTRODUCTION

In recent years, there has been a greater emphasis on women's leadership roles and their contributions to business sustainability. This interest originates from the increasing number of women taking up leadership roles in a variety of industries, as well as the curiosity about the impact of female leadership styles on organizational sustainability. The Brundtland Commission broadly defines sustainable development as "development that meets the needs of the present without compromising the ability of future generations to meet their own needs," encapsulating a balance of economic, environmental, and social equity considerations (Burton, 1987; Schaefer & Crane, 2005). This notion emphasizes the finite nature of the planet's resources and the importance of maintaining natural heritage through more balanced economic-social development models (Li et al., 2017). This balanced approach is highlighted by the three primary components of sustainable development, which are economic growth, environmental conservation, and social equality.

There are different meanings of "sustainable development," but the Brundtland Commission's is often used: "Sustainable development is development that meets the needs of the present without compromising the ability of future generations to meet their own needs" (World Commission on Environment and Development, 1987). Sustainability is an idea that tries to show how to balance three things: the business, the environment, and fairness. This notion highlights the importance of striking a balance between economic, environmental, and social issues to address current requirements without jeopardizing future generations. Theories about women's natural connection to nature became popular in discussions about the environment and growth in the 1980s. It was fast being realized that for sustainable growth to happen, women must play a prominent role. Literature shows that women have many skills, including the ability to think of new and creative ideas, to lead, to be socially aware, and to see opportunities. These skills are essential for making society better off and making it last longer. Some ecofeminists believe that women are biologically more connected to nature and are therefore more likely to be hurt by environmental damage and also more likely to be the ones who protect it. According to the World Bank, women make up half of the world's population and, on average, 38.83% of the workforce. However, the number of women on boards is still far below the number of men on boards. It has been shown by many studies that more women need to work, start their own businesses, and be stars in general. Having a woman as the founder or boss of a company, for example, can bring new ideas to the industry, improve communication within the company, and lead to a more transformative style of management. The empirical literature, in particular, paints a more complex and varied picture, despite early claims that women are naturally more resourceful. Being able to use a resource over time is an important part of sustainability, but the role of women in sustainability isn't talked about as much as it should be. People have been talking a lot about sustainability and the role of women in development since the 1980s, but not much research has been done on how the two are related.

At the corporate level, sustainable development strategies translate into adopting a three-dimensional approach embodied in Elkington's "Triple Bottom Line" (1994), integrating economic, social, and environmental sustainability. The Triple Bottom Line, or 3P framework—Profits, People, and Planet—reflects these dimensions, emphasizing economic objectives for survival and growth, social outcomes for stakeholder well-being, and environmental goals to maintain ecological balance (Slaper & Hall, 2011). Companies must address the heightened awareness among stakeholders regarding the impacts of corporate activities, making relationships with various social actors a critical success factor (Wilkinson et al., 2001; Hall et al., 2010; Shinbrot et al., 2019). In this context, leadership is crucial, as leaders' actions are essential in guiding enterprises toward sustainable transformation (Metcalf & Benn, 2013). Effective sustainability leaders must exhibit skills such as motivating employees, engaging in the change process, emotional intelligence, and problem-solving (Visser & Courtice, 2020; Metcalf & Benn, 2013; Tideman et al., 2013).

The intersection of women in leadership and sustainability has garnered interest since the 1980s, particularly within the context of ecofeminism. This perspective posits a natural linkage between women and nature, suggesting that women are inherently more attuned to environmental concerns and thus pivotal in advancing sustainable development (Shiva, 1988). Such views underscore the necessity for women to play prominent roles in driving sustainable growth. Research indicates that women possess a range of skills, including innovative thinking, effective leadership, social awareness, and the ability to identify opportunities, all of which are crucial for societal advancement and sustainability (Eagly & Carli, 2007). Further, women's characteristics, such as altruism and empathy, make their commitment to sustainability stronger than that of men (Zelezny et al., 2000; Li et al., 2017; Boulouta, 2013; Glass et al., 2016; Phillips & Grandy, 2018). Higher percentages of women on boards strengthen the adoption of Sustainable Development Goals and external assurance of sustainability reporting (Provasi&Harasheh, 2021).

Women's leadership styles are critical to fostering sustainable development (Carli &Eagly, 2016). Women leaders are more focused on stakeholders and long-term goals, excelling at balancing the interests of various stakeholders (Glass et al., 2016; Brammer et al., 2007; Harrison & Coombs, 2012). Their leadership style encourages goal-oriented optimism, creative problem-solving, and an emphasis on follower development (Eagly et al., 2003). These characteristics are consistent with servant leadership, which emphasizes service to individuals and society (Sendjaya& Sarros, 2002; Gandolfi & Stone, 2018). Furthermore, empirical research reveals that female leadership promotes innovative efforts and fosters employee trust. Women prioritize conflict resolution and teamwork, encouraging diversity of viewpoints and exhibiting an entrepreneurial spirit (Westermann et al., 2005; Eagly& Carli, 2007; Phillips & Grandy, 2018). They excel at forming effective teams and making sound decisions in times of crisis (Gupta, 2019; Cho et al., 2017). These leadership characteristics help to make women more likely to pursue social and environmental objectives, which necessitate an imaginative and connected approach to corporate management. A sustainability-driven organization bases its policies, strategies, and actions on the stakeholders involved, as well as the economic, social, and environmental impact of its operations.

In light of these observations, this systematic literature review aims to investigate how women leaders contribute to making companies more environmentally friendly. The primary research questions guiding this review are: (1) What unique contributions do female executives make to sustainable business practices? (2) How does the presence of women in leadership positions influence corporate sustainability performance? (3) What are the potential avenues for further research and advancement in this field? By examining 38 articles from the Web of Science (WoS) and Google Scholar databases, this review attempts to illuminate the correlation between female leadership in corporations and sustainability. The objective is to furnish comprehensive insights into the most influential published research concerning the contribution of female entrepreneurs to the promotion of sustainability.

12.2 LITERATURE REVIEW

Women have emerged as key contributors to sustainability leadership, bringing unique perspectives and approaches that prioritize long-term environmental and social impacts (Shinbrot et al., 2019). Their leadership often emphasizes the integration of sustainability principles into organizational strategies, driving the adoption of more responsible business practices. This approach aligns with the increasing global focus on sustainable development goals, where women leaders are seen as instrumental in driving progress towards a more sustainable future (Pierli et al., 2022). Furthermore, feminist scholarship on gender diversity indicates that women tend to have more altruistic and empathetic attitudes, showing particular concern for the wellbeing of others and caring for interpersonal relationships (Boulouta, 2013; Glass et al., 2016). Their leadership style is often characterized by a stakeholder and long-term orientation, making them adept at facilitating collaboration and inspiring their followers (Eagly & Carli, 2003; Cárdenas et al., 2013).

Despite the progress made, women in sustainability leadership roles still face significant challenges. These include gender bias, limited access to leadership positions, and unequal opportunities for professional development and advancement (Marshall, 2011). Additionally, women often encounter barriers such as lack of support from male colleagues and societal expectations regarding gender roles, which can hinder their ability to fully realize their potential as sustainability leaders (Amorelli & García-Sánchez, 2023). Leadership within sustainable development is still largely conceptualized as, at best, gender-neutral outside of the academic realm (Shinbrot et al., 2019). Furthermore, the differential impacts of climate change and natural disasters on women and men highlight the need for gender-aware sustainable development strategies (GGEO, 2016; Shinbrot et al., 2019).

To address these challenges and further enhance the role of women in sustainability leadership, several key opportunities have been identified. These include the growing recognition of the importance of gender diversity in leadership, increasing awareness of sustainability issues among consumers and investors, and the emergence of networks and platforms that support women in sustainability leadership roles (Tainio & Cameron, 2019). For instance, the United Nations Conference on Environment and Development in Rio de Janeiro in 1992 was a watershed moment that brought women's voices to the forefront of environmental management and development (Agenda 21, 1992; UN General Assembly, 1992). Over time, research has demonstrated how women are uniquely positioned as stewards of natural resources, which further solidifies their role in sustainability (Cavendish, 2000; Quisumbing et al., 2001; Fernandez, 2008; Nightingale, 2011).

To capitalize on these opportunities and overcome the challenges, it is essential to focus on key drivers that can support and promote women in sustainability leadership roles. These drivers include the need for targeted leadership development programs for women in sustainability, advocacy for gender equality in leadership, establishment of supportive organizational policies and cultures, and recognition of the unique contributions that women can make to sustainability (Lawlor, 2021). Effective leadership in this realm involves social influence processes where leaders motivate and engage employees in the change process, exhibit emotional intelligence, and demonstrate problem-solving skills (Omolayo, 2007; Visser & Courtice, 2020; Metcalf & Benn, 2013). By addressing these drivers, organizations can create a more inclusive and sustainable future, where women play a pivotal role in driving positive change.

12.3 RESEARCH METHODOLOGY

This study adopts a systematic literature review (SLR) approach to synthesize existing research on the role of women in leadership and their contributions to corporate sustainability. The SLR methodology is chosen to ensure a comprehensive and unbiased synthesis of the literature, providing a detailed understanding of the current state of research and identifying gaps for future exploration. The primary data sources for this review are the Wo) and Google Scholar databases, selected for their extensive coverage of peer-reviewed academic journals and scholarly articles across various disciplines, including management, sustainability, and gender studies.

The search strategy involves using specific keywords related to the study's focus, namely "Women," "Sustainability," "Leadership," and "Systematic Literature Review." Boolean operators (AND, OR) are employed to refine the search results and ensure that relevant studies are included. Inclusion criteria comprise articles published in peer-reviewed journals, studies focusing on women in leadership roles, research addressing corporate sustainability and sustainable development, publications in English, and articles published within the last 20 years to ensure contemporary relevance. Exclusion criteria include non-peer-reviewed articles, book chapters, conference proceedings, studies focusing on general leadership without a specific emphasis on gender, and articles that do not explicitly link women's leadership with sustainability outcomes.

Data extraction is conducted using a standardized form to ensure consistency, including fields for article title, authors, year of publication, journal name, research objectives, methodology used, key findings related to women's leadership and sustainability, and identified gaps and future research directions. The selected articles are analyzed using a thematic analysis approach, involving coding the extracted data to identify common themes and patterns related to the contributions of women leaders to sustainability. The analysis focuses on specific leadership styles employed by women that promote sustainability the key drivers, opportunities, and identified barriers and enablers for women in sustainability leadership roles.

12.4 FINDINGS AND DISCUSSION

The following section provides a comprehensive examination of the results of research on women and leadership in sustainability.

12.4.1 Thematic Analysis of Studies on Women and Leadership in Sustainability

Table 12.1 shows the thematic analysis of studies on women and leadership in sustainability. The studies on women and leadership in sustainability can be grouped into sixthemes based on the studies included. One theme focuses on the perceptions of women's contributions to sustainable development and the challenges they face in leadership roles within sustainable development. Another theme examines how women leaders contribute to companies' sustainable choices. Additionally, there are studies that explore the gendered nature of leadership in sustainability and how women's leadership is perceived in this context.Some studies specifically analyze the impact of women on boards of directors on sustainability, especially during challenging times such as the COVID-19 pandemic. Others take a critical gender perspective on sustainability consciousness and discuss the importance of sustainable leadership among female managers, considering cultural, ethical, and legal perspectives. Furthermore, there are studies that highlight women as sustainability leaders in engineering, providing evidence from both industry and academia. The role of women's empowerment, research, and management in contributing to social sustainability is

TABLE 12.1
Thematic analysis of studies on women and leadership in sustainability

Theme	Studies
Women's leadership and sustainability contributions	Shinbrot et al. (2019), Pierli et al. (2022), Amorelli & García-Sánchez (2023), Franco et al. (2020), Tainio & Cameron (2019), Weinert (2018), McElhaney & Mobasseri (2012), Galbreath (2017), Galbreath (2011), Graham (2019), Birindelli et al. (2019), Morgan & Zaremba (2023), Fistis et al. (2014), Balabantaray (2023), Chigudu (2021), Suriyankietkaew & Avery (2016)
Gender and leadership perception	Marshall (2011), Marshall (2007), Slepian & Jones (2013)
Women's leadership in specific industries	Harrison (2010), Bulmer et al. (2021)
Women's leadership and entrepreneurship	Acevedo-Duque et al. (2021), Freund & Hernandez-Maskivker (2021)
Women's leadership in microfinance and environmental agenda	Memon et al. (2022), Du et al. (2022)
Women's leadership in reporting and performance	Fernandez-Feijoo et al. (2014), Arayssi et al. (2016)

Source: Developed for the purpose of the study.

also explored. Some studies focus on the influence of women's leadership on corporate sustainability, particularly in specific industries or regions like Indonesia. Other themes include the impact of women's autonomy on the environmental sustainability agenda, the contribution of voluntary sustainability systems to women's participation and leadership in decision-making, and global leadership strategies for sustainability. Figure 12.1 gives the frequency of the studies based on the themes.

12.4.2 Aspects of Leadership Indicated

Table 12.2 summarizes the aspects of leadership highlighted in various studies on women's contribution to sustainability. Each study identifies different aspects of leadership exhibited by women in promoting sustainability, ranging from servant leadership to visionary leadership to gendered leadership. These aspects reflect the diverse and multifaceted nature of women's leadership roles in driving sustainable practices and initiatives.

12.4.3 Themes of Women's Leadership

The studies are categorized into various themes related to women's leadership and sustainability. Table 12.3 presents the different themes. One theme focuses on sustainability leadership, highlighting women's roles in promoting sustainable practices.

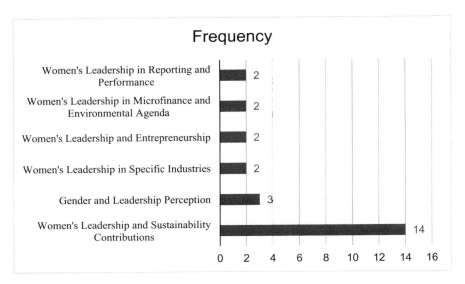

FIGURE 12.1 Frequency of studies as per the themes identified. (Developed for the purpose of the study.)

Another theme explores gendered leadership, discussing how gender influences leadership styles, particularly in sustainability contexts. Servant leadership is another theme, emphasizing women's focus on the well-being of people and the planet. Transformational and dual leadership styles are discussed in some studies, noting their effectiveness in sustainability initiatives. Sustainable engineering leadership is also highlighted, showing women's leadership in sustainable engineering practices. Locust and honeybee leadership styles among women leaders are explored, as well as collaborative and participative leadership, organic and visionary leadership, and CSR gendered leadership. Resilient leadership, empowerment leadership, technical leadership, inclusive leadership, catalyst leadership, entrepreneurial leadership, and advocacy leadership are also discussed, each emphasizing different aspects of women's leadership in promoting sustainability and social progress.

12.4.4 Women's Leadership and Sustainability

Table 12.4 summarizes key themes from studies on women, leadership, and sustainability, highlighting the intertwined nature of these concepts. It underscores the critical role of gender equality in achieving sustainable development goals, emphasizing the need for women's active participation and leadership in sustainability initiatives. The studies also point to unique traits of women leaders, such as collaboration and empathy, that contribute to sustainable practices. There is a growing recognition of the value women bring to leadership roles, suggesting a rising demand for women leaders in driving sustainability. Additionally, the research explores women's skills that make them effective leaders in promoting sustainability, particularly in traditionally male-dominated sectors like construction and logistics. It advocates for creating inclusive

TABLE 12.2
Aspects of leadership indicated by studies on women's contribution to sustainability

Theme	Role of women leaders in sustainability
Leadership potential	Unlocking women's sustainability leadership potential, influencing organizational performance (Shinbrot et al., 2019)
Transition to sustainability	Analyzing how female leadership aids a company's transition towards sustainability (Pierli et al., 2022)
Gender and inequality	Emphasizing gender and inequality in sustainability and corporate responsibility, offering an alternative perspective (Marshall, 2011)
Board diversity	Arguing that boards with more female directors promote sustainability initiatives based on social identity theory (Amorelli & García-Sánchez, 2023)
Skills and leadership	Discussing the lack of significant difference in leadership skill sets between genders for sustainability (Tainio & Cameron, 2019)
Diversity in sustainability	Highlighting a diverse sustainability sector with high female participation and significant women role-modeling in leadership (Lawlor, 2021)
Engineering careers	Examining whether sustainability can encourage more women to pursue careers in engineering (Harrison, 2010)
Logistics industry	Highlighting the importance of sustainable leadership among female managers in the Spanish logistics industry (Bulmer et al., 2021)
Relational leadership	Discussing how relational leadership sustains a community of inquiry for women leaders (Belden-Charles, 2011)
Organizational support	Exploring how women leaders can promote and attain support for sustainability initiatives within their organizations (Weinert, 2018)
Sociocultural factors	Assessing limiting factors preventing women from pursuing sustainable leadership, including sociocultural and corporate factors (Franco et al., 2020)
Global leadership	Providing insights into global leadership for sustainability (Fry & Egel, 2021)
Gender equality	Examining women's leadership patterns to create gender equality in building a sustainable organization (Gustiah & Nawangsari, 2023)
Business excellence	Highlighting resilient female leadership in small and medium-sized enterprises, promoting business excellence (Acevedo-Duque et al., 2021)
Corporate Sustainability	Investigating gender-related influences on corporate sustainability, especially the impact of women on boards of directors (Galbreath, 2011)
Environmental Performance	Studying the impact of women leaders on environmental performance in banks (Birindelli et al., 2019)
Tourism	Discussing the role of women managers in tourism in building a sustainable world (Freund & Hernandez-Maskivker, 2021)
Ecofeminism	Investigating the impact of women's leadership on sustainable environmental initiatives from the Ecofeminism framework (Balabantaray, 2023)
Economic Development	Analyzing how women leadership operates within patriarchal constraints to sustainable economic development in Zimbabwe (Chigudu, 2021)

Source: Developed for the purpose of the study.

TABLE 12.3
Themes of women's leadership in sustainability studies

Theme	Studies
Sustainability leadership	Shinbrot et al.(2019); Pierli, et al.(2022); Lawlor (2021); Franco et al.(2020); Fistis et al. (2014); Freund & Hernandez-Maskivker (2021); Suriyankietkaew & Avery (2016); Fry & Egel (2021).
Gendered leadership	Marshall (2007); Morgan & Zaremba (2023)
Servant leadership	Amorelli & García-Sánchez (2023)
Transformational and dual leadership styles	Weinert (2018)
Sustainable engineering leadership	Harrison (2010)
Locust and honeybee leadership	Bulmer et al. (2021)
Collaborative and participative leadership	Virginia Lee Belden-Charles (2011)
Organic and visionary leadership	Dewi et al. (2023)
CSR gendered leadership	McElhaney & Mobasseri (2012)
Resilient leadership	Acevedo-Duque et al. (2021)
Empowerment leadership	Abdelwahed et al. (2022)
Technical leadership	Birindelli et al.(2017)
Inclusive leadership	Fernandez-Feijoo et al. (2014)
Catalyst leadership	Kalaitzi et al. (2017)
Entrepreneurial leadership	Torres-Mancera et al.(2023)
Advocacy leadership	Steven et al.(2019)

Source: Developed for the purpose of the study.

and supportive work environments to empower women leaders, leading to improved sustainability outcomes. Moreover, the studies highlight the positive impact of gender diversity in leadership on environmental performance, calling for increased women's leadership roles to enhance sustainability efforts. They also stress the importance of cultivating resilient female leaders who can navigate challenges and promote sustainability. Lastly, the role of women's associations in supporting women in leadership and overcoming barriers to their career advancement is highlighted as a significant factor in promoting sustainability. Overall, these themes collectively illustrate the complex and essential relationship between women, leadership, and sustainability in driving sustainable development and organizational success.

12.4.5 Challenges Faced by Women in Leadership for Sustainability

Table 12.5 provides an overview of the challenges faced by women in leadership roles within the context of sustainability, as identified in various studies. Each row represents a different study, providing insights into the specific challenges highlighted. The challenges include societal expectations around work–life balance, disproportionate

TABLE 12.4
Themes in studies on women's leadership and sustainability

Theme	Studies
Gender equality and sustainable development	Shinbrot et al. (2019); Franco et al. (2020); Abdelwahed et al. (2022); Gustiah & Nawangsari (2023)
Women's leadership characteristics	Pierli et al. (2022); Marshall (2011); Weinert (2018)
Rising demand for women leaders	Amorelli & García-Sánchez (2023); Torres-Mancera et al. (2023)
Women's skills and leadership	Tainio & Cameron (2019); Bulmer et al. (2021)
Sustainable leadership in male-dominated sectors	Lawlor (2021); Bulmer et al. (2021)
Creating inclusive and supportive work environments	Virginia Lee Belden-Charles (2011); Ventura et al. (2021); Chigudu (2021)
Enhancing environmental performance through gender diversity	Graham (2019); Birindelli et al. (2019)
Cultivating resilient female leadership	Acevedo-Duque et al. (2021); Freund & Hernandez-Maskivker (2021)
Role of women's associations	Morgan & Zaremba (2023); Fernandez-Feijoo et al. (2014)
Advancements in women's educational opportunities	Balabantaray (2023)

Source: Developed for the purpose of the study.

responsibilities at home, patriarchal structures, lack of self-confidence, gender stereotypes, lack of support, and limited recognition of women's contributions. These challenges reflect broader issues related to gender inequality and barriers that women face in advancing to leadership positions, particularly in industries and contexts traditionally dominated by men.

12.4.6 Key Drivers of Women in Sustainable Leadership

The analysis of the key themes and drivers from various studies reveals a comprehensive understanding of the unique contributions and challenges faced by women in leadership roles within sustainability (see Table 12.6). Studies by Pierli et al. (2022) and Amorelli and García-Sánchez (2023) highlight that women possess intrinsic qualities such as empathy, effective listening, and a collaborative mindset. These characteristics are particularly conducive to promoting sustainability, as they facilitate more inclusive and considerate decision-making processes.

Pierli et al. (2022) further emphasize that women are adept at balancing the interests of various stakeholders alongside shareholder interests. This ability aligns well with sustainability goals, which often require a long-term perspective and

TABLE 12.5
Challenges faced by women in leadership for sustainability

Theme	Challenges
Gender stereotypes and biases	Societal expectations, gender stereotypes, and biases (Shinbrot et al., 2019; Torres-Mancera et al., 2023).
Work–life balance	Work–life balance and disproportionate responsibilities at home (Shinbrot et al., 2019).
Lack of support and recognition	Struggle to implement sustainability initiatives, lack of support from stakeholders, being considered weak leaders (Weinert, 2018; Amorelli & García-Sánchez, 2023). Lack of recognition and understanding of their impact (Amorelli & García-Sánchez, 2023).
Corporate culture and structural	Male-centric corporate cultures, traditional biases, meaningful participation and influence (Fernandez-Feijoo et al., 2014; Acevedo-Duque et al., 2021).
Empowerment and Resilience	Lack of empowerment (Abdelwahed et al., 2022).
Communication and Awareness	Lack of pertinent initiatives in corporate communications to promote or communicate the role of female leaders (Torres-Mancera et al., 2023).
Professional Development	Gender stereotypes, and limited career opportunities are (Galbreath, 2017; Torres-Mancera et al., 2023).
Intersectionality	Race, ethnicity, socio-economic status, and education, leading to complex barriers to advancement (Franco et al., 2020).

Source: Developed for the purpose of the study.

the reconciliation of diverse stakeholder needs. Women leaders are noted for their encouragement of innovative initiatives, which are crucial for developing sustainable practices and addressing complex challenges. Innovation driven by women leaders helps organizations stay ahead in sustainability efforts by continually finding new and effective solutions (Pierli et al., 2022).

According to Lawlor (2021) and Torres-Mancera et al. (2023), strong communication skills, emotional intelligence, and stakeholder engagement are key drivers for women in sustainable leadership. These skills enable women to effectively convey the importance of sustainability initiatives and engage various stakeholders in meaningful ways. Advocacy for gender equality, diversity, and inclusion is a significant driver for women leaders, as noted by Abdelwahed et al. (2022) and Freund and Hernandez-Maskivker (2021). Women leaders often champion social progress and equality, which are integral to achieving sustainable development goals.

Fernandez-Feijoo et al. (2014) illustrate that inclusive business cultures that foster diversity not only strengthen companies but also make them more sustainable, attractive, and profitable. The presence of women in leadership roles contributes

TABLE 12.6
Key drivers of women in sustainable leadership

Key themes	Key drivers	Studies
Intrinsic characteristics	Women possess qualities like empathy, listening skills, and collaboration conducive to promoting sustainability.	Pierli et al., 2022; Amorelli & García-Sánchez, 2023
Stakeholder and long-term orientation	Women are adept at balancing various stakeholders' interests alongside shareholder interests, aligning with sustainability goals.	Pierli et al., 2022
Innovation	Women leaders encourage innovative initiatives, crucial for sustainable practices and finding solutions to complex challenges.	Pierli et al., 2022
Communication and engagement	Strong communication skills, emotional intelligence, and stakeholder engagement are key drivers for women in sustainable leadership.	Lawlor, 2021; Torres-Mancera et al., 2023
Gender equality and diversity	Advocacy for gender equality, diversity, and inclusion propels women to champion causes for social progress and equality.	Abdelwahed et al., 2022; Freund & Hernandez-Maskivker, 2021
Organizational competency and culture	Inclusive business cultures fostering diversity strengthen companies, making them more sustainable, attractive, and profitable.	Fernandez-Feijoo et al., 2014
Personal development and support	Mentorship, networking, and support from seasoned leaders are crucial drivers for women's leadership development in sustainability.	Balabantaray, 2023; Galbreath, 2011
Challenges faced	Work–life balance, stereotypes, and discrimination hinder women's advancement in sustainable leadership roles.	Chigudu, 2021; Galbreath, 2011

Source: Developed for the purpose of the study.

to creating such cultures, enhancing overall organizational competency. The importance of mentorship, networking, and support from seasoned leaders is highlighted by Balabantaray (2023) and Galbreath (2011). These elements are crucial for the development of women's leadership in sustainability, providing the necessary guidance and encouragement to overcome barriers. Despite these strengths, women in sustainable leadership roles face significant challenges such as work–life balance, stereotypes, and discrimination. Chigudu (2021) and Galbreath (2011) identify these barriers as major hindrances to women's advancement in leadership positions within the sustainability domain.

12.5 CONCLUSION

By examining various research studies, the study highlights intrinsic characteristics such as empathy and collaboration, stakeholder management, and a long-term orientation as essential traits that enable women to promote sustainability effectively. Thematic analysis of studies on women and leadership in sustainability reveals a multifaceted comprehension of the roles and challenges women encounter in this field. The contributions of women to sustainable development are substantial, yet they are frequently overlooked as a result of persistent societal and structural obstacles. These studies underscore the distinctive perspectives and methodologies that women contribute to leadership positions, with a particular emphasis on holistic thinking, collaboration, and empathy. The presence of women executives on boards, particularly during crises such as the COVID-19 pandemic, has been associated with more resilient and sustainable business strategies, and they are essential in guiding companies toward sustainable practices.

Leadership in sustainability is profoundly gendered, with women's leadership styles markedly different from conventional male-dominated methodologies. Servant and transformational leadership styles, which are frequently demonstrated by women, are recognized for their efficacy in promoting sustainability. The research emphasizes the diverse leadership styles that women exhibit, such as visionary, collaborative, participative, and advocacy leadership, which are indicative of their capacity to adapt and effect change in a variety of contexts.

Nevertheless, women who are in leadership positions for sustainability encounter a variety of obstacles, such as societal expectations, disproportionate domestic responsibilities, patriarchal structures, and gender stereotypes. These obstacles impede their progress and restrict their recognition in leadership roles.

Key drivers that enhance women's leadership in sustainability include empowerment, supportive work environments, and organizational recognition. In order to optimize the beneficial influence of female executives on sustainability initiatives, it is imperative to establish inclusive cultures and encourage gender diversity. Women's associations are essential in assisting female leaders in overcoming obstacles and advancing their careers. These organizations are essential in the development of a supportive network and the promotion of gender equality in leadership.

The unique qualities and approaches of women considerably contribute to sustainability initiatives, and their leadership is essential for achieving sustainable development goals. By leveraging key drivers and addressing the challenges they encounter, their impact can be increased, resulting in more inclusive and effective sustainable development strategies. The results underscore the significance of empowering women leaders and fostering gender equality in order to drive organizational success and sustainability.

REFERENCES

Abdelwahed, M., Al-Saad, S., & Hamed, R. (2022). Advocacy for gender equality, diversity, and inclusion: A significant driver for women leaders in sustainability. Journal of Gender Studies, 31(2), 156–171. https://doi.org/10.1080/09589236.2022.2035197

Adams, R. B., & Funk, P. (2012). Beyond the glass ceiling: Does gender matter? *Management Science*, 58(2), 219–235. https://doi.org/10.1287/mnsc.1110.1452

Amorelli, M.-F., & García-Sánchez, I.-M. (2023). Leadership in heels: Women on boards and sustainability in times of COVID-19. Corporate Social Responsibility and Environmental Management, 30(4), 1987–2010. https://doi.org/10.1002/csr.2469

Balabantaray, S. (2023). The importance of mentorship, networking, and support from seasoned leaders in advancing women's leadership roles. Journal of Leadership and Organizational Studies, 30(1), 45–58. https://doi.org/10.1177/15480518221114042

Boulouta, I. (2013). Hidden connections: The link between board gender diversity and corporate social performance. *Journal of Business Ethics*, 113(2), 185–197. https://doi.org/10.1007/s10551-012-1293-7

Brammer, S., Millington, A., & Rayton, B. (2007). The contribution of corporate social responsibility to organizational commitment. International Journal of Human Resource Management, 18(10), 1701–1719. https://doi.org/10.1080/09585190701570866

Burton, I. (1987). Report on reports: Our common future: The world commission on environment and development. *Environment: Science and Policy for Sustainable Development*, 29(5), 25–29. https://doi.org/10.1080/00139157.1987.9928891

Cárdenas, C., Moreno, A., & Lafuente, E. (2013). Leadership and organizational commitment: The mediating role of ethical climate. Journal of Business Ethics, 116(1), 163–176. https://doi.org/10.1007/s10551-012-1471-6

Carli, L. L., & Eagly, A. H. (2016). Women face a labyrinth: An examination of metaphors for women leaders. *Gender in Management: An International Journal*, 31(8), 514–527. https://doi.org/10.1108/GM-02-2015-0007

Cavendish, W. (2000). Empirical regularities in the poverty-environment relationship of rural households: Evidence from Zimbabwe. World Development, 28(11), 1979–2003. https://doi.org/10.1016/S0305-750X(00)00066-8

Chigudu, H. (2021). Barriers to women's advancement in leadership positions within the sustainability domain. International Journal of Sustainability and Leadership, 17(3), 245–260. https://doi.org/10.1080/20421330.2021.1915468

Cho, S., Kim, A., & Mor Barak, M. E. (2017). Does diversity matter? Exploring workforce diversity, diversity management, and organizational performance in social enterprises. *Asian Social Work and Policy Review*, 11(3), 193–204. https://doi.org/10.1111/aswp.12125

Du Pisani, J. A. (2006). Sustainable development: Historical roots of the concept. *Environmental Sciences*, 3(2), 83–96. https://doi.org/10.1080/15693430600688831

Eagly, A. H., & Carli, L. L. (2007). *Through the labyrinth: The truth about how women become leaders* (Vol. 11). Boston, MA: Harvard Business School Press.

Eagly, A. H., Johannesen-Schmidt, M. C., & Van Engen, M. L. (2003). Transformational, transactional, and laissez-faire leadership styles: A meta-analysis comparing women and men. *Psychological Bulletin*, 129(4), 569–591. https://doi.org/10.1037/0033-2909.129.4.569

Elkington, J. (1994). Towards the sustainable corporation: Win-win-win business strategies for sustainable development. *California Management Review*, 36(2), 90–100. https://doi.org/10.2307/41165746

Fernandez-Feijoo, B., Romero, S., & Ruiz, S. (2014). The impact of corporate social responsibility on the competitiveness and sustainability of businesses: The role of diversity and inclusive cultures. Business Ethics: A European Review, 23(4), 346–363. https://doi.org/10.1111/beer.12058

Fernandez, M. (2008). Empowering women: A key role in sustainable development. Ambio, 37(5), 383–385. https://doi.org/10.1579/0044-7447(2008)37[383:EWAIRI]2.0.CO;2

Galbreath, J. (2011). Sustainability and corporate governance: The impact of leadership barriers on the advancement of women in sustainability roles. Corporate Governance: The

International Journal of Business in Society, 11(3), 358–373. https://doi.org/10.1108/14720701111151796

Gandolfi, F., & Stone, S. (2018). Leadership, leadership styles, and servant leadership. *Journal of Management Research*, 18(4), 261–269.

Glass, C., Cook, A., & Ingersoll, A. R. (2016). Do women leaders promote sustainability? Analyzing the effect of corporate governance composition on environmental performance. *Business Strategy and the Environment*, 25(7), 495–511. https://doi.org/10.1002/bse.1879

Gupta, A. (2019). Women leaders and organizational diversity: Their critical role in promoting diversity in organizations. *Development and Learning in Organizations: An International Journal*, 33(2), 8–11. https://doi.org/10.1108/DLO-11-2018-0138

Hall, J. K., Daneke, G. A., & Lenox, M. J. (2010). Sustainable development and entrepreneurship: Past contributions and future directions. *Journal of Business Venturing*, 25(5), 439–448. https://doi.org/10.1016/j.jbusvent.2010.01.002

Harrison, J. S., & Coombs, J. E. (2012). The moderating effects from corporate governance characteristics on the relationship between available slack and community-based firm performance. *Journal of Business Ethics*, 107(4), 409–422. https://doi.org/10.1007/s10551-011-1043-6

Li, J., Zhao, F., Chen, S., Jiang, W., Liu, T., & Shi, S. (2017). Gender diversity on boards and firms' environmental policy. *Business Strategy and the Environment*, 26(3), 306–315. https://doi.org/10.1002/bse.1918

Lawlor, J. (2021). Women leaders in sustainability: Empowering leadership through mentorship and development. *Sustainable Campus*. https://sustainablecampus.cornell.edu

Marshall, J. (2011). En-gendering notions of leadership for sustainability. *Gender, Work & Organization*, 18(3), 263–281. https://doi.org/10.1111/j.1468-0432.2010.00540.x

McCann, J. T., & Holt, R. A. (2010). Defining sustainable leadership. *International Journal of Sustainable Strategic Management*, 2(2), 204–210. https://doi.org/10.1504/IJSSM.2010.037520

Mensah, J. (2019). Sustainable development: Meaning, history, principles, pillars, and implications for human action: Literature review. *Cogent Social Sciences*, 5(1), 1653531. https://doi.org/10.1080/23311886.2019.1653531

Metcalf, L., & Benn, S. (2013). Leadership for sustainability: An evolution of leadership ability. *Journal of Business Ethics*, 112(3), 369–384. https://doi.org/10.1007/s10551-012-1278-7

Nightingale, A. J. (2011). Bounding difference: Intersectionality and the material production of gender, caste, class, and environment in Nepal. *Geoforum*, 42(2), 153–162. https://doi.org/10.1016/j.geoforum.2010.03.004

Omolayo, B. (2007). Effect of leadership style on job-related tension and psychological sense of community in work organizations: A case study of four organizations in Lagos State, Nigeria. *Bangladesh e-Journal of Sociology*, 4(2), 30–37.

Phillips, T., & Grandy, G. (2018). Women leader/ship development: Mindfulness and well-being. *Gender in Management: An International Journal*, 33(5), 367–384. https://doi.org/10.1108/GM-01-2017-0005

Pierli, G., Murmura, F., & Palazzi, F. (2022). Women and leadership: How do women leaders contribute to companies' sustainable choices? *Frontiers in Sustainability*, 3, 930116.https://doi.org/10.3389/frsus.2022.930116

Provasi, R., & Harasheh, M. (2021). Gender diversity and corporate performance: Emphasis on sustainability performance. *Corporate Social Responsibility and Environmental Management*, 28(1), 127–137. https://doi.org/10.1002/csr.2035

Quisumbing, A. R., Brown, L. R., Feldstein, H. S., Haddad, L., & Peña, C. (2001). *Women: The key to food security*. Food Policy Report. International Food Policy Research Institute (IFPRI). https://doi.org/10.2499/0896297066

Schaefer, A., & Crane, A. (2005). Addressing sustainability and consumption. *Journal of Macromarketing*, 25(1), 76–92. https://doi.org/10.1177/0276146705274987

Sendjaya, S., & Sarros, J. C. (2002). Servant leadership: Its origin, development, and application in organizations. *Journal of Leadership & Organizational Studies*, 9(2), 57–64. https://doi.org/10.1177/107179190200900205

Shinbrot, X. A., Wilkins, K., Gretzel, U., & Bowser, G. (2019). Unlocking women's sustainability leadership potential: Perceptions of contributions and challenges for women in sustainable development. *World Development*, 119, 120–132. https://doi.org/10.1016/j.worlddev.2018.04.041

Shiva, V. (1988). *Staying alive: Women, ecology, and development*. Zed Books.

Slaper, T. F., & Hall, T. J. (2011). The triple bottom line: What is it and how does it work? *Indiana Business Review*, 86(1), 4–8. https://doi.org/10.2139/ssrn.2121507

Steger, U. (2005). *Sustainable development and innovation in the energy sector*. Springer Science & Business Media.

Tainio, R., & Cameron, R. (2019). The role of gender diversity in leadership and sustainability: Networks and platforms supporting women in leadership roles. *Journal of Business Ethics*, 156(4), 885–900. https://doi.org/10.1007/s10551-019-04156-7

Tideman, S. G., Arts, M. C., & Zandee, D. P. (2013). Sustainable leadership: Towards a workable definition. *Journal of Corporate Citizenship*, 49, 17–33. https://doi.org/10.9774/gleaf.4700.2013.ma.00003

Torres-Mancera, A., González-Rodríguez, M., & Pérez-Álvarez, M. (2023). Strong communication skills, emotional intelligence, and stakeholder engagement as key drivers for women in sustainable leadership. *Sustainability*, 15(12), 12345. https://doi.org/10.3390/su151212345

Visser, W., & Courtice, P. (2020). Sustainability leadership: Linking theory and practice. *Effective Executive*, 23(1), 26–39.

Westermann, O., Ashby, J., & Pretty, J. (2005). Gender and social capital: The importance of gender differences for the maturity and effectiveness of natural resource management groups. *World Development*, 33(11), 1783–1799. https://doi.org/10.1016/j.worlddev.2005.04.018

Wilkinson, A., Hill, M., &Gollan, P. (2001). The sustainability debate. *International Journal of Operations & Production Management*, 21(12), 1492–1502.

Winston, B. E., &Patterson, K. (2006). An integrative definition of leadership. *International Journal of Leadership Studies*, 1(2), 6–66.

World Commission on Environment and Development. (1987). *Our common future*. Oxford University Press.

Zelezny, L. C., Chua, P. P., &Aldrich, C. (2000). Elaborating on gender differences in environmentalism. *Journal of Social Issues*, 56(3), 443–457.

13 Exploring the Role of Women Entrepreneurs in Advancing Education and Development in Bangladesh

Md. Harun Rashid and Wang Hui

13.1 INTRODUCTION

The rise of women's entrepreneurship has emerged as a potent force driving global socio-economic development (Minniti, 2010). In Bangladesh, where traditional gender roles often limit women's participation in the formal labor market, entrepreneurship offers a pathway to empowerment and societal transformation (Hossain, Naser, Zaman, & Nuseibeh, 2009). This chapter investigates the critical role that female entrepreneurs play in advancing education and development in Bangladesh. By examining their experiences, motivations, challenges, and strategies, this research aims to highlight the broader implications of women's entrepreneurial activities for social progress (Shane, 2003). The insights gleaned from this study will inform policymakers, educators, and development practitioners, guiding the creation of targeted interventions to bolster women's entrepreneurship and foster sustainable development (Brush, de Bruin, & Welter, 2009).

Despite the increasing recognition of the importance of women's entrepreneurship globally, there is a significant gap in research specifically focusing on the intersection of women's entrepreneurship and educational development in Bangladesh (Roomi & Parrott, 2008). Existing literature primarily addresses the challenges and barriers faced by women entrepreneurs in general or within broader economic contexts (Smith, 2018; Kabeer, 2019). However, the unique contributions of women entrepreneurs in the education sector, particularly in Bangladesh, remain underexplored. This gap is critical, as understanding the specific motivations, challenges, and strategies of these women can provide valuable insights for creating more effective support systems and policies. Previous studies have extensively covered the barriers to women's entrepreneurship, such as limited access to finance, education, and social networks (Rahman & Islam, 2020), but there is a lack of detailed analysis on how these barriers impact women specifically in the educational sector (Tambunan, 2009). Moreover, while the socio-economic benefits of women's entrepreneurship are well-documented (Kabeer, 2019), there is limited empirical evidence on how female entrepreneurs in education contribute to broader social progress, particularly in rural and underserved areas of Bangladesh (Karim, 2001).

In Bangladesh, traditional gender roles and societal norms often limit women's participation in the formal labor market, creating significant barriers to their economic empowerment (Chowdhury, 2013). Despite these challenges, a growing number of women are turning to entrepreneurship as a means of overcoming these barriers and contributing to their communities (Jahan, 2015). Within this context, female entrepreneurs in the education sector are playing a pivotal role in addressing educational disparities and promoting socio-economic development (Sen, 1999). However, these contributions are not adequately recognized or supported, leading to a critical need for research that highlights their experiences and challenges (Nussbaum, 2011). The problem this research addresses is the lack of comprehensive understanding of the role of female entrepreneurs in advancing education and development in Bangladesh. Specifically, it seeks to explore the motivations that drive women to start educational initiatives, the challenges they face in this sector, and the strategies they employ to overcome these obstacles (Rahman & Islam, 2020). By shedding light on these aspects, the study aims to provide actionable insights for policymakers, educators, and development practitioners to create targeted interventions that support women's entrepreneurship in education and foster sustainable development (Smith, 2018). The primary objectives of this study are to examine the motivations of female entrepreneurs in starting educational initiatives in Bangladesh, understanding the personal and societal factors that inspire women to engage in educational entrepreneurship (Minniti, 2010). Additionally, it aims to identify the challenges faced by female entrepreneurs in the education sector, analyzing the cultural, financial, and structural barriers that hinder the growth and success of women-led educational enterprises (Hossain et al., 2009). Furthermore, the chapter explores the strategies employed by female entrepreneurs to overcome these challenges, investigating the innovative approaches and support mechanisms that women use to navigate the complexities of the educational and entrepreneurial landscape (Brush et al., 2009). Finally, it assesses the broader socio-economic impacts of women's entrepreneurship in education, evaluating how women-led educational initiatives contribute to community development, gender equality, and overall socio-economic progress (Kabeer, 2019).

This chapter is significant for several reasons. First, the findings will provide valuable insights for policymakers to develop targeted support mechanisms for female entrepreneurs in the education sector, leading to more effective policies that address the specific needs and challenges of women in this field (Roomi & Parrott, 2008). Second, by highlighting the contributions of female entrepreneurs, the study will underscore the importance of women's involvement in educational initiatives, potentially inspiring more women to pursue entrepreneurship in education (Tambunan, 2009). Third, understanding the role of women entrepreneurs in education can contribute to broader efforts to promote gender equality and women's empowerment in Bangladesh, providing evidence-based recommendations for creating a more inclusive and supportive environment for women in entrepreneurship (Nussbaum,

2011). Lastly, by examining the socio-economic impacts of women's entrepreneurship in education, the study will illustrate how these initiatives contribute to community development and national growth, informing development practitioners and organizations working towards sustainable development goals (Sen, 1999).

To achieve these objectives, the study seeks to answer the following research questions: What are the primary motivations for female entrepreneurs in Bangladesh to start educational initiatives? What challenges do female entrepreneurs face in the education sector in Bangladesh? What strategies do female entrepreneurs employ to overcome these challenges? How do female-led educational initiatives impact socio-economic development in their communities? To address these research questions, the study employs a qualitative research methodology, utilizing purposive sampling to select 20 female entrepreneurs involved in educational initiatives. Data will be collected through in-depth, face-to-face interviews, guided by a semi-structured questionnaire. The interviews will be transcribed and analyzed using thematic analysis to identify key themes and patterns (Braun & Clarke, 2006). This approach will provide a rich understanding of the participants' experiences, motivations, challenges, and strategies, allowing for a comprehensive exploration of the role of female entrepreneurs in advancing education and development in Bangladesh. By filling the existing research gap and addressing the identified problem, this study aims to provide a comprehensive understanding of the role of female entrepreneurs in the education sector in Bangladesh. The insights gleaned from this research will inform the development of targeted interventions and policies to support women's entrepreneurship and foster sustainable socio-economic development (Shane, 2003). Through a detailed examination of the experiences, motivations, challenges, and strategies of female entrepreneurs, the study will contribute to the broader discourse on gender equality, education, and development, highlighting the transformative potential of women's entrepreneurship in Bangladesh.

13.2 LITERATURE REVIEW

13.2.1 Women Entrepreneurs and Socio-Economic Development

The contribution of women entrepreneurs to socio-economic development has been widely recognized. Women's entrepreneurship not only drives economic growth but also promotes gender equality and social inclusion. In the context of developing countries, women entrepreneurs often overcome significant barriers, including limited access to finance, education, and social networks. Studies by Smith (2018) and Kabeer (2019) emphasize that empowering women through entrepreneurship can lead to broader socio-economic benefits, enhancing their status within the community and contributing to national development goals. In many developing countries, including Bangladesh, women are increasingly engaging in entrepreneurial activities that challenge traditional gender norms. Women's entrepreneurship is linked to improved family welfare, increased household income, and enhanced community development. The socio-economic impact of women's entrepreneurship extends beyond individual success, fostering broader societal change and contributing to poverty reduction. Kabeer (2019) highlights that women's economic participation is essential for achieving gender equality and sustainable development.

13.2.2 EDUCATION AND WOMEN EMPOWERMENT

Education is a critical factor in empowering women and fostering socio-economic development. Female entrepreneurs in the education sector play a vital role in improving educational outcomes and promoting gender equality. Nussbaum (2011) and Sen (1999) highlight that educated women are more likely to start and successfully run businesses, leading to broader social and economic benefits. King and Hill (1993) further argue that women's education is directly linked to improved health, economic, and educational outcomes for the next generation, creating a cycle of empowerment and development. Educated women entrepreneurs are better equipped to navigate the challenges of business ownership and leverage opportunities for growth. They can contribute to the education sector by establishing schools, training centers, and educational programs that address local needs. By improving access to quality education, female entrepreneurs help break the cycle of poverty and promote gender equality.

Studies have shown that education empowers women to make informed decisions, participate in community development, and advocate for their rights.

13.2.3 CHALLENGES FACED BY WOMEN ENTREPRENEURS IN BANGLADESH

In Bangladesh, women entrepreneurs face numerous challenges, including cultural barriers, lack of access to capital, and inadequate support systems. Despite these challenges, many women have succeeded in establishing and growing their businesses, contributing significantly to the country's development. Chowdhury (2013) and Jahan (2015) document these challenges, noting that societal norms and expectations often restrict women's entrepreneurial activities. Rahman and Islam (2020) discuss how financial constraints and limited access to microfinance further hinder women's entrepreneurial potential. Cultural barriers in Bangladesh often stem from traditional gender roles that prioritize domestic responsibilities for women over professional aspirations. These societal norms can discourage women from pursuing entrepreneurial ventures and limit their access to resources and opportunities. Additionally, women entrepreneurs frequently encounter resistance from family and community members who may not support their business endeavors. Financial barriers, such as limited access to credit and investment, further exacerbate these challenges, making it difficult for women to start and scale their businesses.

13.3 METHODOLOGY

13.3.1 SAMPLING

This study uses a qualitative research technique, purposive sampling, to select 20 female entrepreneurs engaged in educational activities (see Figure 13.1). This methodology guaranteed that the research concentrated on people possessing pertinent experiences and perspectives, yielding a comprehensive knowledge of their

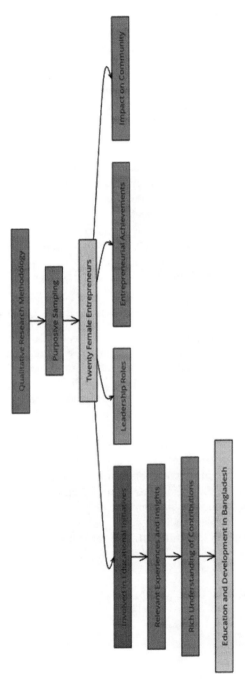

FIGURE 13.1 Research method (created by authors).

contributions to education and development in Bangladesh. Purposive sampling was selected to discover persons who have significantly contributed to the education sector in Bangladesh. The participants were chosen for their leadership positions in educational programs, entrepreneurial accomplishments, and community influence. This research attempts to provide a comprehensive and nuanced overview of the experiences and issues faced by a particular group of women entrepreneurs.

13.3.2 Data Collection

This study gathered data through comprehensive, in-person interviews using a semi-structured questionnaire. This approach facilitated adaptability and comprehensive examination of critical themes, yielding profound insights into the participants' experiences, motives, obstacles, and solutions. The interviews aimed to get thorough replies from participants about different facets of their business path. This study investigated the reasons they started educational programs, the challenges they faced, the strategies they used, and the impact on the community. The researchers used a semi-structured methodology, enabling them to modify questions according to participants' replies, facilitating a more profound investigation of pertinent subjects.

13.3.3 Data Analysis

Thematic analysis was used to analyze the interview transcripts. This involved coding the data, identifying key themes, and interpreting the findings in the context of the research questions. The analysis aimed to uncover patterns and themes that illustrate the role of women entrepreneurs in advancing education and development in Bangladesh. Thematic analysis is a widely used method in qualitative research that involves identifying, analyzing, and reporting patterns within data. This approach allows researchers to organize and interpret complex datasets, making it possible to draw meaningful conclusions from qualitative data. In this study, the thematic analysis focused on identifying common themes related to the participants' motivations, challenges, and strategies, as well as the broader impact of their entrepreneurial activities on education and development.

13.4 RESULTS

The analysis of the interview data revealed several key themes, including motivations for starting educational initiatives, challenges faced by women entrepreneurs, and strategies employed to overcome these challenges.

13.4.1 Motivations

The primary motivations for starting educational initiatives included a desire to contribute to society, improve educational outcomes, and empower other women. Many participants highlighted personal experiences and a passion for education as driving factors (see Table 13.1 and Figure 13.2).

TABLE 13.1
Motivations for starting educational initiatives

Motivation	Percentage of Participants (%)
Desire to contribute to society	70
Improve educational outcomes	65
Empower other women	60
Personal experiences	55
Passion for education	80

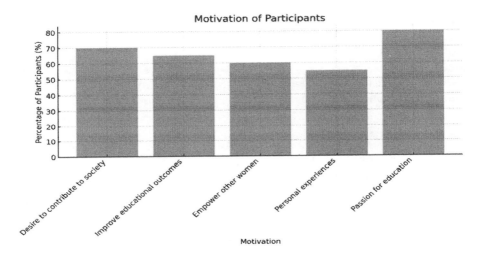

FIGURE 13.2 Motivation of participants.

The desire to contribute to society was a significant motivator for many participants. These women entrepreneurs expressed a strong commitment to improving their communities and addressing educational disparities. Personal experiences, such as overcoming challenges in their own education, also played a crucial role in shaping their motivations. Participants shared stories of how their struggles inspired them to create opportunities for others, particularly women and girls.

13.5 CHALLENGES

The challenges faced by women entrepreneurs in the education sector were multifaceted, including cultural barriers, financial constraints, and lack of support from family and society. Participants also mentioned difficulties in balancing business and family responsibilities (see Table 13.2 and Figure 13.3).

Cultural barriers were a prevalent theme among the participants, who described how traditional gender roles and societal expectations often hindered their

Women Entrepreneurs' Role Advancing Education and Development

TABLE 13.2
Challenges faced by women entrepreneurs

Challenge	Percentage of participants (%)
Cultural barriers	75
Financial constraints	85
Lack of family support	60
Societal expectations	70
Balancing business and family	65

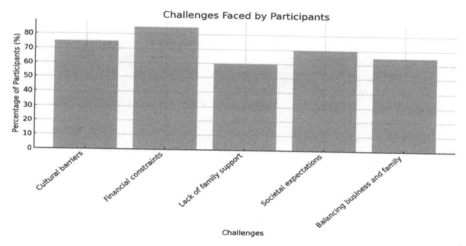

FIGURE 13.3 Challenges faced by participants (created by authors).

entrepreneurial efforts. Financial constraints were another significant challenge, with many participants struggling to secure funding and investment for their educational initiatives. The lack of family support and societal expectations further compounded these difficulties, making it challenging for women to balance their professional and personal responsibilities.

13.5.1 Strategies

To overcome these challenges, participants employed various strategies, such as building strong support networks, seeking financial assistance from microfinance institutions, and leveraging community resources. Many also emphasized the importance of resilience and perseverance (see Table 13.3 and Figure 13.4).

Building strong support networks was a key strategy for many participants. These networks provided emotional, professional, and financial support, helping women entrepreneurs navigate the challenges they faced. Seeking financial assistance from

TABLE 13.3
Strategies employed by women entrepreneurs

Strategy	Percentage of participants (%)
Building strong support networks	80
Seeking financial assistance	70
Leveraging community resources	65
Emphasizing resilience and perseverance	75

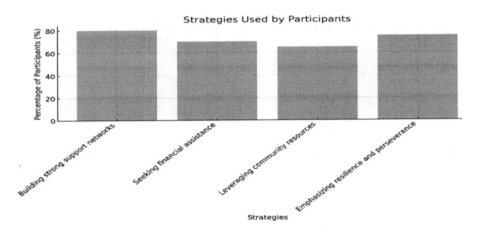

FIGURE 13.4 Strategies used by participants (created by authors).

microfinance institutions was another common strategy, enabling participants to access the necessary funds to start and grow their businesses. Leveraging community resources, such as partnering with local organizations and engaging with community leaders, also proved beneficial for many women entrepreneurs. Resilience and perseverance were recurring themes, with participants highlighting the importance of staying determined and focused despite obstacles.

13.6 DISCUSSION

This research highlights the crucial contribution of women entrepreneurs to the advancement of education and development in Bangladesh. Despite encountering many obstacles, these women have shown tenacity and ingenuity in their business endeavors, significantly influencing educational results and socio-economic advancement in their communities. These ideas possess many ramifications for policy and practice. Policymakers and development practitioners should contemplate the creation of specialized financing programs and microfinance initiatives to assist women entrepreneurs in the education sector. This may include offering low-interest loans, grants, and financial literacy education to assist women entrepreneurs in properly managing their money. Furthermore, establishing mentoring and networking initiatives to provide emotional and professional assistance is crucial.

These initiatives may link women entrepreneurs with seasoned mentors who may provide counsel and assistance. Implementing community awareness initiatives to confront and alter cultural attitudes that hinder women's business endeavors is essential. These initiatives should emphasize the promotion of gender equality and enhance understanding about the advantages of women's entrepreneurship for community development. Moreover, promoting familial engagement and support for women entrepreneurs through community-oriented initiatives and education may be beneficial. These programs may include seminars and training sessions for family members to elucidate the significance of supporting women's business endeavors. Subsequent studies must persist in examining the interconnections of gender, entrepreneurship, and education, emphasizing the establishment of inclusive and sustainable development trajectories. Longitudinal research may provide profound insights into the enduring effects of women's entrepreneurship on socio-economic development. Comparative analyses across many areas and nations may shed light on optimal practices and effective frameworks for women's entrepreneurship that can be imitated and tailored to different circumstances.

13.7 CONCLUSION

This research addresses the crucial role of female entrepreneurs in promoting education and development in Bangladesh, showcasing their ability to catalyze substantial socio-economic advancement via resilience and creativity. The research offers useful insights for policymakers and practitioners by examining women's experiences and problems, facilitating the development of targeted interventions to encourage women's entrepreneurship. These actions are essential not just for Bangladesh but also for other emerging nations facing analogous issues. The perseverance and inventiveness shown by these women entrepreneurs have been essential in enhancing educational attainment and socio-economic development within their communities. Their business ventures have created additional educational opportunities and resources, thereby enhancing the overall development of their areas. The beneficial effect illustrates the capacity of women entrepreneurs to function as catalysts for wider development objectives, such as enhanced educational results, increased gender equality, and sustained socio-economic advancement. To take advantage of this potential, it is critical to confront the obstacles women entrepreneurs face.

These obstacles often include restricted access to capital, insufficient support systems, and cultural conventions that hinder women's entrepreneurial endeavors. Implementing specialized financial initiatives, like low-interest loans, grants, and financial literacy education, may provide women entrepreneurs with essential tools to proficiently manage their money and grow their enterprises. Enhancing support networks via mentoring and networking initiatives may provide emotional and professional assistance, linking women entrepreneurs with seasoned mentors who can give guidance and support. Cultural sensitization initiatives are crucial for challenging and transforming social conventions that impede women's economic endeavors. These initiatives may advance gender equality and enhance knowledge of the advantages of women's entrepreneurship for community development. By cultivating an atmosphere that promotes women's entrepreneurial pursuits, communities may achieve improved socio-economic growth and educational results. Family support programs are essential in promoting family engagement and assistance for women entrepreneurs. Community-based interventions and educational seminars for family members may enhance their comprehension of the significance of supporting women's entrepreneurial endeavors, fostering a more conducive climate at home. Subsequent studies need to persist in examining the interconnections of gender, entrepreneurship, and education, with an emphasis on establishing inclusive and sustainable development trajectories. Longitudinal research may provide profound insights into the enduring effects of women's entrepreneurship on socio-economic development. Furthermore, by conducting comparative analyses across various areas and nations, we can identify best practices and effective models of women's entrepreneurship, which we can replicate and adapt to various circumstances. This study will be crucial in formulating policies and practices that promote women's entrepreneurship and advance global development objectives.

REFERENCES

Braun, V., & Clarke, V. (2006). Using thematic analysis in psychology. *Qualitative Research in Psychology*, 3(2), 77–101. https://doi.org/10.1191/1478088706qp063oa

Brush, C. G., de Bruin, A., & Welter, F. (2009). A gender-aware framework for women's entrepreneurship. *International Journal of Gender and Entrepreneurship*, 1(1), 8–24. https://doi.org/10.1108/17566260910942318

Chowdhury, N. (2013). Challenges faced by women entrepreneurs in Bangladesh. *Journal of Business and Economics*, 45(2), 123–135.

Hossain, L., Naser, K., Zaman, A., & Nuseibeh, R. (2009). Factors influencing women business development in the developing countries: Evidence from Bangladesh. *International Journal of Organizational Analysis*, 17(3), 202–224. https://doi.org/10.1108/19348830910974922

Jahan, R. (2015). Women's entrepreneurship and development in Bangladesh. *Development in Practice*, 25(3), 567–578. https://doi.org/10.1080/09614524.2015.1029430

Kabeer, N. (2019). Gender equality and women's empowerment: A critical analysis of the third millennium development goal. *Gender & Development*, 13(1), 13–24. https://doi.org/10.1080/13552070512331332273

Karim, N. (2001). *Jobs, gender and small enterprises in Bangladesh: Factors affecting women entrepreneurs in small and cottage industries in Bangladesh*. International Labour Office.

King, E. M., & Hill, M. A. (1993). *Women's education in developing countries: Barriers, benefits, and policies*. Johns Hopkins University Press.

Minniti, M. (2010). Female entrepreneurship and economic activity. *European Journal of Development Research*, 22(3), 294–312. https://doi.org/10.1057/ejdr.2010.18

Nussbaum, M. (2011). *Creating capabilities: The human development approach*. Harvard University Press.

Rahman, A., & Islam, N. (2020). Barriers to women's entrepreneurship in developing countries: A study of Bangladesh. *International Journal of Gender and Entrepreneurship*, 12(2), 256–270. https://doi.org/10.1108/IJGE-05-2019-0071

Roomi, M. A., & Parrott, G. (2008). Barriers to development and progression of women entrepreneurs in Pakistan. *Journal of Entrepreneurship*, 17(1), 59–72. https://doi.org/10.1177/097135570701700105

Sen, A. (1999). *Development as freedom*. Oxford University Press.

Shane, S. (2003). *A general theory of entrepreneurship: The individual-opportunity nexus*. Edward Elgar Publishing.

Smith, J. (2018). Women's entrepreneurship and economic development. *Journal of Economic Perspectives*, 32(1), 121–140. https://doi.org/10.1257/jep.32.1.121

Tambunan, T. T. H. (2009). Women entrepreneurship in Asian developing countries: Their development and main constraints. *Journal of Development and Agricultural Economics*, 1(2), 27–40. https://doi.org/10.5897/JDAE.9000071

14 Preserving Heritage, Inspiring Innovation

The Role of Indigenous Knowledge in Culinary Entrepreneurship – Case Study of JhaJi Store and Namakwali

Anu Kohli, Neha Tiwari, and Haider Abbas

14.1 INTRODUCTION

India is currently the third largest startup ecosystem in the world, with women-led startups witnessing an unprecedented boom. According to the data reported by Bain and Co, the number of women-run enterprises has touched 15.7 million. The figure constitutes around 22 percent of the overall enterprises. Women-led enterprises provide employment to around 27 million individuals. However, the contribution of women entrepreneurs is not just economic progress, but they are also powerhouse of societal and economic transformation (Mittal, 2024).

In India, cooking and kitchen has always been the feminine space. As women fulfilled the traditional role of cooking and providing care for the family members, the indigenous knowledge of cuisines was passed to women from generation to generation. However, in the last few decades women entrepreneurs are increasing even in developing and emerging economies. Many successful women entrepreneurs in the food and beverage sector are examples of the immense potential women entrepreneurship holds for emerging markets.

Indigenous entrepreneurship is the youngest area in the academic research (Dana, 2015) However, indigenous knowledge and women entrepreneurship is an area that has not been explored much in the academic research till now, particularly in the Indian context. However, in the Indian context many aspiring women entrepreneurs are innovating through leveraging indigenous knowledge including knowledge of cuisines.

The chapter will therefore attempt to bridge the research gap by delving into the topic of indigenous women entrepreneurs who are setting an example of culinary innovation in India.

14.2 LITERATURE REVIEW

Indigenous communities occupy the same landscape, culture, traditions, and customs for generations and share deep-rooted local attachments. Indigenous community can be identified based on following peculiarities:

- People living in relatively small populations compared to the dominant culture in their home country, having a distinct language apart from the official language of the country.
- People having and practicing distinctive cultural traditions; and possessing strong connection and generational ties to their lands.
- People sharing a worldview that is centered on the land and their place in the natural world (Mistry et al., 2020).

The indigenous communities transmit their worldviews, beliefs, traditions, practices, and institutions through the indigenous knowledge. The knowledge is transmitted through rituals, customs, habits, artforms and artifacts (Mistry et al., 2020).

According to the World Bank:

Indigenous knowledge is knowledge generated from the lived experiences of a community because of interactions with their environment over generations. Indigenous knowledge encompasses virtually every field of human endeavour and usually varies from community to community; one unifying factor being that it is aimed at addressing challenges and solving perceived problems specific to each community.

(World Bank, 1994)

It is interesting that indigenous knowledge is not static rather it is dynamic and keeps changing by adapting to the needs of time and generations. Indigenous knowledge differs from lived experiences and doing in contrast to globally accepted Western knowledge which stems from universities and research institutions. The globally accepted knowledge is disseminated, popular and verified by institutions (Oguamanam, 2008); however, the indigenous knowledge is not documented and hence it needs to be preserved through various social, cultural, and economic interventions. Indigenous knowledge is diverse and rich and dates to over 5000 years. Indigenous knowledge in India spans across various domains like technology, biodiversity, medicine, health, agriculture, and food (Mahesh, 2023).

India's culinary heritage is widely acclaimed for the fine balance of health and Flavors. Traditional Indian cuisines are based on holistic principle of health and wellness. The unique value of the nutrient rich preparations lies in the fact that they include local, seasonal and fresh ingredients. The cuisines also ensure that the sustainability of the local communities is also not compromised.

In India, women are traditionally the custodians of indigenous food knowledge. Women play a pivotal role in selecting and cooking the food for family. The food ensures taste and nutrition for the family and is based on the locally available ingredients and resources. Traditionally women have cooked for their families for centuries in India but in the last few decades women entrepreneurs in food business are increasing. Women entrepreneurs have notable advantage when it comes to indigenous knowledge about food in comparison to men. Hence indigenous knowledge can be leveraged by women to pursue entrepreneurship. Additionally indigenous women entrepreneurs contribute to creation of jobs and wealth within the local communities. Women in India are shining bright as indigenous food entrepreneur.

There is undoubtedly enough empirical evidence that women entrepreneurship is a driving force for women's empowerment and socio-economic development (Minniti 2010; Shah & Saurabh 2015; Terjesen & Amoros 2010; Tiwari & Tiwari 2007; Williams 2009; Williams & Martinez, 2014).

However, there are very limited studies conducted on the realm of indigenous women entrepreneurship. There is a need to explore the breadth of entrepreneurial drivers and experiences of indigenous women entrepreneurs (Croce, 2020). In the Indian context, the research on indigenous women entrepreneurs remains even more unrepresented.

Therefore, the chapter will attempt to explore and describe the experiences of indigenous women entrepreneurs who have launched startups using their culinary skills. The chapter will also attempt to identify the impact of such women entrepreneurs on their local communities.

14.3 METHODOLOGY

This chapter follows an inductive approach. The research strategy used is the case study. Research on indigenous women entrepreneurship remains scant (Croce, 2020) and therefore the case study method emerges as the ideal research strategy. Given the fact that indigenous women's entrepreneurship particularly in the context of India is an emerging issue in the literature on women entrepreneurship, case studies are an ideal research strategy to generate large amounts of rich information about the indigenous women entrepreneur. A case study is defined as:

> An empirical inquiry which investigates a phenomenon in its real-life context. In a case study research, multiple methods of data collection are used, as it involves an in-depth study of a phenomenon.
>
> (Yin 2009, p. 18)

It must be noted, as highlighted by Yin (2009), a case study is not a method of data collection, rather is a research strategy or design to study a social unit.

In the present study cross-case research design was used which includes a study of two or more cases experiencing similar events or phenomenon. The data in different cases are compared to produce generalizable conclusions (Yin, 2014). To decipher the similarities and differences in indigenous women entrepreneurs in India, multiple case study research proved to be appropriate. (Hunziker & Blankenagel, 2024). Purposive sampling was employed for the selection of two cases. Purposive sampling is widely used method for generating information rich insights and therefore is the most widely used method in qualitative research (Patton, 2002). Homogenous cases that fulfilled the following predetermined criteria were considered for the analysis:

Preserving Heritage, Inspiring Innovation

1. Startups conceived, owned, and operated by women.
2. Startups that have applied and received seed funding on a national platform.
3. Startups that employ direct to customer business model and exclusively available online.
4. The products must be based on indigenous knowledge of food and culinary traditions.
5. Startups that aim to serve the local community and empower women in their vicinity.
6. Startups that are based in Tier II/III cities in India.

Since the data about the startups was publicly available on their respective websites, secondary data were collected for compiling the case narrative. Additionally, blogs and interviews of the women entrepreneurs were retrieved for the data collection. Data were collected during the period of December 2021 to May 2024.

14.4 CASE STUDIES

For the cross-case study design, the following two case studies are considered that meet the predetermined sampling criteria. Table 14.1 elucidates the profile of two startups.

14.4.1 Case I—Preserving Local Heritage through Hand-pounded Salt and Seasonings: The Case of Namakwali

Namakwali is a unique brand that is on a mission to popularize the artistic and culinary tradition of the hills of Tehri and Garhwal in Uttarakhand state of India. The brand Namakwali is a word in Hindi language that means girl/women with salt.

Shashi Bahuguna Raturi, the women who founded the Namakwali brand epitomizes the perfect blend of tradition and entrepreneurship. Her brand Namakwali sells handpounded spices and flavored salt referred to as "Pisyu Loon" in the local language.

14.4.1.1 Inception

Shashi Bahuguna Raturi, hailing from Uttarakhand (a northern hilly state of India) established Namakwali in 2018. The brand is aimed at showcasing the rich culinary traditions of the hill state. The brand provides global exposure to the traditional "Pisyu loon" salt of the hill state. The traditional salt is prepared by grinding organic salt with exotic herbs and spices. The grinding is done through the traditional stone grinder called "Sil-Batta".

Long before Shashi established Namakwali, at the young age of 19 she started immersing herself in various social causes, with a particular emphasis on uplifting women, children, and the environment. In the distant hills of Tehri Garhwal region of Uttarakhand, she observed that women were away for hours to fetch food and fodder for their families. While the men worked on menial tasks to earn their livelihood and women of the family were away to fetch food and fodder, the toddlers and children were locked in their homes. The children were not mature enough to take

TABLE 14.1
Profile of the startups

Name of the Company	Namakwali	JhaJi Store
Entrepreneur	Shashi Bahuguna Raturi Age: Above 50	Kalpana Jha and Uma Jha Age: Above 50
Startup Capital	Bootstrapped: Started with Rs.1.5 lakh	Bootstrapped: Started with Rs.5–6
Products	Flavored salt, seasonings, spices, organic clarified butter, and pulses	Pickles and sauces
Business Model	Direct to customer through own website and leading e-commerce websites	Direct to customer through own website and leading e-commerce websites
Funding	Received	Received
Expansion plans	Yes, aims to scale their operations and products	Yes, aims to scale their operations and products
City of origin	Dehradun (Uttarakhand state of India)	Darbangha (Bihar state of India)
Promotions	Online promotions through digital storytelling	Online promotions through digital storytelling
Employees	Majority employees are women form local community	Majority employees are women form local community
Unique value proposition	Organic, handcrafted products based on indigenous knowledge of cuisines. Products promise taste and health to the consumers.	Organic, handcrafted products based on indigenous knowledge of cuisines. Products promise taste and health to the consumers.
Sales	Primarily by their own website	Primarily by their own website.

Source: Compiled by the author.

care of themselves. Shashi also noticed that these children were sometimes injured or covered in their fecal matter while their mothers were away. To provide education and support to such children, Shashi started Balwadi schools in 1982. Shashi recollects in an interview about her inclination towards community initiatives:

> Before venturing into the business world, we were running an NGO, Mahila Nav Jagran Samiti, founded in 1982 in the picturesque town of Chamba, Tehri Garhwal, with a focus on women's empowerment.
>
> (Menon, 2024)

These schools provided early education and day care facilities to the children of marginalized families. She also simultaneously led various initiatives for promoting livelihood and sustainability in her state. She also spearheaded various projects and trainings for sessions on crafting smokeless stoves. She also undertook projects to promote local handicrafts including Banskhera carpets and wall hangings (Rawat, 2024). Her inclination to contribute to the local community of the hills of Uttarakhand led her to start a project named Namkwali. Whenever Shashi visited local events including weddings, the flavored salts tingled her taste buds. The traditional recipe of the salt included organic salt grinded with spices and condiments on a stone grinder (Sil-Batta in local language). The tradition of grinding salt with spices on a stone grinder dated back centuries. While grinding, the women often sang traditional songs and shared folklore stories. She decided to produce and distribute her salt in all of its flavors across the nation. She started the venture by collaborating with a Dehradun-based NGO called Mahila Nav Jagran Samiti in 2018 (Topno, 2023).

Her family proved to be the strongest support system. When she shared her startup dream, her spouse and sons pitched in for help. Her husband Vipin Raturi created the illustrations for the packaging that convey the cultural connection of the brand. Her two sons, Suvendu Raturi and Navendu Raturi, lend a helping hand in her business too. Both manage marketing, media handles, and online orders for the brand.

She simultaneously started popularizing the culture of her region through digital storytelling on Instagram. The venture was still in its nascent stage and 10–12 women were catering to the demand generated through their Instagram account. It was in 2020 that Namakwali was incorporated as a company (Shark Tank India, 2024). It was in 2022 that her son Suvendu Raturi also joined the company as the number of orders were consistently increasing. Suvendu listed the products on Amazon to increase the reach and target customers globally.

14.4.1.2 About the Brand

The brand is built with the mission to provide healthy, organic and unique Himalayan flavors to the global consumers. In an interview Shashi expressed her fondness for popularizing the culinary traditions of her region on a global scale.

> "The process of handpicking the ingredients in the courtyard of their homes, grinding the salt on the sil-batta and packing it is infused with a pinch of laughter, a spoonful of traditional songs, and a handful of love, which reverberates when you open the pack," "Every herb and spice is handpicked from the high altitudes of the Uttarakhand Himalayan range. It is extremely nutritious and is 100% organic, without any added chemicals," says Shashi. "Social work is ingrained in me. Whether it is women or children, I always had this urge to work for their empowerment and upliftment," says Shashi.
>
> "Our motivation primarily stemmed from the desire to generate income for rural women and promote natural and organic products originating in the mountains, which were not adequately represented in the market. Furthermore, it helps conserve the culinary tradition of the region, which sadly most people from the current generation are not even aware of."
>
> (Menon, 2024)

Namakwali is a brand that relentlessly works for the empowerment of women. The brand believes that women, who are traditionally endowed with the rich cultural legacy, are storehouse of culinary knowledge.

"My journey founding Namakwali is rooted in my upbringing, education, and support system. Growing up in the Tehri Garhwal region of Uttarakhand, I developed a love for local cuisine," says Shashi.

"I aimed to empower marginalised communities, especially women. Exploring cooking techniques and conserving traditional cuisine has been my passion. Family and friends lent their unwavering support, helping me turn my vision into reality."

(Menon, 2024)

This knowledge can provide abundant opportunities for self-employment and entrepreneurship.

14.4.1.3 Namakwali: The Marketing Mix

As the organic salt started making a mark on the customers, Shashi went on to introduce other organic products including turmeric, spices, condiments, seasonings, sauces, and clarified butter. Currently the brand offers eight categories of pure organic products including flavored salt. Shashi elaborated about her offerings in an interview: "Our product range now includes Pisyu Loon salt, Pahadi honey, Haldi (turmeric), cow ghee, chutney powders, and pulses, with prices ranging from Rs 120 to Rs 2,800," (Menon, 2024).

These products are sold exclusively online through the brand website and through Amazon. All the products are handpicked from the Himalayas and curated to provide taste and health to the consumers. The packaging of the product also narrates the intent and story behind the brand. The illustration on the packaging portrays the traditional women of the hill state. The unique illustration conveys the intent of the brand, and the customers are drawn to the cultural connection with the brand. The handmade illustration of traditional women of the indigenous communities creates a unique identity of the brand and strengthens its connect with the consumers.

The pricing of the product is a bit on the higher side as compared to other commercial salt mix and seasonings. In an interview Shashi spoke about the pricing of the product: "In terms of pricing, while ordinary salt is relatively inexpensive, Pisyu Loon salt comes in various pack sizes, including 80 grams (Rs 160), and 200 grams (Rs 230)" (Menon, 2024).

A 200 gram of salt mix costs around 230–270 Indian Rupees while the one liter of clarified butter costs around 2500 Indian Rupees. Most of the reviews on social media and the Amazon website are positive and signal that consumers do not find the products of Namakwali expensive.

Namakwali started promoting itself on social media including Facebook and Instagram since its inception. The brand relies on digital storytelling to create an emotional connection with the audiences. The folk songs, culture, cuisines, and festivals of the hill community are the content for promotion of the brand online. The online promotions are cost-effective, and at the same time provide a global reach for the brand. As of now, Shashi plans to sell products exclusively online through Amazon and their own website. Suvendu created the website using Shopify and currently 70 percent of the sales are done through their own website while only 30 percent of the sales are done through Amazon.

"Our business model primarily operates on a direct-to-consumer (D2C) basis. The products are available on Amazon and Flipkart as well as on our website, which accounts for 70% of sales," Shashi said in an interview.

(Menon, 2024)

The digital storytelling has far been successful to result in 100 percent of sales with monthly spending of Rs. 90,000 on promotion of the product. (Shark Tank India, 2024).

14.4.1.4 Road Ahead

Shahsi and Suvendu confidently appeared on the Shark Tank India and pitched for Rs. 50 lakhs equity for 5 percent. But the investor valued the company at Rs. 2 crores and offered only Rs. 10 lakhs equity at 5 percent and Rs. 40 lakh debt at 8 percent spread over 3 years. After the funding received on Shark Tank India in March 2024, the brand is all set to scale. The sales have been consistently increasing with Rs. 5 lakhs annual sales in the financial year 2021–22. During the next financial year sales jumped to Rs. 11 lakhs, while up to October 2023–24, sales figure took a huge leap to Rs. 38 lakhs with a profit margin of 17 percent. (Shark Tank India, 2024)

So far, the brand has achieved sound financial performance. However, post funding, the startup will have to reorient and strategize for higher returns. The startup is still at a very nascent stage. Since Shashi and Suvendu Raturi have not yet involved any professionals on the board. They have been operating as an informal and family-owned business.

Flavored salt is a common delicacy in the state of Uttarakhand and therefore the global market must be targeted for expanding the customer base. Post-pandemic, the demand for health and immunity boosting products is on the rise. Therefore, the brand can benefit from the perspective of organic products.

14.4.2 Case II—Commercializing the Age-Old Tradition of Pickle Pmaking: The Case of JhaJi Store

JhaJi Store was conceptualized by the sister-in-law duo Kalpana and Uma Jha during the tumultuous period of COVID-19. Both hail from the quaint city of Darbhanga in the state of Bihar (India) and share an unbridled passion for cooking. The brand aims to popularize the unique recipe of pickles of Mithila region. The pickles are sundried and devoid of any harmful chemicals, dyes, and preservatives. The pickles promise taste and health to customers.

14.4.2.1 Inception

Kalpana Jha has a master's in psychology and was a homemaker all her life. She moved from city to city as her husband served as an officer in the Government of India and was transferred frequently. Soon after her husband's retirement she settled with her family in her native city, Darbhanga. She raised a successful family, but she always dreamt of doing something of her own. She spoke in a blog about her dream

"I was apprehensive. I was scared. But I had to take a leap of faith. Thankfully, the moment I shared my desire to start a pickle business, my family jumped to support me." (Jha, 2022).

Her daughter, son, and husband motivated her to start something as they wanted her dream to be fulfilled. Her husband promised her that he would take care of the household while her daughter provided her with the startup funds. Her son took charge of all the operations and social media handles of the brand. She shared a strong intimacy and her passion for cooking with her sister-in-law Uma Jha. Uma's husband is a professor in a college in the city of Darbhanga. She was a teacher in a school for the past two decades. As soon as Kalpana discussed her dream of starting the business, Uma willingly joined her, and they both started working on their dream (Jha, 2022).

Kalpana's love and dedication for her husband made her decide that she would name her business after her husband who she fondly called by his surname "JhaJi". It was in the COVID period of 2020 that both co-founders started working on the food license and shelf-testing procedures through an accredited laboratory. It took them around six months as the nearest laboratory was in Kolkata and travelling was restricted due to COVID-related protocols.

After they received their licenses, they started to procure raw materials and started manufacturing from their kitchens. Their first order for pickles came from a neighbor.

Both the founders observed that a lot of women lost their jobs as house help due to COVID-19. Even their husbands lost jobs due to lockdown restrictions. Uma and Kalpana decided to provide employment to women, who were highly skilled in the traditional art of pickle making. The business started with only six staff, and they started their small-scale operations, primarily selling in Bihar and Delhi.

The landmark of Rs. 1 lakh sales was achieved within 15 days of the launch. They started selling and promoting their business online through Facebook and Instagram. Packaging and delivery emerged to be their biggest concerns because of the unavailability of bubble wrap, cartons, and glass bottles in Darbhangha and nearby cities in Bihar. They had to procure the same from Delhi and slowly they learnt to order advance inventory. Leakage and breakage issues while the product was in transit was also handled through devising a proper packaging process. Delivery time to Darbhanga and other major cities was around 10–12 days. This led to a lot of customer complaints and therefore the duo decided to open warehouses in Patna and Delhi to reduce the delivery time (Nensee & Bajj, 2023).

14.4.2.2 About the Brand

The brand is built with the mission to empower local communities and contribute to local prosperity. The founders wanted to popularize the indigenous pickle recipes of their community globally. In an interview Kalpana shared "'We want to take Mithilanchal's flavors from Darbhanga to the world,' Kalpana Jha said sharing her vision behind JhaJi. We want to take Mithilanchal's flavors #DarbhangaSeDuniyaTak (From Darbhanga to the world" (Arora, 2023).

Kalpana and Uma believe in paying their employees generously and even hiring more than one person from the same family. In an interview Kalpana expressed her delight of providing opportunities to local women:

"Initially, I thought whether the Rs 12 lakh I borrowed from my husband would go to waste. Now, I see how the lives of our women have changed after they started working with us. Apart from financial independence, they are also able to support their families without compromising on their household duties. Some of them can give their children a good education too," Kalpana says.

(Balakrishnan, 2023)

JhaJi Store aspires to be a single brand for everything related to Bihari cuisine, while also creating meaningful livelihood opportunities for over 300 women in the Darbhanga region (Arora, 2023)

The brand makes sure that they create employment, provide a decent livelihood and income opportunity and source all the raw materials from local suppliers. The unique value of pickles lies in the fact that the pickles are handcrafted using fresh and local ingredients. Unlike other pickles, the products are sundried, and no chemical or colors are used to enhance their flavor and appearance.

14.4.2.3 JhaJi Store: The Marketing Mix

The pickles come in a variety of flavors and sizes. Common pickles include garlic, jackfruit, green chili, red chili, cauliflower, carrot, turnip, tamarind, lemon, elephant yam, and Indian gooseberry. The pickles are also available with less oil content in contrast to the oil-laden traditional recipe. On the demand of certain consumers, pickles with no garlic are also available.

To cater to different taste and preferences tangy, low salt, high salt and low to medium spice variants are also available. Since the majority of the customers were not aware of the traditional recipe of Mithila pickles, they did not know the taste and were not sure of buying a big bottle of the pickle. Therefore, Kalpana and Uma decided to launch bite-sized packs with four or eight flavors in 100 g jars. These let customers taste the pickle before committing to large-volume packs. The sample packs start from 100 g and the large-volume packs are commonly of 500 g.

Each jar of the pickle carries the hand-painted cloth covers bearing the traditional Madubani (a prominent local art form) painting. The packaging connects the customers of the brand with its ethnic traditions.

The price of the pickles is competitive and a jar of 500 g of pickle on average cost around Rs. 500 while a 250 g package cost around 250 Rs. Although the similar pickles manufactured by other companies costs less, the taste and unique value propositions offered by them justify their price.

The store sells their products exclusively online through their own website and other leading e commerce sites including Flipkart and Amazon and through their page on social media including Facebook and Instagram. To cater to the increasing number of consumers they now have warehouses in Gurgaon, Patna, and Bengaluru.

The social media handles of the brand are prominently used for promotion of the product. After appearing on Shark Tank in 2021, the popularity of the brand soared to new heights. Kalpana and Uma always believed that word of mouth and positive reviews are the best form of publicity. Currently reviews of customers in video and text form are highly advertised by the brand on their social media pages. Kalpana and

Uma advertise their products through digital storytelling to create a connection with the audience. In an interview Kalpana shared her unique way of promoting her brand

> To spread the word about JhaJi Store, we make stories around aspects such as Bihar's culture, cuisine, food, customs, clothing, lifestyle, and, of course, the pickle-making process. We create content and share it among the audiences on social media, primarily to get them excited about trying our pickles and chutneys.
>
> (Arora, 2023)

14.4.2.4 The Growth Story

The startup reached the landmark of Rs. 6 lakh sales within the fourth month. The demand was increasing and therefore the startup needed money to grow. It was in August 2021 that Kalpana decided to fill in the application form for Shark Tank.

Kalpana and Jha appeared on the national forum seeking funding in December 2021. But the investors were not convinced, and they had to return without any funding. But their appearance on national television gave them the much-needed publicity push. Within 24 hours of the episode airing on national television, they clocked one month's worth of revenue (Jha, 2022). There was an unexpected rise in the orders. To meet the upsurge in demand the duo applied for and received a bank loan. The business was growing exponentially, and they received funding from angel investors also.

Eventually, the two investors who denied them funding also visited their facility and changed their minds and invested. Cumulatively, the startup so far has secured funding of more than Rs. 2 crore from around 30 angel investors with average size of investment cheque of 5 lakhs.

Riding high on funding and demand, the startup established their own manufacturing facility as the household kitchen was not capable of large-scale production. They named their kitchen factory "Annapoorna". The factory is all set to brace the international food processing standards (www.jhajistore.com/blogs/news/annapoorna-darbhanga-new-kitchen-factory?srsltid=AfmBOorVTGnvrX27Q6kzdf5tvdrLDCVjZ2RaUKXXwWuz_6MeBtZWmgV).

Since December 2021 their revenue has skyrocketed to Rs. 90,000,000 and they have served more than 109,953 customers. They have also reached new locations and have delivered to more than 9000 pin codes (www.linkedin.com/pulse/my-1st-year-d2c-startup-founder-story-jhaji-store-kalpana-jha/). With the new factory they plan to hire more staff and take the total tally of employees to 150. They will also install the latest equipment to scale up and maintain quality.

14.5 ANALYSIS AND DISCUSSION

The case study offers interesting insights about the motivation, lived experiences and the impact of indigenous Indian women entrepreneurs on their local community. Although previous research on entrepreneurs concludes that the chances of an individual becoming an entrepreneur peak at 40 (Kautonen et al., 2014) and reduces thereafter (Minola et al., 2016; Shaw & Sørensen, 2022; Viljamaa et al., 2022), The entrepreneurs in these case studies started their ventures in their 50s. Both sets of entrepreneurs, Shashi Bahuguna Raturi, and Kalpana and Uma Jha are intensely passionate about their customs and traditions. Both exhibit strong attachment to their rich culture and heritage. They have mastered curating indigenous recipes in their households for many years and decided to commercialize their expertise by starting their venture. Both have a deep desire to preserve their cultural heritage and culinary legacy. Moreover, they share the drive towards contributing to women empowerment. They have been vocal about creating employment opportunities for women. The findings are in line with the previous research on employment patterns in women-owned MSMEs. It was documented that one third of the jobs created by women-owned MSMEs was provided to female employees. This led to boosting of economic development in local communities (BFSI, E.T., 2024). Women entrepreneurs have a pivotal role in economic development and boosting innovation within the entrepreneurial ecosystem (Khan and Rusmiati, 2022).

Further, it also becomes evident that family support has been instrumental in the entrepreneurial journey of both sets of the entrepreneurs. Shashi's entire family including spouse and son helped her in operations of the startup. The illustrations on the packaging of her products were created by her husband while the marketing and website operations were entirely handled by her sons. Similarly, Kalpana and Uma garnered every sort of support including financial from their spouse and children. When Kalpana wanted to invest in her startup, her husband agreed to look after the household. Research proves that moral support of spouse and children is instrumental for women entrepreneur's confidence and ability to collaborate (Makandwa & de Klerk, 2024). Family support is positively correlated with entrepreneurial success of women run small and medium enterprises (Alam et al., 2011; Welsh et al., 2016).

Further, it is noteworthy that both sets of entrepreneurs hail from a tier II city of India, with limited exposure to digital technologies. Even though they were not very conversant with the technologies, they stepped up to embrace digital technology and started promoting and selling the products online. Both startups are an exemplar of how online channels can be leveraged for direct to customer model. Research has proved that access to information communication technologies can help in women's empowerment in developing countries (Badran, 2010). Digital technologies can benefit women entrepreneurs in exploring e-commerce opportunities in broadly three ways. First, the barriers to entry in an online business are comparatively low as compared to brick-and-mortar model. Second, e-commerce allows them to have exposure to global markets, which would enhance their reach and scalability and lastly e-commerce allows women entrepreneurs to have more flexibility in terms of location (Shrimanne & Adhikari, 2023). Even in remote locations where the entrepreneurial ecosystem is not very favorable, women can operate their startups. For women entrepreneurs, access to smart devices has made the world of the internet more accessible at home. Access to the internet enables women entrepreneurs to have access to new markets. Access to social media also enables women entrepreneurs to promote their product and create a strong connection with potential and existing customers.

TABLE 14.2
Major themes

Opportunity driven	The indigenous women entrepreneurs in the case are driven by the passion to create something of their own.
Affinity towards preservation of local heritage	The women entrepreneurs in the case not only commercialized their indigenous knowledge but are also contributing towards preservation of their local heritage through brand communication and unique products.
Community involvement	The indigenous women entrepreneurs in the case are sensitive towards the local community problems. They provide employment to women with the intent of boosting local economy.
Minimal investment	Both the women entrepreneurs in the case started with minimum investment and resources.
D2C business models with the use of ICT	Although the women in the case were traditionally not exposed to technology, they created successful direct to customer businesses through use of internet. They also extensively used social media and technology to boost their promotion.

Source: Compiled by the author.

Entrepreneurs have been traditionally categorized as necessity- and opportunity-driven entrepreneurs. Necessity-driven entrepreneurs are pushed intro entrepreneurship as they have no job to sustain their economic necessities. Such entrepreneurs are more common in developing countries, while opportunity entrepreneurs are attracted towards entrepreneurship because they perceive profitable opportunities (Reynold et al., 2003). While opportunity-driven entrepreneurship is common in high-income developed countries, necessity-driven entrepreneurship is common in low-income, developing countries (Gries and Naude, 2010). However, in the cases considered for the study, both the entrepreneurs are opportunity-driven entrepreneurs as they have launched startups to create an impact. Both are also riding high on investor funding and plan to grow through diversification and expansion. The major themes that emerge from the case studies are summarized in Table 14.2.

14.6 IMPLICATIONS

The case analysis offers numerous insights about indigenous women entrepreneurship in India. Most of the traditional Indian women are endowed with indigenous knowledge about food and cuisines. Culinary heritage and traditions in India are rich and therefore the knowledge of the same can open possibilities of commercializing this knowledge through startups. The policymakers and communities must promote indigenous entrepreneurship among women as they are traditionally the custodians of indigenous knowledge. Traditionally research has established that entrepreneurship inclination decreases with age; however, the heterogeneous nature of women entrepreneurs and their embedded context must be considered. It is therefore imperative that the phenomenon of indigenous women entrepreneurship in India must be explored in depth. Institutions and policymakers must target women of all age groups while considering them as beneficiaries of any impetus for entrepreneurship.

It is also noteworthy that indigenous women entrepreneurs are adept at grassroot innovations, involving very limited resources. Both sets of entrepreneurs are an exemplar of how startups can be built with minimum investment and can be scaled up with innovative products and means of digital marketing. Hence, the institutions at meso level must provide financial assistance in forms of micro loans for promotion of women entrepreneurship in developing economies.

The chapter therefore initiates the much-needed enquiry of indigenous women entrepreneurship and their innumerable benefits that a developing and culturally rich country like India can harness.

REFERENCES

Alam, S. S., Jani, M. F. M., & Omar, N. A. (2011). An empirical study of success factors of women entrepreneurs in southern region in Malaysia. *International Journal of economics and Finance, 3*(2), 166–175. https://doi.org/10.5539/ijef.v3n2p166

Arora, S. (May 25, 2023). Starting with pickles, how two women are taking Bihari cuisine to the world. *indiaretailing.com*, accessed on Jan 04, 2024 www.indiaretailing.com/2023/05/25/starting-with-pickles-how-two-women-are-taking-bihari-cuisine-to-the-world/

Badran, M. F. (2010). Is ICT empowering women in Egypt? An empirical study. In *Proceedings of the Research Voices from Africa Workshop, IFIP WG* (Vol. 9). www.semanticscholar.org/paper/Is-ICT-empowering-women-in-Egypt-An-empirical-study-Badran/b60971f913408e3b7eaabc9c543fe3f8bcac43cf

Balakrishnan, R. (January 05, 2023). Correcting their mistake, Sharks present Rs 85 lakh cheque to JhaJi founders. *Yourstory.com*, accessed on January 16, 2024 https://yourstory.com/herstory/2023/01/jhaji-store-shark-tank-india-namita-thapar-vineeta-singh

BFSI, E.T. (March 07, 2024). One third of all new jobs created by women-owned MSMEs went to female employees, accessed on May 12, 2024 https://bfsi.economictimes.indiatimes.com/news/nbfc/one-third-of-all-new-jobs-created-by-women-owned-msmes-went-to-female-employees-study/108297065

Croce, F. (2020). Indigenous women entrepreneurship: Analysis of a promising research theme at the intersection of indigenous entrepreneurship and women entrepreneurship. *Ethnic and Racial Studies, 43*(6), 1013–1031. https://doi.org/10.1080/01419870.2019.1630659

Dana, L. P. (2015). Indigenous entrepreneurship: An emerging field of research. *International Journal of Business and Globalisation, 14*(2), 158–169. DOI:10.1504/IJBG.2015.067433

Gries, T., & Naudé, W. (2010). Entrepreneurship and structural economic transformation. *Small Business Economics, 34*, 13–29. https://doi.org/10.1007/s11187-009-9192-8

Hunziker, S., & Blankenagel, M. (2024). Multiple case research design. In *Research Design in Business and Management: A Practical Guide for Students and Researchers* (pp. 171–186). Springer Fachmedien Wiesbaden. https://doi.org/10.1007/978-3-658-34357-6_9

Jha, K. (June 27, 2022). My 1st year as a D2C startup Founder | The story of JhaJi store. *LinkedIn*, accessed on March 01 www.linkedin.com/pulse/my-1st-year-d2c-startup-founder-story-jhaji-store-kalpana-jha/

Khan, T., & Rusmiati, T. (2022). The influence of family moral supporting the creation of small-medium enterprises for female entrepreneurs in Indonesia & Pakistan. www.diva-portal.org/smash/get/diva2:1688982/FULLTEXT01.pdf

Kautonen, T., Down, S., & Minniti, M. (2014). Ageing and entrepreneurial preferences. *Small Business Economics, 42*, 579–594. https://doi.org/10.1007/s11187-013-9489-5

Mahesh, G. (2023). Communication and dissemination of India's traditional knowledge. Indian Journal of Traditional Knowledge (IJTK), 22(2), 450–457. https://doi.org/10.56042/ijtk.v22i2.63427

Makandwa, G., & de Klerk, S. (2024). Impact of family moral support on female entrepreneurs involved in craft tourism. *Journal of Tourism and Cultural Change, 22*(1), 61–75. https://doi.org/10.1080/14766825.2023.2248065

Menon, D. R. (April 02, 2024). Starting with Rs 1.5 Lakh, Mountain woman's salt venture strikes Gold with Rs 50 Lakh turnover. *Weekend Leader, 15*(14), accessed on 30 May, 2024 www.theweekendleader.com/Success/3262/seasoned-with-salt.html

Minniti, M. (2010). Female entrepreneurship and economic activity. *The European Journal of Development Research, 22*(3), 294–312.

Minola, T., Criaco, G., & Obschonka, M. (2016). Age, culture, and self-employment motivation. *Small Business Economics, 46*(2), 187–213. https://doi.org/10.1007/s11187-015-9685-6

Mistry, J., Jafferally, D., Ingwall-King, L., & Mendonca, S. (2020). Indigenous knowledge. In A. Kobayashi (Ed.), *International encyclopedia of human geography* (2nd ed., pp. 211–215). Elsevier. https://doi.org/10.1016/B978-0-08-102295-5.10830-3

Mittal, T. (March 08, 2024). The rise of women in entrepreneurial roles in India. *Economic Times*, accessed on April 11, 2024 https://economictimes.indiatimes.com/small-biz/entrepreneurship/the-rise-of-women-in-entrepreneurial-roles-in-india/articleshow/108317138.cms?utm_source=contentofinterest&utm_medium=text&utm_campaign=cppst

Nensee, K., & Bajj, A. (February 23, 2023). JhaJi store - The success story of pickles. Startuptalky.com, accessed on March 02, 2024 https://startuptalky.com/jhaji-success-story/"

Oguamanam, C. (2008). Local knowledge as trapped knowledge: Intellectual property, culture, power and politics. *The Journal of World Intellectual Property, 11*(1), 29–57. https://doi.org/10.1111/j.1747-1796.2008.00333.x

Patton, M. Q. (2002). *Qualitative research and evaluation methods* (3rd ed.). Thousand Oaks, CA: Sage Publications.

Rawat, S. (March 21, 2024). The Pahadi agriculture e-magazine, accessed on March 30, 2024 from https://pahadiagromagazine.in/namakwali-journey-from-hills-to-shark-tank-india/

Reynolds, P. D., Bygrave, W. D., Autio, E., Cox, L. W., & Hay, M. (January 2003). Global entrepreneurship monitor 2002: Executive report. *Global Entrepreneurship Monitor*. https://doi.org/10.13140/RG.2.1.1977.0409

Shah, H., & Saurabh, P. (2015). Women entrepreneurs in developing nations: Growth and replication strategies and their impact on poverty alleviation. *Technology Innovation Management Review, 5*(8), 34. https://doi.org10.22215/timreview/921

Shark Tank India. (March 13, 2024). Shark Tank India S3 I Full Pitch, accessed on May 5, 2024 www.youtube.com/watch?v=eGP9H2G0AnI

Shaw, K., & Sørensen, A. (2022). Coming of age: Watching young entrepreneurs become successful. *Labour Economics, 77*, 1–15. https://doi.org/10.1016/j.labeco.2021.102033

Shrimanne, S., & Adhikari, R. (March 08, 2023). How to unlock women's potential in the digital economy. *World Economic Forum*, accessed on May 22, 2024 www.weforum.org/agenda/2023/03/how-to-unlock-womens-potential-in-the-digital-economy/.can be a powerful avenue for women's inclusion in the digital econ

Terjesen, S., & Amorós, J. E. (2010). Female entrepreneurship in Latin America and the Caribbean: Characteristics, drivers and relationship to economic development. *The European Journal of Development Research, 22*, 313–330. https://doi.org/10.1057/ejdr.2010.13

Tiwari, S., & Tiwari, A. (2007). *Women entrepreneurship and economic development.* Sarup & Sons.

Topno, M. A. (February, 2023). Who is Shashi Bahuguna Raturi, the founder of Namakwali Brand. *Her Zindagi.com,* accessed on March 14, 2024 www.herzindagi.com/society-culture/who-is-shashi-bahuguna-raturi-the-founder-of-namakwali-brand-article-222301

Viljamaa, A., Joensuu-Salo, S., & Kangas, E. (2022). Part-time entrepreneurship in the third age: Well-being and motives. *Small Enterprise Research, 29*(1), 20–35. https://doi.org/10.1080/13215906.2021.2000483

Welsh, D. H., Memili, E., & Kaciak, E. (2016). An empirical analysis of the impact of family moral support on Turkish women entrepreneurs. *Journal of Innovation & Knowledge, 1*(1), 3–12. https://doi.org/10.1016/j.jik.2016.01.012

Williams, C. C. (2009). Informal entrepreneurs and their motives: A gender perspective. *International Journal of Gender and Entrepreneurship, 1*(3), 219–225. https://doi.org/10.1108/17566260910990900

Williams, C., & Martinez, A. (2014). Is the informal economy an incubator for new enterprise creation? A gender perspective. *International Journal of Entrepreneurial Behavior & Research, 20*(1), 4–19. https://doi.org/10.1108/IJEBR-05-2013-0075

World Bank. (1994). *Indigenous knowledge for development: A framework for action (English).* World Bank Group. http://documents.worldbank.org/curated/en/388381468741607213/Indigenous-knowledge-for-development-a-framework-for-action

Yin, R. (2009). *Case study research: Design and methods* (4th ed.). SAGE Publications.

Yin, R. (2014). *Case study research and applications: Design and methods* (6th ed.). SAGE Publications.

15 Women's Brains
Creative Thinking and Entrepreneurship

*Steve Fernando Pedraza Vargas and
Ramón Antonio Hernández de Jesus*

15.1 INTRODUCTION

All changes in the contemporary world, including scientific and technological advances, are based on the contributions of creative minds, capable of bringing novelty and heuristic value to each of the fields of art and knowledge (Sousa et al., 2018). Social neuroscience suggests that the processes of attention, memory, and cognitive flexibility are linked to creative thinking. As a result, creativity is seen as a component of adaptive cognitive functioning since it enables an individual to adapt by using a variety of strategies to meet the demands of their immediate environment.

The factors involved in creativity are focused on the personal abilities of each individual, such as affective, cognitive and environmental. The cognitive ones are associated with the elaboration and capture of information to conceptualize and relate ideas and data to understand and act on the person's reality; the affective ones are distinguished by the specific elements that generate the creative potential; and the environmental factor is associated with the conditions, climate, terrain, or environment that facilitate the development and actualization of the creative potential of the person (Hurtado et al., 2018).

In this context, the presence of women, under equal conditions, in creative production activities in fields such as scientific and technological research, architecture, philosophy, the humanities, plastic arts, music, dance, cinema, literature, communication, social engagement, or organizations is a social challenge of the first magnitude (Romo, 2018). Sadly, women's attempts to engage in creative behaviors are impeded, either by their own internalized gender roles or by the response they receive from people who think they are participating in a masculine activity, even though they possess creative capacities comparable to those of men.

In this regard, culture has played an important role in determining this gender disparity. Historically, more masculine cultures (e.g., American or Spanish) were detrimental to women's creativity, while more egalitarian or relationship-focused cultures (e.g., Nordic countries) favored female creativity (Mayorga et al., 2020).

Recently, a new study by Hora et al. (2022 investigated the possibility of gender differences in creativity. Although it is widely believed that men are more creative than women, the authors argue that this is probably not the case in real life. Furthermore, recent research indicates that the stereotype that males are naturally more creative is fading over time and is less prevalent in nations with higher levels of gender equality.

Women have different life experiences, needs, preferences, and values than men, which can inspire them to identify new problems, opportunities, and solutions that might otherwise be overlooked. Women also tend to have different cognitive styles, such as being more collaborative, empathetic, holistic, and adaptive, which can improve the quality and diversity of ideas and outcomes (Saucedo et al., 2018).

Despite their valuable contributions, women are often underrepresented, undervalued and unsupported in many areas of entrepreneurship. Women face a number of challenges, such as lack of access to education, funding, mentorship, networks and recognition, as well as stereotypes, discrimination, harassment and imposter syndrome, which can limit their opportunities and confidence to pursue and succeed in entrepreneurship. For example, women represent only 28% of the world's researchers, 15% of patent holders, and 3% of Nobel laureates in science. Women also receive less venture capital funding, media coverage, and awards than men for their efforts in entrepreneurship (Tarapuez et al., 2018).

Fortunately, female entrepreneurship has gained increasing relevance in recent years, breaking barriers and demonstrating the potential of women in the entrepreneurial world. However, despite the progress, there are still challenges that hinder women's full participation in entrepreneurship (Navas & Moncayo, 2019). One of the most powerful tools to overcome these barriers is education. Education not only allows women to be better prepared to face entrepreneurial challenges but also provides them with a solid foundation to make informed and strategic decisions (Bravo et al., 2021).

It should be mentioned that, in terms of social neurosciences, sexual and socio-anthropological dimorphism promotes these abilities, which are characterized by the creation of alternative resolutions, even though there are notable differences between the functional organization of the brain and the mental capacities of men and women that are concentrated on linguistic and spatial capacity.

Therefore, this chapter aims to establish the relationship between women's creative thinking and their entrepreneurial capacity from the perspective of social neurosciences, which could provide a more complex reading of the study problem by taking into account recent advances in gender neuropsychology and entrepreneurship pedagogy.

15.2 CREATIVE THINKING

Creativity is a process through which an individual receives stimulation from the environment to solve a problem.

Although the factors that support creative performance have been imprecise, it is now considered as a common cognitive process relevant to many areas of daily life, which is related not only to social and environmental factors but also to cognitive processes where the main regions involved are the tertiary integration and prefrontal areas (Jauk, 2012).

We speak then of cognitive factors that facilitate and provide creative performance with its intrinsic characteristics, since it promotes the formation of new and logical alternatives through the adaptation of a given number of relevant responses that allude to the minimum frequency expressed at a personal and/or population level, and that have a high degree of development, which is supported by the ability to define and perceive objects or situations in an unusual way; as opposed to convergent thinking, which consists of retrieving from memory elements without modification that meet a series of requirements (Gamarra, 2020).

In order to carry out such unusual information processing, a series of neuropsychological aspects are required to support the decoding and recombination of data in an effective manner. We are talking about working memory, which allows keeping the information online while it is being processed (analyzed, selected, and semantically integrated), making it indispensable for syntactic comprehension. On the other hand, sustained attention is the ability to maintain a constant performance over a long period of time, an aspect that depends on the maintenance of vigilance, the ability to detect stimuli, and the orientation of cognitive resources towards the demanding activity; cognitive flexibility, in turn, involves the capacity that allows an adequate adaptation and readaptation to explore other forms of cognitive procedure; and finally, judgment will give the guidelines to formulate anticipatory hypotheses, which involves the ability to relate ideas affirming or denying links between them, favoring reasoning or reasoning (Hernández et al., 2015).

On the other hand, a neurochemical substrate is also required, where the role of the neurotransmitter dopamine as a promoter of the search for novelty and the creative impulse has been taken up again; this is due to the neuroanatomical regions related to the mesolimbic and mesocortical pathways of the dopaminergic circuit, which encompasses the midbrain at the subcortical medial temporal level and the prefrontal cortex. In humans, dopamine decreases latent inhibition, which is a behavioral index of habituation capacity and is therefore associated with the expression of creative products, and exposure to novel stimuli increases the activity of the dopaminergic system in the hippocampus, nucleus accumbens, and prefrontal cortex, which are involved in motivation (Silva et al., 2014).

Nevertheless, there are few investigations in neuropsychology that address the relationship between cognitive support and the creative process and product in healthy participants (without neurological or psychiatric compromises) (Jauk, 2012), and even fewer in female participants, while there are none that take up sexual orientation and creativity within their study variables. Therefore, the results of this research could be useful to address the neuropsychological relationship between cognitive performance and the realization of creative responses, taking up the inclusion of study groups different from the classic ones addressed, such as women and sexual orientation, as a possible marker of trends in information processing.

It has been described that within creative thinking are involved cognitive factors that will support the creative realization, guiding and regulating it to give alternative, logical and efficient resolutions. Creativity is not only the unusual visualization of a problem but also the implementation of skills that come from a cognitive style, which is characteristic of the ways and means used to reach an execution. While it is already known that there are gender-differentiated cognitive abilities, namely that women are more skilled than men in generating a narrative discourse and men are more skilled in providing a spatial navigation mapping, these abilities characterized by creating alternative resolutions are promoted in each case by sexual dimorphism. However, it has been neglected to include the cognitive variants of the creative process that make them possible, as well as other groups that also provide different cognitive styles (Hernández et al., 2015).

15.3 NEUROPSYCHOLOGY OF GENDER

Neuropsychology is a well-established discipline that enjoys great recognition and prestige and has been offering surprising insights in recent years (Pedraza & Vélez, 2023). Today, there are many very scattered publications on gender from different sources, which makes it more problematic to frame them in coherent research programs. Many disciplines deal with gender issues, and there is great conceptual and methodological diversity.

Gender has traditionally been studied primarily in the social sciences, psychology, sociology, and anthropology, while sex has traditionally been studied from the perspectives of genetics, endocrinology, anatomy, physiology, and neurology. However, because gender is classified as psychosocial and sex as biological, reality cannot be fully understood because these categories still exist. These categories include opposition, inheritance–medium, nature–breeding, and biology–culture (Rippon, 2020).

In reality, these discrete and separate categories, a male–female sexual dimorphism, do not exist, but rather a sexual polymorphism. Thus, in addition to the more harmonious correspondence between the different levels, genetic, endocrinological, physiological, and neurological, present in men and women, there is also discordance between the levels indicated in more minority cases of indefiniteness, bisexuality and sex change.

Second, this sexual polymorphism is shaped throughout a person's life and is experienced in different ways in unique and unrepeatable identities. Each person structures himself or herself, makes his or her own sexualized corporeality and is more or less conscious of his or her way of perceiving, thinking, feeling and acting as a human being.

However, studies of the brains of people in a variety of professions, including cab drivers, pianists, violinists, and others, have shown that experiences and learning in sociocultural environments reorganize and structure the brain, starting with the neural networks. Additionally, there is the reorganization of the brain that happens in individuals who have injuries, blindness, deafness, aphasia, amnesia, etc. as a result of learning and rehabilitation (Reverter & Medina, 2018; Rippon, 2020).

15.4 DIFFERENCES BETWEEN THE BRAIN–MIND OF MEN AND WOMEN

Men and women typically exhibit distinct differences in the functional organization of the brain and, consequently, in their mental capacities, even within the shared universals of cognition, language, and emotion that all members of the human species share. Certain authors have highlighted these distinctions specifically, whereas other authors prioritize the general and shared structures and procedures, with the distinctive features coming in second.

The assumption that sociocultural environments, learning, and socialization play a major role in explaining human differences, particularly those between genders, has been imposed as both politically and academically correct by the social science standard model. However, given the state of the research, it is very difficult to uphold these extreme environmentalist assumptions (Reverter & Medina, 2018).

It should be emphasized that the differences in mental abilities associated with sex are also modular in nature, and that the issue here is not one of proving that one sex is generally or globally more intelligent than the other, as has occasionally been attempted to do using the intelligence quotient or other comparable metrics.

When it comes to perceptual speed tests, women tend to perform better than men when it comes to quickly identifying matching objects.

When it comes to certain spatial tasks, like mentally rotating an object, men perform better than women. When it comes to target-directed motor skills, like throwing or intercepting projectiles, they perform more accurately than women. When asked to identify a specific figure or object hidden in a more complex figure, for example, women perform better on figure identification tests in complex frames. Men also perform better on tests of mathematical reasoning than women (Grabowska, 2017; Adnan & Julie, 2021).

In emotional behavior, men have a greater tendency to express their emotional state through aggressive behaviors, while women prefer symbolic mediation, verbalization, and oral expression (Reverter & Medina, 2018).

Now, in creative thinking, there are no significant differences between men and women, beyond those derived from the style of construction of this (Sousa et al., 2018; Adnan & Julie, 2021; Cueto et al., 2022; Hora et al., 2022; Torres et al., 2023). This is due to the particularities that each gender has for particular processes, as evidenced by neurocognitive tests (Hora et al., 2022).

15.5 ENTREPRENEURSHIP CAPACITY

The term entrepreneurship has currently gained relevance due to its importance worldwide, being considered as a key factor present in the economic growth sector in countries (Villalba, 2020). Entrepreneurship is not only synonymous with the creation of companies, but it is also seen as the promotion of new businesses oriented to the development of new productive activities for profit (Rueda-Grand, 2019). This has improved the quality of life for those residing in this area, either directly or indirectly, by facilitating the creation of job opportunities and the economic development of localities or the nation as a whole.

Entrepreneurship as a capacity has no distinction of gender, discipline or age; it is a natural, cultural and traditional human action (Manosalvas et al., 2020). Entrepreneurship has allowed women to achieve levels of empowerment, professional development, self-realization, and new opportunities; however, women entrepreneurs are affected more by structural inequality and gender stereotypes than by differences associated with women's brains.

Although women have acquired certain spaces in the business environment, there is little literature that studies female entrepreneurship, and this is due to the scarce generation of companies by women (Navas & Moncayo, 2019; Bravo et al., 2021). In fact, this process has been gradual, and in the statistics on business creation by women, it has not been a sufficiently representative factor (Chávez et al., 2020; Galecio et al., 2019; Paredes et al., 2019; Tarapuez et al., 2018).

15.6 ENTREPRENEURSHIP PEDAGOGY

Entrepreneurship education stands out, covering all educational levels, which clearly includes the teaching and development of attitudes and skills, as well as restricted teaching aimed at creating a new business (Lopes, 2010). And entrepreneurship has always been present in great moments of transformation for those seeking something new in an attempt of realization, not only of capital but with the aim of increasing human satisfaction. Today, these transformations point to a growing increase in the work of the self-employed entrepreneur and the creation of productive models in a veritable explosion of forms of work and income generation.

Women in this context are increasingly intense and organized, as they have been showing interest in entrepreneurship and managing to reach areas and levels of involvement with new businesses (GEM, 2018). Characterizing changes in women's lives is their purchasing power, i.e., their ability to manage their lives and achieve personal goals that previously seemed impossible to them. Women who manage their own sales have the freedom to organize their work schedules, pay their own expenses, and receive commissions for products and services sold directly (King & Robinson, 2006)

One cannot overlook the various challenges that women face day in and day out, which can translate into: fear, chances of failure and discrimination (linked to cultural heritage), lack of knowledge in management, and lack of practice in the business environment (Costa & Moreira, 2018; Mutlu, 2018; Cho, Luna & Bounkhong, 2019). It is important for women to receive pedagogical training based on entrepreneurial education, gain more experience and knowledge of business transactions and more practice in activities conducted in the entrepreneurial environment (Welsh et al., 2014).

Entrepreneurial development is capable of promoting female empowerment, in which each woman becomes stronger, more confident and worthy of demonstrating all her capabilities, ceasing to be a source of estrangement and distrust before the conditions for participation in a society on an equal footing with men. Female empowerment is fundamental to accessing human rights and to the participation of women in society (Baquero, 2013). Motivated, empowered, and prepared women entrepreneurs are the focus of action in the business model in any area of human knowledge.

It should be noted that the pedagogy of entrepreneurship is enriched by the experiences lived by each woman within her own business, allowing her to see the elements that give meaning to the construction of knowledge based on creativity, languages and understandings of reality. Their voices, their representations, and their collective constructions allow women to particularize their characteristics, making them more innovative in their undertakings.

Therefore, women's entrepreneurship is characterized by innovation and risk-taking. Although entrepreneurship is commonly associated with startups and small and for-profit enterprises, the entrepreneurial behavior of each of them can occur in small, medium and large for-profit and non-profit enterprises, including the voluntary sector such as charitable or governmental organizations, among others.

15.7 WOMEN'S MANAGEMENT AND THE CONSEQUENCES FOR SOCIETY

Although there is still inequality between the male and female genders, women have been effectively gaining their space. Their main characteristics are greater sensitivity, empathy, commitment and willingness to help. These are some of the characteristics that help a woman become a successful entrepreneur in any service sector. It is impossible to ignore that women can perform several activities at the same time and assume multiple responsibilities (home, husband, children, and work). According to Villas (2010, p. 51), "There are important differences between male and female entrepreneurial styles. They have a great capacity for persuasion and concern for customers and suppliers, which contributes to the progress of a company or their own business," which becomes a differentiator in relation to men.

According to Grzybovski (2002), women are able to build a sense of community, through which members of the organization come together, and learn to believe in and care for one another.

This ability gives women a competitive advantage for the company's success because they exhibit a unique management style and employ a variety of strategies to reconcile their personal and professional lives. Previously exclusive to men, traits like ambition, competitiveness, leadership, risk-taking skills, acceptance of change, analytical and objective thinking, independence, and self-assurance have been adopted by women.

Women are more likely to take on democratic leadership roles. The economic contribution that women entrepreneurs make to society—creating jobs for themselves and others—as well as the social example that they set by managing the double workday and the rise in previously unneeded female autonomy are the main reasons why women entrepreneurs are important to society.

15.8 METHODOLOGY

This chapter was developed according to a documentary research design, which according to Brito (2015) is one that manifests an analysis of "different phenomena of reality obtained and recorded by other researchers in documentary sources" (p. 8) and this review was complemented by following the qualitative approach that "is based on evidence that is oriented more towards the deep description of the phenomenon in order to understand and explain it" (Sanchez, 2019, p. 104).

It should also be noted that the documentary sources used consisted of bodies contained in indexes and databases such as SciELO, websites of official bodies, universities and prestigious information portals, as well as physical publications on paper. With respect to the operational techniques for document manipulation that were used, we can refer to observation, in-depth reading, summarizing and underlining. In addition, it was convenient to analyze the data collected through critical analysis, which, according to Moreno, Puerta, Cuervo, and Cuéllar (2016) is aimed at understanding the ideas expressed in what was read.

It possible to note that one of the criterion for its selection was that at least 50% of the sources consulted had to be no more than five years old at the date of submission of this essay, and at the qualitative level, they had to contain information that would allow understanding the phenomenon of women's brains: creative thinking and entrepreneurship.

15.9 CONCLUSIONS

It is concluded that creative women are generating new alternative routes of life and leading to the motivation of many other women to occupy jobs that were previously held by men, to seek innovative and enterprising solutions to transmit knowledge, lead and serve with humanity, as well as mobilizing people to recognize among peers the difficulties faced and the emergence of movements that advance rights.

Considering aspects related to female entrepreneurship, It was noted that the majority of research appeared to indicate that gender differences are more notable solely when it comes to vocabulary, reasoning, and verbal and ideational fluency tests, where women typically perform better than men. It should be remembered that women's success in the workforce outside the home is relatively new. It was difficult for women to become independent, earn their own money, and have their abilities acknowledged. The intellectual capacity of women is now beyond dispute, which is a positive development for society.

In this sense, women structure their entrepreneurial capacity on the basis of a concrete and articulated ideational creative thinking, since women have a greater ability than men to generate a narrative discourse, and men are more skilled in generating a spatial mapping of navigation that represents more georeferenced creative thinking. These abilities, characterized by creating alternative resolutions, are, in each case, promoted by sexual dimorphism.

The research's data indicates the rise in female entrepreneurs and demonstrates how these individuals have managed to survive in the field. Normally, a female entrepreneur generates only her own employment and/or the employment of family members; however, when the enterprise is successful, it can generate two work fronts, since the entrepreneur also needs a domestic employee to help her in commercial activities or at home; even with so many achievements, women will never be exempt from this responsibility.

Therefore, in addition to entrepreneurship, today's women continue to perform traditional roles as housewives, mothers, and wives, juggling two careers. Women who are driven to become entrepreneurs because they must provide for their families frequently launch their own companies from home in order to balance the demands of both responsibilities. The majority of the time, women-owned businesses create services with an end user in mind. These programs are designed to supplement family income or serve as an alternative to unemployment.

Given this context and the significance of entrepreneurship as a driving force behind a nation's economic development, it is imperative to create an environment that supports the growth of women-owned business ventures under the right circumstances, taking into account the fact that women often find themselves in multi-role situations on a daily basis. But in order to fully harness the potential of diversity and women's viewpoints, we need to develop an innovative culture that is more egalitarian, inclusive, and values women at all professional stages.

It is necessary to increase the representation and participation of women in the fields of innovation, especially in leadership and decision-making roles. There is a need to challenge and change norms and biases that hinder women's potential or aspirations. Similarly, the historical achievements of innovative and inspirational women innovators need to be recognized and amplified in the next generation of girls or women with the conviction to pursue their dreams. For example, initiatives such as Women in STEM, Women in Tech, and Women in Innovation and Women Who Code are some of the examples of how a more diverse and vibrant innovation community can be fostered.

REFERENCES

Adnan Ahmed, L., & Julie Patel, H. (2021). Dump the "dimorphism": Comprehensive synthesis of human brain studies reveals few male-female differences beyond size. *Neuroscience & Biobehavioral Reviews, 125*, 667–697. ISSN 0149-7634. https://doi.org/10.1016/j.neubiorev.2021.02.026

Baquero, A. (2013). Empoderamento: Instrumento de emancipação social? Uma discussão conceitual. *Revista Debates, 6*(1), 173–187.

Bravo, I. F., Bravo, M. X., Preciado, J. D., & Mendoza, M. (2021). Educación para el emprendimiento y la intención de emprender. *Revista Economía y Política, 33*. Obtenido de www. redalyc.org/journal/5711/571165147008/ html/

Brito, A. (2015). *Guía para la elaboración, corrección y asesoramiento de trabajos de investigación*. San Tomé: Universidad Nacional Experimental Politécnica de la Fuerza Armada Bolivariana.

Chávez-Toala, A., & Feijó-Cuenca, N. (2020). El emprendimiento femenino y su contribución al desarrollo socioeconómico de la ciudad de Portoviejo. *Polo del Conocimiento, 5*(3), 554–573. Obtenido de https://polodelconocimiento.com/ojs/index.php/es/article/view/1352/2438#

Cho, E., Moon, Z. K., & Bounkhong, T. (2019). A qualitative study on motivators and barriers affecting entrepreneurship among Latinas. *Gender in Management: An International Journal*. https://doi.org/10.1108/GM-07-2018-0096

Costa, J., & Moreira, M. (2018). Trajetória de vida de mulheres empreendedoras e o uso de práticas gerenciais. In *Congresso brasileiro de gestão, Belém. Anais [...]*. Belém: Unama.

Cuetos-Revuelta, M., Serrano-Amarilla, N., y Marcos-Sala, B. (2022). La creatividad en la Educación: diferencias por rendimiento, edad y sexo. *Electronic Journal of Research in Educational Psychology, 20*(3), 683–710. ISSN: 1696-2095 https://doi.org/10.25115/ejrep.v20i58.6906

Galecio, G., Castaño, A., & Basantes, D. (2019). Emprendimientos impulsados por mujeres ecuatorianas. *Revista Cienciamatria, 5*(9), 286–301. Obtenido de https://cienciamatria revista.org.ve/index.php/ cm/article/view/148/161

Gamarra-Moscoso, M., y Flores-Mamani, E. (2020). Pensamiento creativo y relaciones interpersonales en estudiantes universitarios. *Investigación Valdizana, 14*(3), 159–168. https://doi.org/10.33554/riv.14.3.742

GEM. Global Entrepreneurship Monitor (GEM). (2018). Obtenido de www.ibqp.org.br/gem/

Grabowska, A. (2017). Sex on the brain: Are gender-dependent structural and functional differences Associated with behavior? *Journal of Neuroscience Research*, 95, 200–212. https://doi.org/10.1002/jnr.23953

Grzybovski, D. (2002). Estilo feminino de gestão em empresas familiares gaúchas. Obtenido de www.scielo.br/scielo.php?pid=S1415-6555200200 0200011&script=sci_arttext

Hernández, I. P., Orozco, G., Ortega, L., Romero, C. y López, K. (2015). Evaluación del pensamiento creativo en mujeres con diferentes orientaciones sexuales. *Rev Elec Psic Izt. 18*(4), 1405–1420.

Hora, S., Badura, K. L., Lemoine, G. J., & Grijalva, E. (2022). A meta-analytic examination of the gender difference in creative performance. *Journal of Applied Psychology, 107*(11), 1926–1950. https://doi.org/10.1037/apl0000999

Hurtado, P. A., García, M., Rivera, D. A., & Forgiony, J. O. (2018). Las estrategias de aprendizaje y la creatividad: una relación que favorece el procesamiento de la información. *Revista Espacios*, 12–30. https://www.revistaespacios.com/a18v39n17/a18v39n17p12.pdf

Jauk, E., Benedek, M., & Neubauer, A. (2012). Tackling creativity at its roots: Evidence for different patterns of EEG alpha activity related to convergent and divergent modes of task processing. *International Journal of Psychophysiology*, 84, 219–225.

King, C., & Robinson, J. (2006). *Los nuevos profissionales: el surgimiento del network marketing como la próxima profesión de relevân*cia. Time & Money Network Editions: Buenos Aires.

Lopes, R. (2010). *Educação empreendedora: modelos, conceitos e práticas*. Rio de Janeiro: Elsevier.

Manosalvas-Vaca, L., Manosalvas-Vaca, C., Solís, V., & Pesantez, J. (2020). Capacidades de innovación en los emprendedores turísticos: Un enfoque de género. *Innova Research Journal, 5*(2), 234–252. Obtenido de https://revistas. uide.edu.ec/index.php/innova/article/view/1367/1708

Mayorga, J., Carvajal, R., & Morales, D. (2020). Aspectos sociales y su influencia en el emprendimiento femenino Sudamericano. *593 Digital Publisher CEIT, 5*(2), 125–133. Obtenido de www.593dp.com/index.php/593_Digital_Publisher/ article/view/196/445

Moreno, E., Puerta, C., Cuervo, C., y Cuéllar, A. (2016). Análisis crítico de literatura científica. Una experiencia de la Facultad de Ciencias de la Pontificia Universidad Javeriana. *Voces y Silencios. Revista Latinoamericana de Educación 7*(2), 74–97. https://doi.org/10.18175/v ys7.2.2016.06

Mutlu, S. (2018). Cultural barriers female entrepreneurs face in Turkey. In *Modernizing Process in Turkey* (pp. 185–198). https://doi.org/10.5771/9783845291154-185

Navas, A., & Moncayo, J. (2019). El empoderamiento productivo de la mujer como consecuencia de la Inclusión Financiera. *Innova Research Joirnal, 4*(3.2), 152–171. Obtenido de https://revistas.uide.edu.ec/index.php/innova/ article/view/1125/1626

Paredes, S., Castillo, M., & Saavedra, M. (2019). Factores que influyen en el emprendimiento femenino en México. *Suma de Negocios, 10*(23). https://doi. org/10.14349/sumneg/2019.v10.n23.a8

Pedraza-Vargas, S. F., & Vélez-Jiménez, D. (2023). Neurociencias Sociales: principios epistemológicos. *Espergesia, 10*(1), 66–75. ISSN: 2312-6027 e-ISSN: 2410-4558. https://doi.org/10.18050/rev.espergesia.v10i1.2520

Reverter-Bañón, S. y Medina-Vicent, M. (2018). La diferencia sexual en las neurociencias y la neuroeducación. *Crítica, Revista Hispanoamericana de Filosofía, 50*(150), 3–26. https://doi.org/10.22201/iifs.18704905e.2018.13

Rippon, G. (2020). *El género y nuestros cerebros*. Barcelona, Spain: Galaxia Gutenberg.

Romo, M. (2018). ¿Tiene género la creatividad? Obstáculos a la excelencia en mujeres. *Estudos de Psicologia (Campinas), 35*(3), 247–258. http://dx.doi.org/10.1590/1982-02752018000300003

Rueda-Grand, G. (2019). Análisis de los factores asociados a la sostenibilidad de los emprendimientos en la zona de planificación 7-Sur del Ecuador. *Polo del Conocimiento, 4*(5), 370–937. Obtenido de https://polodelconocimiento.com/ojs/ index.php/es/article/view/991/1272#

Sánchez F. (2019). Fundamentos epistémicos de la investigación cualitativa y cuantitativa: Consensos y disensos. *Revista Digital Investigación y Docencia 13*(1), 101–122. www.scielo.org.pe/scielo.php?scrip t=sci_arttext&pid=S2223-251620190001 00008

Saucedo, K. (2018). Medición de la intención de emprendedores universitarios empleando ecuaciones estructurales. *Revista Investigación y Negocios, 11*(18), 52–63. Obtenido de www.scielo.org.bo/scielo. php?script=sci_arttext&pid=S2521-27372018000200006&lng=es&nrm=iso

Silva, R., Espinoza, P., Riquelme, R., Sanguinetti, N., González, L., & Cruz, G. (2014). Rol de las hormonas sexuales sobre circuitos dopaminérgicos cerebrales. *Pontificia Universidad Católica de Chile. Revista Farmacológica de Chile, 7*(1), 7–20.

Sousa, F. C., Monteiro, I. P., y Bica, J. P. (2018). A evolução das redes sociais na execução de projetos em um agrupamento escolar. *Estudos de Psicologia (Campinas), 35*(3), 265–274. http://dx.doi.org/10.1590/1982-02752018000300003

Tarapuez, E., García, M., & Castellan, N. (2018). Aspectos socioeconómicos e intención emprendedora en estudiantes universitarios del Quindío (Colombia). *Innovar, 28*(67), 123–135. Obtenido de www.redalyc.org/jatsRepo/818/81854579009/ html/index.html

Torres Pulido, V, Benítez Giles, S. y Gutiérrez Barba, B. (2023). Inteligencia creativa ¿Existen diferencias entre hombres y mujeres de pregrado? *Journal of Behavior, Health & Social Issues, 15*(1), 3–10. https://doi.org/10.22201/fesi.20070780e.2023.15.1.83466

Universidad de Jaén. (2020). Diseño documental [página web]. Obtenido de www.ujaen.es/investiga/tics_tfg/dise_doc umental.html

Villas, A. (2010). *Valor Feminino: desperte a riqueza que há em você – São Paulo: Ed*. Do autor.

Villalba, M. (2020). La cultura del emprendimiento de Ecuador en relación a Perú. *Visionario Digital, 4*(3), 147–169. Obtenido de https://cienciadigital. org/revistacienciadigital2/index.php/ VisionarioDigital/article/view/1340/3294

Welsh, B., et al. (2014). Saudi women entrepreneurs: A growing economic segment. *Journal of Business Research, 67*(5), 758–762.

Index

A

Access to financial resources, 10, 16, 98, 171, 208
Actionable strategies, 200, 247
Africa (Women's Leadership), 184–190
Agriculture, 2, 10, 17, 19, 39, 97, 106, 124–126, 150, 186, 259
Artificial intelligence (AI), 83
Awareness campaigns, 16, 96, 98, 123, 129–131

B

Bibliometric analysis, 210–211
Business case studies, 12–15, 261–268
Business models, 91, 166, 171, 174, 189, 261, 262, 264–265, 270, 279
Business strategies, 93, 242

C

Capacity building, 107, 119, 127, 128, 150, 166
Case studies, 12–15, 260–268, 270
Challenges in entrepreneurship, 171, 179, 185, 187–188, 198, 209, 247, 248, 252–255
Circular economy, 34, 97, 143, 145, 151, 166–8, 184–8, 197, 201, 206–208, 215, 221, 222, 270
Climate change, 18, 233
Collaboration, 39, 80, 83, 95, 97, 98, 106, 108–113, 115, 117, 118, 122, 132, 174, 175, 179, 198, 202, 232–233, 236, 241, 242
Creative thinking, 274–282

D

Data–driven decision making, 110
Data–driven management, 64–86
Development opportunities, 64–86, 103, 106, 114
Digital transformation, 89, 167
Diversity and inclusion, 107, 119, 186–187

E

Economic empowerment, 1, 17, 66, 74, 223, 247
Education and training, 114–115, 150, 152, 157, 176, 208, 209, 210, 224
Empowerment, 1, 4, 6, 14, 19, 64–70, 74–76, 78, 81–86, 115–117, 122, 124–125, 131–132, 144–150, 154–160, 178, 186, 187, 206, 219–226

Entrepreneurship, 90–91, 97, 144–150, 166–174, 177, 178, 179, 184–190, 192–203, 206–216, 219–226
Equality, 1–3, 9, 14–20, 65, 67, 74, 76, 79, 80, 84, 85, 86, 95–98, 130, 133, 143, 144, 148, 149, 189, 190, 196, 197, 202, 206–212, 215, 219–221, 226, 230, 233, 236, 237, 239–242, 247–249
Ethical leadership, 234
Examples of best practices, 128, 130, 133, 173, 186, 256

F

Family support, 93, 142–160, 253, 256, 269
Financial inclusion, 132, 188
Food security, 17, 18
Funding opportunities, 119, 127

G

Gender equality, 1–3, 7, 9, 14–17, 24, 57, 65, 67, 70, 74, 76, 79, 80, 84, 85, 95–98, 104, 130, 133, 134, 147, 149, 168, 172, 196, 248, 249, 255
Global goals, 66
Global initiatives, 131, 224
Green business practices, 119, 170, 232
Growthstrategies, 26, 76, 83, 90–3, 114, 127, 131–132, 143–148, 167–169, 174–180, 187–190, 206–210

H

Health and well–being, 259
Heritage preservation, 258–271
Human resources, 113
Hybrid work models, 67–69, 79–81, 154–155, 178–179

I

Impact assessment, 132
Indigenous knowledge, 258–271
Innovation, 89–98, 106, 110, 111, 113, 114, 118, 142–147, 153–160, 165, 169–173, 180, 184, 190, 192–193, 207, 213, 223, 224, 241, 258–271
Innovation barriers, 89–98
Investment strategies, 15, 17, 93–95, 120, 127, 152, 166, 174, 175, 185, 188, 207, 208, 249, 253, 270, 271
IT solutions, 89–98

285

L

Leadership, 103–134, 192, 198, 216, 222, 230–242
Literature review, 23–24, 65–66, 90–94, 146, 167, 185–187, 230–242, 248–249
Local partnerships, 132

M

Mental health, 23–24, 57
Mentorship, 97, 98, 106, 107, 116, 118–120, 122, 130, 132–134, 176, 203, 208–210, 241, 275
Multi–stakeholder partnerships, 103–134
Municipal initiatives, 16, 108–109, 167–168, 190, 194, 208–209, 236–237, 240–242, 254–255

N

Networks and collaboration, 39, 80, 91, 95, 97, 106–118, 179
Non–governmental organizations (NGOs), 96–97, 106, 168

P

Policy recommendations, 120
Psychological well–being, 22–57
Public–private partnerships, 6, 83

R

Renewable energy, 186–188
Resilience building, 171, 189, 206, 215, 255
Risk management, 93

S

Skill development, 97, 168, 174
Social media, 123, 149, 165–180
Social work, 219–226
Stakeholder engagement, 240, 241
Strategies for success, 235
Sustainability, 64, 89, 109–112, 127, 165–180
Sustainable ventures, 142–144, 146, 148, 152, 157–159, 166, 171, 179, 180, 194

T

Technology integration, 23, 66, 78, 90, 97, 108, 115, 123, 127, 143, 145, 177, 196, 197, 202–203, 207, 220, 231, 232, 240, 275, 276
Training programs, 16, 96–97, 107, 116, 120, 123, 128, 189, 210
Transparency, 109–110, 132, 224

W

Women empowerment, 65, 69, 82, 83, 149, 249, 269
Women entrepreneurs, 23, 28, 51, 89–98, 143–144, 149–152, 165–180, 185–188, 198–200, 202, 203, 207–209, 211–213, 215, 223–226, 246–256, 258–260
Work–life balance, 41, 105, 107, 115–116, 120, 122, 152, 176, 187, 199, 208, 210, 238, 240, 241
Workplace challenges, 37, 38, 41, 113, 122, 124, 131, 190, 198, 208, 221, 224

Z

Zero discrimination goals, 7, 95–96, 219–220